The Development and Evolution of *Butterfly* Wing Patterns

Smithsonian Series in Comparative Evolutionary Biology
V. A. Funk and Peter F. Cannell, Series Editors

The intent of this series is to publish innovative studies in the field of comparative evolutionary biology, especially by authors willing to introduce new ideas or to challenge or expand views now accepted. Within this context, and with some preference toward the organismic level, a diversity of viewpoints is sought.

Advisory Board

The Development and Evolution of *Butterfly* Wing Patterns

H. Frederik Nijhout

Smithsonian Institution Press

Washington and London

Library of Congress Cataloging-in-Publication
Data
Nijhout, H. Frederik.
  The development and evolution of butterfly
wing patterns / H. Frederik Nijhout.
      p. cm.—(Smithsonian series in compara-
tive evolutionary biology)
  Includes bibliographical references and
index.
    ISBN 0-87474-921-2
    —ISBN 0-87474-917-4 (pbk.)
    1. Butterflies—Anatomy. 2. Wings—
Anatomy. 3. Butterflies—Development. I.
Title. II. Series.
QL562.N54   1991
595.78'0447—dc20   90-24312

British Library Cataloguing-in-Publication Data
is available.

Cover: *Stichophthalma camadeva* (p. 27 *infra*).

Plate 7A: Photograph by Samuel W. Woo, courtesy A. M. Shapiro.

⊗ The paper used in this publication meets the
minimum requirements of the American National Standard for Permanence of Paper for
Printed Library Materials Z39.48-1984.

Printed in the United States of America
5   4   3   2
95   94   93

For permission to reproduce individual illustrations appearing in this book, please correspond
directly with the owners of the images, as stated
in the picture captions. The Smithsonian Institution Press does not retain reproduction rights for
these illustrations individually or maintain a file
of addresses for photo sources.

To Mary and Monique,

without whom the world would be a

much duller place,

in spite of the butterflies

Table of Contents

Most if not all of the 12,000 or so species of butterflies can be told from one another on the basis of their color pattern alone. Moreover, many species are sexually, geographically, and seasonally polymorphic as well. This extraordinary diversity of patterns is further augmented by the fact that dorsal and ventral patterns are seldom alike and have evolved quite independently of one another. Each element of the color pattern, however, is not just the random dot or stripe seen in the coat colors of mammals but belongs to a system of developmentally and evolutionarily cohesive units whose homologies can be traced across nearly all of the species that exist today. The system of homologies in butterfly wing patterns is as consistent and fundamental as that of the segmental appendages of arthropods or the skull bones and limbs of vertebrates. Butterfly wing patterns illustrate the great richness of diversity that can be achieved by permutation and recombination of a relatively small number of basic units.

The order beneath the exuberant diversity of color patterns was first seen by B. N. Schwanwitsch and F. Süffert in the 1920s. They and their collaborators produced a substantial body of work on the comparative morphology of pattern evolution and the experimental analysis of pattern development in the 1930s. Only one of the early protagonists in this small but active field of study, Schwanwitsch in the Soviet Union, continued a sparse record of publication on the comparative morphology of color patterns after World War II. Unfortunately, with the exception of a few general reviews by Sondhi (1963), Wigglesworth (1965), and Kühn (1971), this work has been largely forgotten. These early workers on wing patterns discovered something important, however: They saw that nearly all the color patterns of butterflies and moths were permutations of a

Preface

It may be said, therefore, that on these expanded membranes Nature writes, as on a tablet, the story of the modification of species, so truly do all changes of the organisation register themselves thereon. Moreover, the same colour-patterns of the wings generally show, with great regularity, the degrees of blood-relationship of the species. As the laws of Nature must be the same for all beings, the conclusions furnished by this group of insects must be applicable to the whole organic world; therefore, the study of butterflies—creatures selected as the types of airiness and frivolity—instead of being despised, will some day be valued as one of the most important branches of Biological science.

—Henry Walter Bates,
The Naturalist on the River Amazons, 1864

single general theme, the nymphalid ground plan.

I became interested in the problem of color pattern formation in the early seventies, when I chanced upon the papers by Schwanwitsch, Süffert, and Kühn and saw that they contained the key to an absolutely first-rate puzzle. Here was a system in which one could study the developmental mechanisms behind a simple morphology and in which one could also study the evolutionary radiation of that morphology on a large scale, in a taxon of which all species are essentially known. The early literature offered many glimpses of the underlying integration, but they were not at all well worked out. For the past 18 years I have been trying to extend the work of these early students of wing patterns. During that time I have been able to fill in a number of the blanks they left, and I have uncovered several relationships among pattern elements that they never saw. This book provides a summary account of that work and places it in the context of that which others have done to elucidate the development, genetics, and evolution of wing patterns. It is also a guide on how to read patterns, provides a window on what can be done to understand their evolutionary and developmental origins once one has seen the wonderful regularity of these structures, and suggests a number of paths for future research.

The organizational principle that runs throughout this book is the nymphalid ground plan, the collection of homologies among pattern elements first elucidated by Schwanwitsch and Süffert. In the course of the past decade, I rejected this ground plan as a working model on two separate occasions and attempted to approach the relationships among pattern elements and patterns in fresh and different ways. The morphoclines in Chapter 4 and the theoretical models in Chapter 7 represent some of the results of

these attempts. In each case, however, I eventually re-created arrays of elements very close to those represented by the nymphalid ground plan. I have come to believe that the nymphalid ground plan, in the way it is developed in Chapter 2, indeed represents an accurate summary of the homologies in butterfly color patterns. Perhaps the most important result of my excursions has been the realization that elements of the nymphalid ground plan develop independently in each wing cell and that they form a developmental continuum with the so-called intervenous stripes. Both of these insights have proved to be crucial for the interpretation of nearly every aspect of pattern evolution.

The book is organized into eight chapters. Chapter 1 provides a general introduction to the structure and development of butterfly wings, their venation, and their pigmentation. Chapters 2, 3, and 4 deal with the comparative morphology of patterns. Chapter 2 describes the homologies of the nymphalid ground plan. Chapter 3 shows how a variety of simple and complex patterns evolved as variants of that ground plan, and Chapter 4 develops a new method for analyzing the structure and diversity of patterns at a much finer level of resolution than that provided by the nymphalid ground plan. Chapter 5 gives an account of the experimental studies that have been done to elucidate the mechanisms whereby color patterns develop. Chapter 6 is an analysis of genetic polymorphisms and seasonal polyphenisms. Much of the literature of genetic polymorphisms that are involved in mimicry systems is reviewed and, for the first time, placed in the context of what we now know about pattern homologies and pattern development. Chapter 7 integrates all the preceding information in the construction of a theoretical developmental model for pattern formation on butterfly wings and discusses a number of features of pattern structure and variation that present particular difficulties in the interpretation of the evolution of pattern development. In Chapter 8 we step back and look at the evolution of color patterns in terms of the evolution of the process that gives rise to the pattern, and we examine some of the general features of color pattern evolution that emerge from the insight developed in the preceding chapters. Diagrams and photographs of patterns are provided throughout the book. Although each of the figures is intended to illustrate a particular point, many also provide supplementary documentation for features of pattern morphology discussed elsewhere. Readers who would like to try their hand at the analysis of additional color patterns should also consult Lewis (1987) and Smart (1975), which provide an inexpensive way of obtaining reasonably good color photographs of thousands of species of butterflies.

A phylogeny of butterflies is given in Appendix A. Appendix B presents an important original contribution by Don Harvey of the U.S. National Museum of Natural History (Smithsonian Institution): an updated higher classification of the family Nymphalidae, including the positions of all genera worldwide. The higher-level systematics of this large and diverse family has been in a state of disarray for a long time, and Dr. Harvey provides an authoritative and critical summary of nomenclatural priorities and hierarchical groupings, with extensive notes that describe the rational basis for this arrangement. The taxonomic arrangement given here represents the most sensible classification scheme I have seen in terms of color pattern morphologies within groups. Monophyletic taxa are identified whenever possible, and the hierarchical groups presented in this section can thus be used to develop phylogenetic hypotheses for this large and diverse family of butterflies.

Acknowledgments

I would like to thank the many friends and colleagues who have contributed to my thoughts on color pattern formation. Some helped by prodding me to go in one direction or another, others by revealing new ways of looking at things. Still others helped by asking pointed questions or by being patient sounding boards as the ideas presented here germinated and developed. They are Greg Wray, Louise Roth, John Lundberg, Susan Paulsen, Diana Wheeler, Larry E. Gilbert, Art Shapiro, Jim Truman, Leah Edelstein, Jim Murray, Don Harvey, Paul Koch, Stan Caveney, and Vernon French. Very special thanks go to Greg Wray for his collaboration in working out the difficult patterns of *Charaxes* and *Heliconius;* to Larry E. Gilbert for sharing his many insights into the structure, genetics, and evolution of *Heliconius* wing patterns; to Laura Grunert for helping in so many ways to make my life and laboratory run smoothly and for doing nearly all the drawings for Chapter 4; and most of all to my wife, Mary, for being a partner, a critic, and a supporter during the alternately difficult and exhilarating episodes that have punctuated these studies over the years. I am also grateful to Greg Wray, Susan Paulsen, Mary Nijhout, Paula Mabee, Art Shapiro, and Joel Kingsolver for their many perceptive comments and critiques of the manuscript, which have much improved this book. I also want to take the opportunity here to extend a special thanks to Don Harvey for providing the taxonomy of the Nymphalidae, which is published here for the first time.

Much of the comparative work presented here was done with materials from the collections at the Museum of Comparative Zoology (Harvard University), Cornell University, and the U.S. National Museum of Natural History (Smithsonian Institution), and I would like to thank the curators of Lepidoptera at these institutions for access to their collections. My work on pattern formation has been supported by grants from the Duke University Research Council and the National Science Foundation.

The color patterns on butterfly wings provide one of the most dramatic examples of evolutionary radiation and morphological diversification in the animal kingdom. They are composed of arrays of pattern elements whose shape, color, and position on the wing are characteristic for each species and whose homologs can be traced across nearly all of the 12,000 species of butterflies in existence today. The development and evolution of butterfly wing color patterns are in large measure the story of the development and evolution of these pattern elements.

The processes that determine the characteristic morphology of the color pattern begin long before metamorphosis, while the butterfly is still a larva and while the wings are still undifferentiated internal imaginal disks. Color pattern determination begins during the early portion of the last larval stage at about the time that the venation pattern of the developing wing is established. As we examine the comparative morphology, development, and genetics of color patterns, we will see that the processes that give rise to the wing pattern depend critically on the wing veins. The veins serve as sources of the inducers that organize the pattern, and as barriers between those portions of the wing whose pattern elements develop and evolve independently of one another. It is therefore appropriate that we start our exploration of the development and evolution of wing patterns with a description of the developmental origin of the wings and their venation pattern. This will occupy the first portion of the present chapter.

The pigments that make up the color pattern reside not in the cells of the wing but in the scales, which are thin flattened cuticular evaginations of specialized cells in the wing epithelium. Each scale bears only a single color, and the overall pattern is a finely tiled mosaic of these colored scales (Plate 1). In ad-

Chapter 1

The Material Basis of Wing Color Patterns

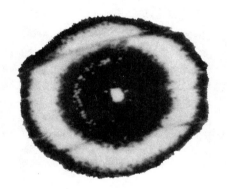

dition to containing chemical pigments, many scales have structural colors as well. Such scales bear several kinds of regular structures whose arrangement and spacing are such that reflected light undergoes constructive interference at certain wavelengths, producing some of the most brilliant colors known in nature. The development of the wing scales and the origin and diversity of their structural and pigmentary colors will be dealt with later in this chapter.

Development and Morphology of the Wings

The wings of a butterfly begin their development as slight enlargements of the epidermal cells in the lateral regions of the mesothoracic and metathoracic segments, the second and third thoracic segments, respectively. These placode-like thickenings first become visible late in embryonic development or during the first larval instar, depending on the species (Comstock, 1918; Köhler, 1932; Kuntze, 1935; Eassa, 1953). Each placode invaginates to form a flattened pouch. The proximal face of the pouch thickens further and becomes the wing bud, or imaginal disk, while the distal face remains thin and becomes a thin cellular envelope for the imaginal disk, the peripodial membrane (Fig. 1.1). It appears that upon invagination, cell division in the wing anlage becomes independent of that of the remainder of the epidermis (Kremen, 1987). Although epidermal cells divide only during a brief period in each molting cycle, the cells of the wing anlage and subsequent imaginal disk divide more or less continuously at a low rate throughout the rest of larval life. The tip of the invagination becomes attached to a tracheal loop that branches off the main lateral thoracic trachea. This tracheal loop soon proliferates into a thick tangle of fine tracheae that will later be the source of tracheation for the developing wing disk. During the penultimate larval instar (in the moth *Samia cynthia*) or the antepenultimate instar (in the moth *Ephestia kuehniella*), the pouch folds upon itself and forms a four-layered structure (Fig. 1.1D,E). The inner two layers constitute the wing imaginal disk and correspond to the future dorsal and ventral surfaces of the adult wing; the outer two layers that envelop the growing disk are now referred to as the peripodial membrane. Throughout development, each surface of the wing disk remains a monolayer of cells, continuous with the peripodial membrane and epidermis.

The venation and tracheation patterns of the wing become established during the last larval instar. At this time the imaginal disk consists of a flattened, crescent-shaped lamina. The basement membranes of the dorsal and ventral epidermal layers are fused over the entire inner surface of the disk. The sequence of events in the development of the initial venation pattern differs somewhat among species. In *S. cynthia* a branching pattern of tubular lacunae forms within the lamina of the disk by local separation of the basement membranes. These lacunae are then invaded by fine tracheae that emerge from the large tracheal cluster at the base of the disk (Fig. 1.1F; Kuntze, 1935). In *E. kuehniella* this sequence appears to be reversed, in that fine tracheae that emerge from the basal tracheal cluster penetrate between the basal laminae of the wing disk before lacunae are evident (Köhler, 1932). The standard elements of the Comstock-Needham venation system (Comstock, 1918) are clearly evident from the earliest stages of lacuna formation and tracheation of the wing disk (Fig. 1.2). In the forewing of *S. cynthia* (Fig.

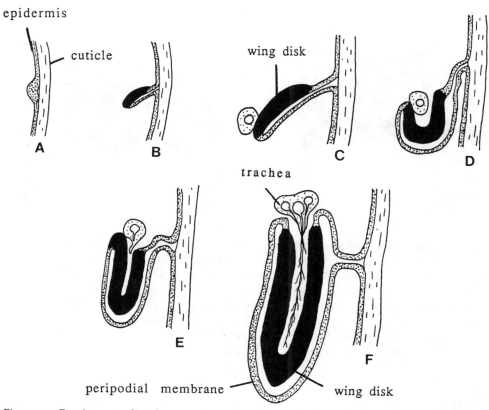

Figure 1.1. Development of lepidopteran wings, shown in cross sections of successive stages in the development of a wing imaginal disk within the caterpillar. The wing starts as a small thickening of the epidermis (*A*), which invaginates (*B*), thickens, and folds (*C–E*) to form a two-layered structure: the wing imaginal disk surrounded by a thin epithelial peripodial membrane. Midway through the last larval instar a bundle of tracheae invades the wing through the lacunae of the presumptive venation system (*F*). (After Kuntze, 1935)

1.2A), each of the main veins arises from an independent stem and branches into a pattern that is virtually identical across most butterflies and moths. The costa and the subcosta are each unbranched, the radius is four-branched, the media and the cubitus are both two-branched, and the two anal veins are each unbranched. The initial venation of the hind wing is similar except that the radius has only three branches and its base is split so that the trachea supplying the anterior branch (R_1) lies in the same lacuna as the subcosta (Fig. 1.2B). Details of the tracheation pattern in the lacunae of the imaginal disks are somewhat variable, particularly in the areas where crossveins develop between media and radius and between media and cubitus. The initial tracheation pattern that develops late in larval life is referred to as the primary tracheation of the wing disk; it remains functional until about the middle of the pupal stage.

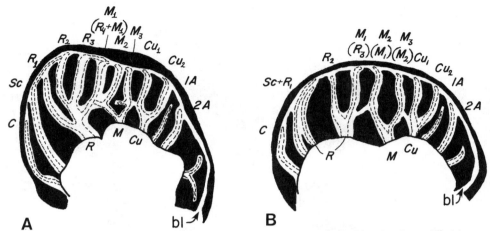

Figure 1.2. Semidiagrammatic view of the wing imaginal disks of *Philosamia cynthia* (Saturniidae) in about the middle of the last larval instar: *A*, forewing; *B*, hind wing. The lacunae of the venation system are white, and the tracheae that invade the lacunae are shown by dashed lines. Labels indicate the standard nomenclature of the wing veins: *C*, costa; *Sc*, subcosta; *R*, radius; *M*, media; *Cu*, cubitus; *A*, anal vein; *bl*, bordering lacuna. It can be seen that the main tracheal branches do not correspond perfectly to the main lacunar branches. This leads to an ambiguity in the nomenclature of the veins, depending on whether one takes the lacunae or the tracheae as the primary determinants of homology. Labels in parentheses refer to the tracheal homologies. (Redrawn from Kuntze, 1935)

The radial lacunae terminate not on the margin of the wing disk but on a bordering lacuna that runs well within the body of the disk and roughly parallel to its margin. This bordering lacuna marks the outline of the future adult wing (Süffert, 1929a). The cells distal to the bordering lacuna die during the pupal stage; most probably they undergo a programmed cell death (Dohrmann and Nijhout, 1990), so that the overall shape and proportions of the adult wing are in effect carved out of a larger imaginal disk much as one would use a cookie cutter on a piece of dough. Details of the wing shape, such as the tails on the hind wings of swallowtail butterflies and some saturniid moths, arise as loops in the bordering lacuna (Fig. 1.3; see Süffert, 1929a, for additional examples).

The wing disks undergo a period of accelerated growth at the end of larval life and during the first half of the pupal stage. The disks grow and expand rapidly during the prepupal stage to form the pupal wings. In most species the pupal wings are between one-third and one-half the size of the adult wing in linear dimensions. A period of cell division during the early part of the pupal stage enlarges the surface area of the wing to adult size, but because of its confinement in the pupal cuticle, the wing does not expand at this time. Instead, its surface is thrown into numerous fine folds that will be stretched out when the wing expands after emergence of the adult.

The two most important events that take place in the wing during the pupal stage are the development of the final wing venation pattern and the development of the scales that will bear the color pattern. At about the middle of the pupal stage a new set of tracheae migrates into the lacunae of the wing. This constitutes the secondary tracheal sys-

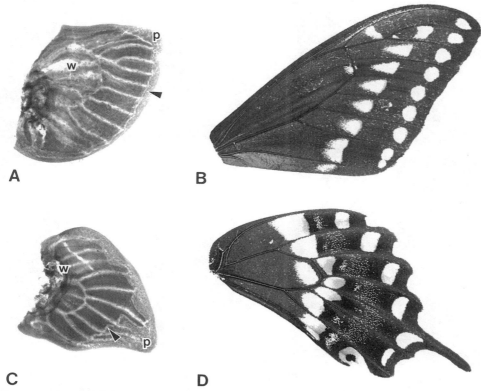

A   B

C   D

Figure 1.3. Imaginal disks and adult wings of *Papilio polyxenes* (Papilionidae). *A* and *C*, imaginal disks of forewing and hind wing, respectively, during late larval life, showing the wing disk proper (*w*) and the peripheral tissue (*p*) that will degenerate during the pupal stage. Arrows indicate the bordering lacuna; the outline of the presumptive tail of the hind wing is clearly visible in *C*. *B* and *D*, ventral surfaces of the fully expanded adult forewing and hind wing, respectively. The imaginal disk is a foreshortened version of the adult wing.

tem, which will provide the air supply for the adult wing (Behrends, 1935; Kuntze, 1935). In *E. kuehniella* and *S. cynthia* the pattern of the secondary tracheae is similar to that of the primary tracheation except that no tracheae enter the base of the media or the first anal vein. The basal portion of the media lacuna and the entire first anal lacuna become constricted during the late larval and prepupal period, and they atrophy entirely during the first half of the pupal stage. The distal portions of the lacunae of the first and second media receive their secondary tracheation

from the radius, and those of the third media from the cubitus, as shown in Figures 1.2 and 1.4 (Behrends, 1935; Kuntze, 1935). There is also a slight lateral displacement of portions of the lacunae of the radius and the distal media so that a large wing cell, the discal cell,* is formed in the region formerly occupied by the base of the media. In some spe-

---

*The term *cell* can have two meanings. Here it refers to an area of the wing bordered by wing veins, not to a cytological cell. To avoid confusion throughout this book, these areas of the wing will be referred to as wing

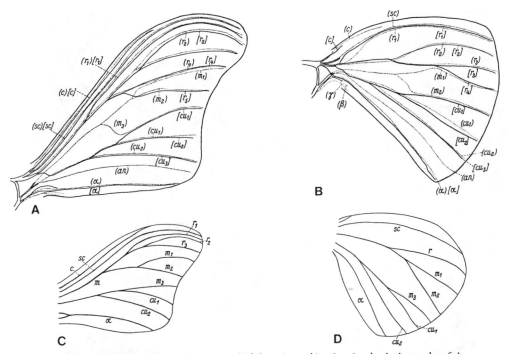

Figure 1.4. Venation system of the adult wing of *Philosamia cynthia*, showing both the tracks of the tracheae and the standard venation system. *A* and *B*, tracheation of pupal forewing and hind wing, respectively. Labels in parentheses refer to nomenclature of the primary tracheal system that develops during larval life; labels in brackets refer to nomenclature of the secondary tracheal system, which develops during pupal life and corresponds to the adult venation. *C* and *D*, venation system of fully mature adult forewing and hind wing, respectively, with conventional nomenclature of secondary tracheal system shown. Note losses of wing veins. (From Kuntze, 1935)

cies a few crossveins develop that close off the discal cell distally, but these crossveins never become tracheated (Köhler, 1932; Behrends, 1935). The adult venation pattern thus differs from that of the imaginal disk by the loss of the base of the media and the first anal vein. In most species it is possible to detect the traces of the atrophied veins in cleared wings, and in many species their former location is revealed by discontinuities in the color pattern (e.g., Fig. 2.10A,B).

_____
cells. The discal cell is the only wing cell with a special name.

Development and Morphology of the Scales

Wings of adult butterflies are covered with a dense array of partially overlapping scales. Each scale is a long, flattened projection of cuticle from a single specialized epidermal cell and emerges from the cell during the first few days of the pupal stage (Fig. 1.5B). The pigments that make up the color pattern occur only within these scales. Scale cell differentiation occurs during the first few days of adult development and coincides with a period of mitotic activity in the pupal wing epidermis. In nondiapausing pupae, extensive mitoses occur in the wing epidermis during the first few days after pupation. These mitoses increase both the cell number and the surface area of the developing wing. The increase in surface area is accommodated within the confines of the pupal cuticle by an array of closely spaced parallel folds perpendicular to the long axis of the wing. Köhler (1932) has presented evidence that at a certain point in development (60 to 72 hours after pupation) in *Ephestia,* the density (or frequency) of mitoses is higher in areas of the wing where dark pigment will subsequently develop. There is, however, no detectable relation between cell density and color pattern. It is possible that Köhler's finding is due to the fact that the cells in presumptive dark areas of the wing tend to undergo mitosis at slightly different times from those in presumptive light areas.

A subpopulation of the dividing epidermal cells undergoes two successive differentiation divisions that yield scale cells (trichogens) and socket cells (tormogens). In *E. kuehniella* these differentiation divisions are readily distinguished from proliferative mitoses in the remaining epidermal cells. The first differentiation division is oriented perpendicular to the surface of the wing, whereas normal epidermal cell divisions always occur in the plane of the wing. The daughter cell that is produced toward the inside of the wing soon degenerates, and the cell that is produced toward the surface of the wing undergoes a second differentiation division. This second division is inclined approximately 45 degrees to the surface of the wing and is oriented parallel to the long axis of the wing (Köhler, 1932; Braun, 1936; Stossberg, 1938). The mechanism that controls this orderly sequence of mitotic orientations is not understood, but the proximo-distal orientation of the spindle axis of the second division suggests that some kind of polarizing signal exists across the wing. The proximal daughter

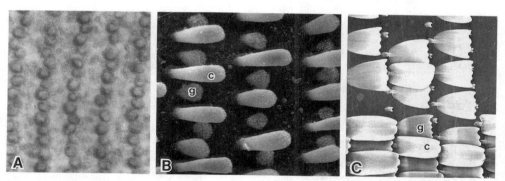

Figure 1.5. Wing scales occur in rows of alternating cover scales and (smaller) ground scales. *A,* whole mount of wing epidermis of *Precis coenia* (Nymphalidae) at about 36 hours after pupation, showing rows of enlarged scale-forming cells. Nuclei of smaller epithelial cells lie outside the plane of focus. *B,* scanning electron micrograph of wing surface at about 36 hours after pupation, showing scale buds just beginning to emerge. Alternation of large cover scales (*c*) and smaller ground scales (*g*) is evident. *C,* scanning electron micrograph of adult wing of dorsal forewing of *Archaeoprepona demophon* (Nymphalidae) with many scales removed to show rows of scale sockets and alternation of cover scales (*c*) and ground scales (*g*).

cell of the second differentiation division becomes the scale-building cell, and the distal cell becomes the socket cell. The scale-building cell then begins to send out a finger-like process, also oriented parallel to the long axis of the wing. This process soon flattens and continues to enlarge, forming the scale. In *E. kuehniella,* about 1 in every 4 epidermal cells differentiates into a scale-building cell (Pohley, 1959), whereas in *Precis coenia* about 1 in every 15 epidermal cells becomes a scale-building cell. Before forming the projection that will become the scale, the scale-building cells become polyploid and greatly enlarged (Fig. 1.5A). Henke (1946) and Henke and Pohley (1952) have shown that in moths, the size of the scale that will be produced is proportional to the degree of ploidy of the scale-building cell.

The scale-building cells of many butter-

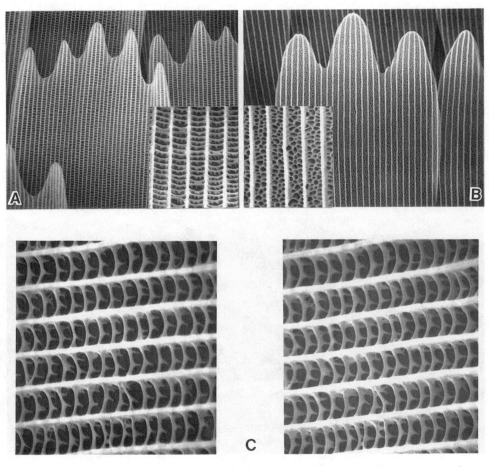

Figure 1.6. Scanning electron micrographs of scale types typical of the Nymphalidae, with square fenestration (*A*, *Precis coenia*), and of the Papilionidae, with a reticulate fenestration (*B*, *Papilio aristolochia*). *C*, stereo pair close-up of fenestration and supporting trabeculae in a scale of *Precis coenia*.

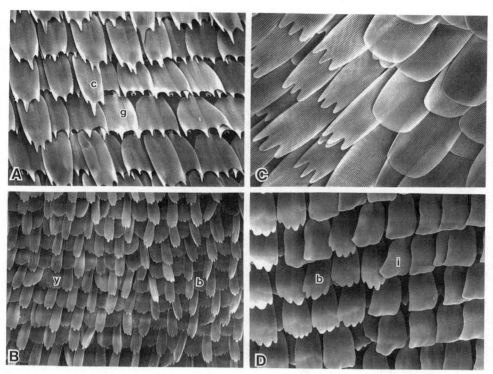

Figure 1.7. Scanning electron micrographs illustrating the diversity of scale morphologies within and among species. *A*, ventral hind wing of *Archaeoprepona demophon* (Nymphalidae) with different ground (*g*) and cover (*c*) scales (compare with forewing scales in Fig. 1.5C). *B*, wing of *Heliconius charitonia* (Nymphalidae). In this species, both the cover scales (thin, long) and the ground scales (short, broad) bear a bright pigment pattern. This photo is taken at the interface of a yellow (*y*) and a black (*b*) region, each occupying exactly half of the picture. Slight differences in the morphology of these two scale colors can be detected. *C*, ventral hind wing of *Colias eurytheme* (Pieridae) at the interface of the brown ocellus (rounded scales) and the yellow background (scalloped scales). *D*, forewing of *Trogonoptera brookiana* (Papilionidae) at the interface between an iridescent green area (*i*, smooth scales) and a velvety black area (*b*, scalloped scales).

flies are arranged in straight parallel rows perpendicular to the long axis of the wing, and the scales likewise form straight rows overlapping like roof tiles (Figs. 1.6 and 1.7). The rows are not entirely regular; there is occasionally some branching, as expected in a system that is attempting to maintain a constant distance between straight rows in a fan-shaped field. In a few taxa the scales are arranged in apparent random order. Among

these are the primitive Papilionidae (*Baronia, Parnassius, Luehdorfia, Zerynthia*), the species of *Acraea* (Nymphalidae), most members of the tribe Satyrini, and the genus *Melanargia* (Satyrinae). Because these taxa are phylogenetically heterogeneous, it is not clear whether the lack of scale rows is a shared primitive (symplesiomorphic) or convergent character among them. Yet, a random arrangement of scales is almost universal among the moths (and thus likely to be the primitive state of this character for the butterflies), suggesting that this character may be of some phylogenetic significance.

Within each row of scale-building cells there is a regular alternation of cell size (Süffert, 1937; Nijhout, 1980a). Two distinct size classes are usually visible that presumably differ in their degree of ploidy. The larger cells give rise to large cover scales; the smaller cells, to the ground scales or under scales (Figs. 1.5 and 1.7). On the adult wing the cover scales are often considerably longer than the ground scales and usually cover them completely. The cover scales are usually much more brightly colored than the ground scales, and in most species that have iridescent patterns, only the cover scales are specialized for iridescence, whereas the ground scales are dull (Süffert, 1924b; Ghiradella, 1985; Nijhout, 1985b).

The cell interaction mechanisms whereby scale-building cells come to be arranged in rows and whereby the nearly perfect alternation of cell size within rows is accomplished are not yet understood. Nardi and Magee-Adams (1986) have shown that epidermal cells of the pupal wing of *Manduca sexta* have long basal processes (the "epidermal feet" of Locke and Huie, 1981a,b) that extend several cell diameters and allow for direct cell-cell interactions with nonadjacent cells. These epidermal filopodia can span distances at least as long as the spacing between rows of scale-building cells. In *Manduca* these filopodia appear at exactly the time that the spacing pattern of scale-building cells is being established, and Nardi and Magee-Adams (1986) suggest that the filopodia may provide the morphogenetic forces for cell rearrangement during row formation. Thus, at the time of scale row determination the means exists for direct long-range and short-range cell-cell interactions of the type that is traditionally believed to be necessary for stable pattern formation (Meinhardt, 1982). Some years ago I investigated the presence of similar epidermal feet in *Precis coenia* and found none at the time of scale cell differentiation or during the period immediately preceding it. In *P. coenia,* scale-building cells do not migrate as they appear to do in *Manduca,* but they differentiate in rows (Nijhout, 1980a, and unpub. data). Thus at least two mechanisms may exist for establishing the straight rows of scale-building cells: random differentiation followed by migration into rows, and direct differentiation in rows possibly caused or guided by a chemical prepattern.

A few days after the initiation of adult development, each scale-building cell begins to send out a fingerlike protrusion directed toward the distal margin of the wing. This process gradually lengthens and then flattens into a spatula-shaped lamina, connected by a narrow neck to the scale-building cell (Köhler and Feldotto, 1937; Stossberg, 1937). This cytoplasmic lamina then acquires an ornately sculptured extracellular cuticle. The upper surface of each scale develops a system of closely spaced ridges or vanes parallel to the long axis of the scale. The undersurface of a scale is usually flat and featureless. The upper surface and the undersurface are held together by a system of trusslike trabeculae under the vanes (Fig. 1.6).

The complex extracellular architecture of a scale is manufactured by the cytoplasm of the scale cell and results from the behavior of cytoskeletal elements immediately below the plasma membrane. Much of the surface sculpturing of the scale cuticle occurs in the epicuticle, the outermost of the three layers of the cuticle. While the nascent scale is still a flattened sac, a system of microfilament bundles becomes arranged parallel to the long axis of the scale and immediately under the plasma membrane. The microfilament bundles become evenly spaced with a center-to-center spacing equal to that of the presumptive longitudinal vanes of the scale (Overton, 1966; Ghiradella, 1974). An extensive system of longitudinal microtubules develops, and although these are concentrated near the plasma membrane, they do not exhibit the highly ordered organization of the microfilaments (Overton, 1966). The longitudinal vanes arise as folds and subsequently develop into ridges in the cuticulin layer exactly halfway between the bundles of microfilaments. The cause of this buckling of the cell surface is not known. The buckling is presumed to be due to mechanical stress that is either caused by contraction of the microfilament bundles or somehow constrained by them (Ghiradella, 1974). Ghiradella and Radigan (1976) have presented evidence that the forces produced by surface tension could be responsible for generating the finer striations of the surface of the scale. In many species these striations are arranged so as to form an interference reflector that gives some scales a metallic or iridescent appearance (see "Structural Colors," below).

In a mature scale the cytoplasm of the scale-building cells dies and the scale becomes filled with air. The flat areas between the lamellae develop a complex fenestration (Fig. 1.6). The fenestration pattern, the spacing of the vanes, and the overall shape of the scale can be quite different among various taxa and in different locations on the wing (Süffert, 1924b; Downey and Allyn, 1975; Ghiradella, 1985). Of particular interest is that in some species, scales that differ in color also differ in these structural details (Fig. 1.7; Süffert, 1924b; Köhler and Feldotto, 1937; Descimon, 1965; Burgeff and Schnei-

der, 1979; Gilbert et al., 1988). The functional significance of this correlation is not clear at present, but the correlation does indicate that scales of different colors can differ in more than just the pigments they synthesize. Scales that are determined to become different colors may also differ in some of the variables responsible for the development of their shape and microarchitecture. Gilbert et al. (1988) have developed a genetic switching hypothesis that accounts for the particular correlation between scale color and structure in *Heliconius* butterflies (Fig. 1.8). Scales of different presumptive colors also develop at different rates. Presumptive pale-colored scales develop a thick cuticle more rapidly than presumptive dark-colored scales (Goldschmidt, 1938; Nijhout, 1980a).

Pigment synthesis in the scales takes place one or two days before the emergence of the adult butterfly and occurs in a stereotyped sequence. In all species that have been examined so far, red and yellow pigments are synthesized first, and brown and black pigments second (Goldschmidt, 1938; Nijhout 1980a, and unpub. data). The reds in all cases appear to be ommochromes; and the blacks and browns, melanins. Thus scales that have been determined to become different colors also come to differ in many structural and chemical features as well as in the rate of their development.

The scales at the wing margin are generally much longer and narrower than those that clothe the rest of the wing, and they are generally of a different color. In the vicinity of the wing margin in *P. coenia* a gradient of scale size exists in which scales become progressively longer as they are located closer to the margin. This fringe of marginal scales appears to be induced by a special signal that originates from the bordering lacuna that de-

Figure 1.8. Hypothetical switching diagram for the control of five scale colors in *Heliconius*. Starting from the left with an undetermined scale, the diagram shows the types of developmental "decisions" that must be made to arrive at the various colors of scales that could occur in a single wing. Each scale contains only a single kind of pigment. Slight differences in the spacing of vanes and in the details of the fenestrations distinguish three morphological types of scales among the five color types. 3-OH-K = 3-hydroxykyn-urenine, which can serve as a pigment in its own right (yellow) or as a precursor for ommochrome synthesis. (After Gilbert et al., 1988)

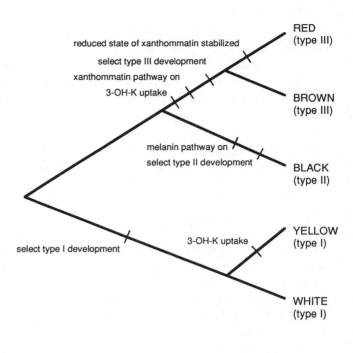

fines the wing margin (Nijhout and Grunert, 1988; Dohrmann and Nijhout, 1990).

Sources of Color

The colors on the wings of butterflies are entirely due to the color of their scales. The cuticle of the wing itself is usually colorless or brown (in *Graphium* and a few species of *Charaxes* the wing membrane is colored, but this coloration is more or less homogeneous and does not correspond to the color pattern), and epidermal cells are absent from the adult wing. The color pattern is in effect a finely tiled mosaic of monochrome scales. The color of a scale may be due to the presence of chemical pigments, or it may be a structural color that comes about when light interacts with regularly spaced physical structures in the scale.

Structural Colors

The work of Mason (1926, 1927a,b) remains one of the most thorough accounts of the origin and diversity of structural color in insects. White is perhaps the most common structural color and also one of the most difficult to characterize properly. Mason (1926) pointed out that white is observed when a colorless cuticle has many small and irregular surfaces that reflect light. In the absence of pigment, scales will appear white because of the reflection of light from vanes and trapped air bubbles. Whether the white of a scale is chalky, pearly, or silvery depends on the intensity of the scattered and reflected light and on the regularity of the scattering structures. Chemical whites are also in a sense structural colors, because their white appearance also comes about through the reflection of light from essentially colorless grains of pigment.

That whites are structural can be demonstrated by wetting a white portion of a wing with xylene (which has approximately the same refractive index as cuticle), in which case the white disappears and only the pale brown color of the cuticle remains.

All iridescent colors and almost all blues and greens are structural colors. Ghiradella (1984, 1985) surveyed the diversity of mechanisms used in the production of structural colors in butterfly scales. She recognized at least six distinct structural causes of color. Most families of Lepidoptera have evolved only one or two of these mechanisms, but the Papilionidae are apparently more versatile. Within this single family are found the whole diversity of structurally colored scale types reported previously in all other families of Lepidoptera. Ghiradella (1984, 1985) noted that three different structural components of a scale can become modified and elaborated into arrays of repeating elements that give rise to structural colors: the vanes (or longitudinal ridges), the microribs on the vanes, and the lamina, the main body of the scale (Fig. 1.9). This diversity and complexity of

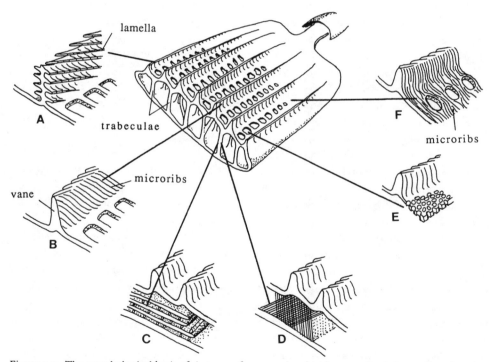

Figure 1.9. The morphological basis of six types of structural and iridescent color in butterfly scales. *A*, lamellar thin-film iridescence (the *Morpho* type), due to constructive interference of light reflected from lamellae on the vanes of a scale. *B*, microrib thin-film iridescence, due to an accentuation of the microribs on the vanes. *C*, laminar thin-film iridescence (the *Urania* type), due to constructive interference of light reflected from multiple thin films that are elaborated from the trabeculae. *D*, diffraction lattice iridescence, due to reflection from a tightly packed and evenly spaced lattice of rodlike trabeculae. *E*, Tyndall blue, due to reflection of light from small tubular holes in the scale surface. *F*, microrib satin iridescence, a soft iridescence due to an elaboration of closely spaced microribs across the body of the scale. (After Ghiradella, 1984, 1985)

fine structure places butterfly scales among the most complicated extracellular structures manufactured by a single cell.

In lamellar thin-film iridescent scales (Fig. 1.9A), raised parallel ridges (lamellae) form on the vanes and are arrayed to create a multiple ridge interference reflector (Plate 2). This type of scale is also known as the *Morpho* type and is found in species of Pieridae, Morphinae, Brassolinae, Amathusiini, Danainae, Nymphalinae, and Riodininae and, among the Papilionidae, in *Parides polyzelus* (Mason, 1927a,b; Lippert and Gentil, 1959; Ghiradella et al., 1972; Ghiradella, 1985). When the lamellar ridge spacing is very fine, these scales support constructive interference in the ultraviolet region of the spectrum and are responsible for the UV reflectance patterns of *Colias* and *Eurema* (Ghiradella et al., 1972).

In microrib thin-film iridescent scales (Fig. 1.9B), the ribs on the vanes are increased in height and angled to form a partially overlapping array that acts as a multiple thin-film interference reflector. This type of scale has been found in the Brassolinae, Hesperiidae, and some Nymphalinae. The blue scales of *Trogonoptera brookiana* (Papilionidae) are of this type.

A different modification of the interlamellar ribs gives rise to microrib satin scales (Fig. 1.9F). Here the microribs are not lengthened but have become closely spaced and extend across much of the body of the scale as well. This structure gives the scale a satin appearance that is best known from the ventral hind wing of *Battus philenor* (Papilionidae). Satin scales are also known to occur in some Nymphalinae (Ghiradella, 1985).

In laminar thin-film iridescent scales (Fig. 1.9C), the interior of the scale contains stacks of evenly spaced sheetlike laminae held apart by irregular ridges that act as spacers. This type is also known as the *Urania* type and,

among the butterflies, is found in the Lycaenidae (*Arhopala* sp.), in *Ornithoptera priamus*, and in several members of the genus *Papilio*—*P. ulysses*, *P. hoppo*, *P. paris*, and *P. palinurus* (Mason, 1927a,b; Lippert and Gentil, 1959; Huxley, 1975; Ghiradella, 1985).

In diffraction lattice scales (Fig. 1.9D), the interior of the scale is completely filled with a very regular crystalline lattice of chitinous scale material, perhaps derived from the trabeculae that normally provide the strutwork between upper and lower laminae. This lattice produces diffraction colors, and this type of scale has been found in the lycaenids *Callophrys rubi* and *Mitoura siva* (Morris, 1975; Allyn and Downey, 1976) and in the papilionid *Parides sesostris* (Ghiradella, 1985).

Finally, the perforations in the lamina between the ridges can become diminished in size and arranged into a closely spaced array of short tubules (Fig. 1.9E) that scatter light and cause a Tyndall blue (Huxley, 1976). Such pale Tyndall blue scales have so far been found only in the Papilionidae—for example, the bluish scales of *P. zalmoxis* (Ghiradella, 1985).

That all six types of scales are widespread among the butterflies and that they can also be found in members of a single family, the Papilionidae, suggest that this diversity of structure is not likely to be of use in reconstructing the phylogenetic relations among butterflies (Ghiradella, 1985). The repeated and apparently independent evolution of each of these scale types indicates that the structural specializations may come about through relatively simple changes in scale development. In support of this suggestion, Ghiradella (1974) noted that the architecture of many reflective scales differs only in degree from that of nonreflective scales. In most cases a simple accentuation of existing folds and striations, or an alteration in their spacing and inclination, is all that distinguishes an iridescent scale from a noniridescent one. In *Colias eurytheme* and *C. philodice,* differences at a single genetic locus are responsible for differences in the UV iridescence of the scales of these species (Silberglied and Taylor, 1973). Changes in the thickness of the epicuticle and in the timing and magnitude of the stress patterns that set up the fine surface sculpturing of the scale might be all that is required to accomplish a switch between iridescent and noniridescent scales.

Most scales that are iridescent are also melanized. The melanin in the background absorbs much of the light that is not reflected, and this causes the iridescence to appear especially brilliant. This enhancing effect of an absorbing background pigment is most readily observed in the white *Morpho* species, where the otherwise bright metallic blue is difficult to see on the bright reflective background. In a few cases, iridescent scales also bear a pigment, and the structural and pigmentary colors interact to give a particular hue. Metallic violets, for instance, are due to a blue iridescence combined with a red pigment (Nijhout, 1985b).

Pigments

The major pigments in the wings of butterflies belong to four chemical categories: melanins, ommochromes (and their precursors), pterins, and flavonoids. A few species have pigments that do not belong to these categories; some of these are bile pigments, and others have not yet been identified (Ford, 1942, 1944a; Umebachi and Takahashi, 1956; Choussy and Barbier, 1973; Vane-Wright, 1979). Here we will deal only with the origin, distribution, and characterization of pigments that belong to the four major categories.

Melanins

These are by far the most common pigments and are responsible for most of the detailed patterns found among the butterflies. Melanins are mostly black (eumelanin) and brown, tan, or reddish brown (phaeomelanin). Eumelanin is a complex polymer of o-diphenols or indoles that are formed by the action of phenoloxidases (tyrosinases). Tyrosine and dopamine are the most common precursors for eumelanin synthesis (Fig. 1.10), but a variety of o-diphenols may be used (Needham, 1974; Nijhout, 1985b). Eumelanins have a nearly flat absorption spectrum throughout the visible range, which is diagnostic. The structure of phaeomelanins is poorly understood. It is believed that the synthesis of nonblack melanins requires the incorporation of additional kinds of molecules in the polymer. Mixtures of sulfur-containing amino acids and tyrosine and of 3-OH-kynurenine and tyrosine are said to be converted to reddish melanins by phenoloxidase (Inagami, 1954; Prota and Thomson, 1976; Riley, 1977), as is N-acetyldopamine (Nijhout, 1980a). The principal difficulty met in the analysis of melanin structure is that melanins are soluble only in hot alkali and some strong acids. Because these solvents also destroy the native configuration of melanin and some of its component parts, it is virtually impossible to reconstruct the original structure from the solubilized form.

In principle the synthesis of melanin requires only the action of a single enzyme (phenoloxidase) on a single substrate (an o-diphenol). The control of melanin synthesis, however, can be rather complex. Phenoloxidases are synthesized in an inactive form

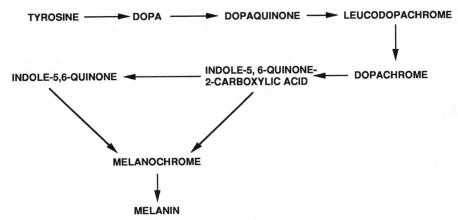

Figure 1.10. Pathway of melanin synthesis. (After Needham, 1974, and Kayser, 1985)

(prophenoloxidase) that must be activated by proteolysis. Specific proteases exist for this function, and these in turn may be part of a multistep activation cascade. A three-step cascade has been described in *Drosophila* (Seybold et al., 1975), and there is evidence for a two-step cascade in Lepidoptera (Saul and Sugumaran, 1987). In addition, the activation of hemolymph phenoloxidase in Lepidoptera appears to be under inhibition by a serine protease inhibitor (Sugumaran et al., 1985; Saul and Sugumaran, 1986, 1987). Finally, in most systems (though not in butterfly scales), melanins occur only in membrane-bound granules, and details of granule size and distribution have considerable influence on the actual phenotype produced by melanization. In practice, therefore, melanization can be subject to a complex system of enzymatic and structural controls requiring the timely expression of many gene products.

The melanins in butterfly scales are diffusely distributed throughout the cuticle of the scale (Nijhout, unpub. data). In the wings of *Precis coenia*, phenoloxidase is active at least one day before the initiation of melanin synthesis. Phenoloxidase activity in the wing is patterned in perfect correspondence with the melanin pattern that develops; when an unpigmented wing (taken 12 to 24 hours before the initiation of normal pigment synthesis) is incubated in a solution of dopamine, a normal melanin pattern appears within a few hours (Nijhout, 1980a). Whether the pattern of phenoloxidase activity is due to patterned synthesis of an active form of phenoloxidase or to patterned activation of the enzyme is unknown.

Ommochromes

Ommochromes are red-to-brown pigments that are derived from tryptophan via the kynurenine pathway (Fig. 1.11). An intermediate in this pathway, 3-OH-kynurenine, is yellow and is also used as a pigment in butterfly wings. The diversity and distribution of ommochromes among the butterflies have not yet been systematically investigated, but so far ommochromes appear to be restricted to the Nymphalinae (Linzen, 1974; Needham, 1974). Rhodommatin and ommatin D have been found in *Vanessa, Aglais, Inachis,* and *Argynnis*. In *Heliconius,* reds and some orange-browns are the oxidized and reduced states, respectively, of xanthommatin, and yellows are 3-OH-kynurenine (Brown, 1981; L. E. Gilbert, pers. com.). Ommochromes are characterized by their solubility in acidified methanol and may be identified by paper chromatography (Linzen, 1974) or cellulose thin-layer chromatography. Perhaps the most accurate way to identify the presence and distribution of ommochromes in a wing is to feed larvae (or inject pupae) with radiolabeled tryptophan and subsequently perform autoradiography on the adult wing. Radiolabel should be concentrated in scales that contain ommochromes and 3-OH-kynurenine, because these are the only products of the kynurenine pathway that are normally present in wings. The red pigment in the wings of *P. coenia* had been tentatively identified as a phaeomelanin, based on its insolubility in all but the harshest solvents (Nijhout, 1980a). We (P. B. Koch and Nijhout, unpub. data) have recently demonstrated by autoradiogra-

phy that this red pigment is, in fact, an ommochrome (Fig. 1.12).

A distinct class of pigments, the papiliochromes, is derived from kynurenine by combination with dopa (dihydroxyphenylalanine) and related phenolic amino acid derivatives (Umebachi, 1980; Kayser, 1985). These pale yellow to reddish brown pigments occur throughout the Papilionidae. Most of the tests that have been done for papiliochromes, however, do not clearly distinguish between them and other kynurenine derivatives in the ommochrome pathway.

Pterins

Pterins are white and yellow-to-red pigments derived from guanosine triphosphate (Needham, 1974; Descimon, 1977). The general pathway of pterin synthesis is shown in Figure 1.13. Pterins are widespread among the

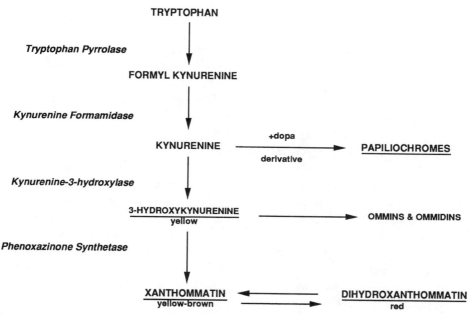

Figure 1.11. Pathway of ommochrome and papiliochrome synthesis, with the enzymes that catalyze each reaction. (After Needham, 1974; Fuzeau-Braesch, 1985; and Kayser, 1985)

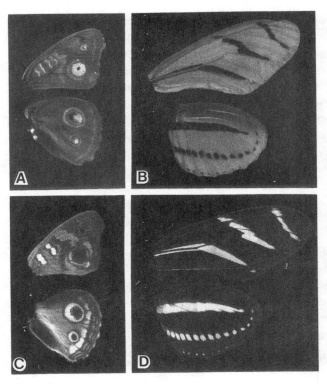

Figure 1.12. Negative autoradiograms of the wings of *Precis coenia* (*A* and *C*) and *Heliconius charitonia* (*B* and *D*) after injection with radiolabeled (^{14}C) tyrosine (*A* and *B*), which is incorporated into melanin, and radiolabeled (^{14}C) tryptophan (*C* and *D*), which is incorporated into 3-OH-kynurenine and ommochromes. In each case the bright portions of the picture are the sites of radiolabel incorporation. In *H. charitonia* there is a clear separation of areas in which only one or the other label is incorporated. In *P. coenia* the tryptophan label is found in all the red, orange, and buff areas of the wing, indicating that the pigment there is an ommochrome, probably xanthommatin. The strongest incorporation of the tyrosine label is in the black areas of the pattern, the eyespots and the lines that flank the bars in the discal cell.

insects, but as wing pigments they have been found only in the Pieridae (Watt, 1964, 1967, 1974; Descimon, 1975a,b, 1977; Kayser, 1985), in the Libytheinae (Shields, 1987), and in the nymphalid genus *Heliconius* (Baust, 1967). The latter finding has been questioned by Descimon (1975a), who was unable to identify pterins in *Heliconius* and who suggested that an ommochrome may have been mistaken for erythropterin. Pterins are easily extracted from wings in aqueous buffers of high and low pH and can be identified by their fluorescence after thin-layer chromatography using samples from small areas of a wing (Watt, 1964, 1967; Descimon, 1973, 1975a). A nice in situ demonstration of pterins was developed by Ford

(1947a). To do this test, it is necessary to expose the wing to molecular chlorine (most readily obtained by treating potassium permanganate with hydrochloric acid—in a fume hood, of course!) for about 20 minutes, followed by fuming with ammonia vapor for a few minutes. Shortly after this procedure, murexide is formed, producing a purple color that intensifies over a period of hours. This test is specific to pterins and allows their identification in single scales. Autoradiography of wings or chromatograms after injection of radiolabeled guanosine triphosphate also provides a sensitive and accurate way of characterizing pterins and pterin synthesis pathways in situ (Watt, 1967; Descimon, 1975a; Kayser, 1985).

Flavonoids

Flavonoids are derivatives of 2-phenylbenzopyrone (flavone). These pigments are widespread among the plants; they are diverse and colorful and include, among others, the yellow, red, and blue anthoxanthins and anthocyanins. Butterflies, like all other animals, are unable to synthesize flavonoids and must obtain them from their food plants (Ford, 1941, 1947b; Needham, 1974; Kayser, 1985). White and yellow flavonoids are fairly widespread among the butterflies; they have been described in a few Satyrinae, Riodininae, and Lycaeninae (Feltwell and Valadon, 1970; Kayser, 1985) and in several species of Dismorphiinae (Pieridae), and they are characteristic of the papilionids Graphiini, *Parnassius,* and Old World *Parides* (Ford, 1941, 1942, 1944a,b, 1947b). The whitish flavonoids in the wings of *Melanargia* (Satyrinae) have been identified as quercitin and tricin (Thomson, 1926a,b; Morris and Thomson, 1963). Tricin has also been identified in *Coenonympha pamphilus* (Morris and Thomson, 1964), and quercitin has been shown to be present as a wing pigment in the lycaenid *Polyommatus icarus* (Feltwell and Va-

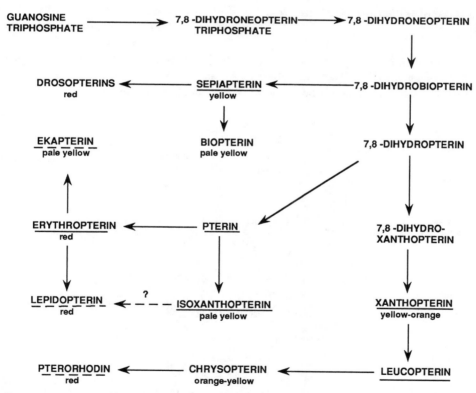

Figure 1.13. Pathway of pterin biosynthesis. Compounds that are underlined have been found as pigments in butterfly wings; those with a dashed underline have been found in butterflies, but not as wing pigments. Unless noted otherwise, the compounds are colorless, and it is possible that some are used as structural whites in some species. (After Needham, 1974, and Descimon, 1977)

ladon, 1970). The presence of flavonoids in a wing can be demonstrated by a simple non-destructive procedure developed by Ford (1941). When wings are fumed with strong ammonia vapors, flavonoids turn a bright yellow. This reaction is specific and diagnostic, can be used to detect minute quantities of flavonoids, and works wonderfully well as a demonstration in *Melanargia*. If the flavonoid is itself yellow or if it is masked by other pigments, it can be extracted with ethyl acetate and reacted with sodium carbonate, which intensifies the yellow color (Ford, 1941).

Pattern Formation in a Cellular Monolayer

Before considering the development and evolution of wing patterns, it will be useful to discuss the constraints imposed by the anat-omy of the wing on the process of pattern formation. The problem in pattern formation is to explain the origin of organized structure within an initially homogeneous and unstructured system. In the butterfly wing, three major pattern formation processes are of interest: the determination of wing shape, the determination of venation pattern, and the determination of color pattern. All three processes take place in the imaginal disk sometime during larval life (Fig. 1.14).

Wing shape is determined by the path of the bordering lacuna (Figs. 1.2 and 1.3), which defines the boundary between the wing proper and the peripheral area of programmed cell death. No work has yet been done on determination of the position of the bordering lacuna. In *Papilio dardanus* and *P. memnon* the tailed condition of the hind wing is determined genetically (see Chapter 6), and it appears that changes at a single gene can affect the tail loop in the bordering

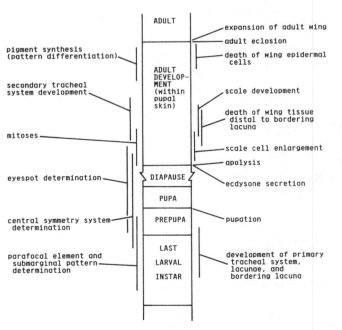

Figure 1.14. Timetable of events during pattern development.

lacuna. Determination of the bordering lacuna occurs at the same time and presumably by the same processes that control determination of the venation system. As noted above, the venation system arises as a local separation of the basal laminae, but what controls its pattern is yet unknown. Surgical ablation of parts of the wing disk during larval life often results in abnormal venation patterns (Henke, 1933a; Nijhout and Grunert, 1988). No systematic analysis has yet been done on such aberrant venation patterns; on first sight they appear to be poorly reproducible and erratic in their expression, but a black-box approach to analysis (Ashby, 1966) might yield some insights here. Knowledge of how the venation system is determined is important for understanding the development and evolution of wing patterns because, as will be seen below, wing veins (and the bordering lacuna) serve as boundaries and sources of inductive signals in the process of color pattern determination.

Determination of the color pattern appears to start before the lacunae of the venation pattern develop, but most color pattern determination takes place after the formation of the venation pattern. Because the color pattern consists of a mosaic of colored scales, pattern formation must consist of a process whereby scale cells become specified to synthesize or produce a particular array of pigments or structural colors. Because the wing disk lies free in the hemocoel of the larva and has no known interaction with its immediate environment, color pattern formation must be a process intrinsic to the cells of the disk itself.

What are the cellular processes that could play a role in pattern organization? Each surface of the wing is an epithelium that is only one cell thick. The two surfaces are separated from each other by the basal laminae of each sheet and, during some portions of development, by a substantial hemolymph space. At the time of color pattern determination the two epithelia are tightly apposed at their basal laminae except at the locations of the lacunae where hemolymph circulates. During the period of pattern determination the wing disk grows through cell division. There is no evidence of cell migration or cell rearrangement during this period, so we must assume that pattern determination takes place in an essentially static sheet of cells and that patterns are not due to patterned cell migration. This means that direct cell-to-cell communication and communication through the hemolymph provide the only means of signal transmission for pattern organization.

The most obvious means of cell-to-cell communication in an epithelium is through gap junctions, which allow the passage of small molecules between cells. Gap junctions are common in insect epidermis, and I have found them in the wing imaginal disks of *Precis coenia*. Insect gap junctions have a molecular weight cutoff of about 1,400 daltons, which means that they should allow the passage of almost all nonpolymeric organic molecules and some very small polypeptides. Passive diffusion through the cytoplasm of the epidermal cells can serve as a cell-to-cell communication mechanism and also as a means of long-range communication across a disk (see Chapter 7). Epidermal feet (Locke and Huie, 1981a,b; Nardi and Magee-Adams, 1986) provide another means for short-range communication by direct cell contact. Epidermal feet may extend several cell diameters and are believed to be involved in the development of the rows of scale cells. It is unlikely, however, that epidermal feet provide a pathway or mechanism for long-range cell-to-cell communication during color pattern determination: They are not present until pattern determination is over, and in any event, they extend only a few cell diameters and are thus very short relative to the scale of the color pattern.

Pattern formation in an initially featureless system must start with the establishment of a discontinuity that can act as an organizing center. In some pattern determination systems the initial discontinuities can be random (Meinhardt, 1982), but more often than not, a physical structure provides a landmark around which pattern organization takes place. For color pattern formation, discontinuities of interest are provided by the margins of the disk and by the venation system. The most obvious distinguishing property of the venation system is that it provides the principal avenue for the supply of extrinsic materials, nutrients, and oxygen (after tracheation becomes established) to the developing wing disk. (Transfer of material across the surface of the disk is inhibited by the enveloping peripodial membrane, Fig. 1.1F.)

There is no reason to believe that the flow pattern of hemolymph in the wing provides an organizing signal for pattern formation. The pattern of circulation in adult wings has been studied by Wasserthal (1982). He showed that the hemolymph is periodically sucked out of all veins simultaneously by the accessory pulsatory organs of the thorax. The hemolymph then flows back into the wing passively. Wasserthal suggested that this oscillating pattern may be an adaptation to prevent the loss of hemolymph in the event of peripheral damage to the wing, which inevitably occurs in adult Lepidoptera. The pattern of circulation in the imaginal disk is unknown. It is possible that in the wing disk the flow pattern is unidirectional, with the return path provided by the bordering lacuna

and by the interstitial spaces between cells. If flow in the imaginal disks is indeed unidirectional, then the principal flow must be in different directions in different wing veins. Because there are no differences in the polarization of the color pattern along different wing veins, the direction of flow of hemolymph probably does not affect the developmental signals responsible for pattern formation.

The lacunae of the wing venation system do, however, play an important role in color pattern determination because they divide the wing disk into compartments—the wing cells—that are the functional units of pattern formation. As will be seen in Chapter 2, the primitive and simplest wing pattern consists of a serial repetition of identical patterns from wing cell to wing cell. In addition, distinctive pigments often develop on or next to the course of these lacunae, so that the venation pattern is also expressed in the color pattern. Compartmentalization of the pattern and vein-dependent patterns both argue for a special role of the wing veins in color pattern determination.

Color pattern formation must ultimately be explicable by processes that use the venation system and bordering lacuna as the initial organizing system and that use cell-cell interactions in a static monolayer as the only means of communication. After reviewing the comparative morphology, development, and genetics of color patterns in the chapters that follow, I will show that it is indeed possible to develop a theoretical model that starts with only these simple structural principles and can generate a broad diversity (perhaps the entire diversity) of butterfly color patterns.

Elements of the Nymphalid Ground Plan

Even the most complicated wing patterns can be dissected into a relatively small number of pattern elements. These pattern elements form a system of homologies that can be identified across thousands of species and are every bit as consistent as the homologies of bones in the tetrapod limb. We owe the elucidation of this homology system to the pioneering work of B. N. Schwanwitsch and F. Süffert, who in the 1920s independently and nearly simultaneously discovered the organizing principles known as the nymphalid ground plan. Priority in this discovery must go to Schwanwitsch, whose publication of a scheme of homologies in the Nymphalidae preceded Süffert's by about three years. Süffert's (1927) system, almost identical to that of Schwanwitsch (1924), is more memorable, however, because Süffert provided a nomenclature for the pattern elements that is more sensible and conveys a much deeper insight into the structure of pattern. His nomenclature was especially powerful in that it led directly to hypotheses about the developmental origin of patterns that served as the basis for extensive experimental studies by A. Kühn, K. Henke, and their co-workers, discussed below.

Both Schwanwitsch and Süffert visualized the nymphalid ground plan as being based on a system of bands and spots that run from the anterior to the posterior margin of each wing. The general scheme, with a nomenclature largely derived from that of Süffert, is illustrated in Figure 2.1. Table 2.1 compares the nomenclature of Süffert with that of Schwanwitsch and with the two systems of nomenclature that are used in this book. For reasons that will become clear as we proceed, I have found it necessary to depart occasionally from

Pattern Elements and Homologies

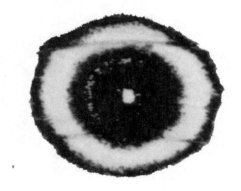

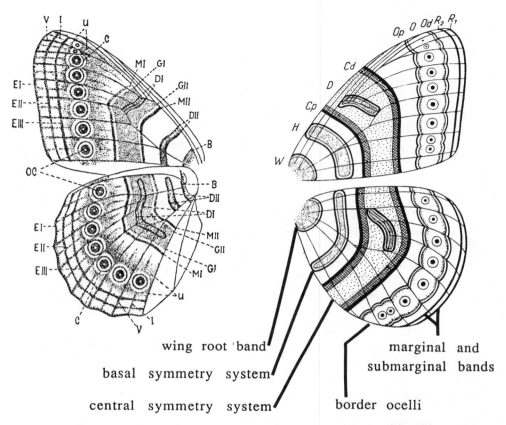

wing root band

basal symmetry system

central symmetry system

marginal and submarginal bands

border ocelli

Figure 2.1. The nymphalid ground plan according to Schwanwitsch (left) and Süffert (right). Nomenclature used by each author, corresponding to the labels in these figures, is given in Table 2.1. (From Schwanwitsch, 1924; Süffert, 1927)

the descriptive terms given in Figure 2.1 and to designate pattern elements simply by a letter code. As Table 2.1 shows, this letter code is strictly synonymous with the nomenclatures in Figure 2.1, and I will use one or the other system in various parts of this book, depending on which system is better suited for clarity of presentation.

The pattern elements of the nymphalid ground plan are the following:

1. The *discal spot* usually consists of a simple dark spot or bar on the crossveins that close off the discal cell. Species that have an open discal cell nevertheless always

have a normal discal spot in the place where the crossveins would otherwise be found. In some species the discal spot is elaborated into an eyespotlike pattern that can be the dominant feature of the wing pattern. Eyespotlike discal spots are common in the Saturniidae, and among the butterflies they are notable in *Mesosemia* species (Riodininae). In some species all or part of the discal spot is devoid of scales, forming a clear "window" in the wing.

2. The *central symmetry system* consists of a pair of bands, usually broad and prominent, that run anterior to posterior in the middle of the wing on either side of the discal spot.

Table 2.1

Nomenclature of the pattern elements of the nymphalid ground plan

Schwanwitsch (1924)	Süffert (1927)	Nijhout (1978, 1985b); this work	Nijhout and Wray (1986); this work
Basalis (B)	Wurzelbinde (W)	Wing root band	a[a]
Discalis II (DII)	Hohlbinde (H)	Basal symmetry system	b, c
Media II (MII); Granulata II (GII)	Innenbinde des centralen; Symmetriesystems (Cp)	Proximal band of central symmetry system	d
Discalis I (DI)	Discoidalfleck (D)	Discal spot	e
Media I (MI); Granulata I (GI)	Aussenbinde des centralen; Symmetriesystems (Cd)	Distal band of central symmetry system	f
(Not recognized)	Innere Binde des ocellaren Symmetriesystems (Op)	(Not named)	g
Circulus (C); Ocellata (OC)	Ocellenreihe (O)	Border ocelli	h
Externa III (EIII)	Aussere Binde des ocellaren Symmetriesystems (Od)	Parafocal element	i
Externa II (EII); Externa I (EI)	Randbinde 2 (R₂); Randbinde 1 (R₁)	Submarginal and marginal bands	j, k
Venosa (V)	(Not recognized)	Venous stripe	(No code)
Intervenosa (I)	(Not recognized)	Intervenous stripe	(No code)

Note: The morphology and disposition of these pattern elements are illustrated in Figures 2.1 and 2.17. The letter codes in parentheses refer to the labels used by Schwanwitsch and Süffert in Figure 2.1.

[a]This pattern element occurs in some moths but is rare or absent in the butterflies.

The term "symmetry system" is derived from the pigment composition of the two bands and does not refer to their shape, which is, as a rule, quite asymmetrical. The bands of the central symmetry system are usually composite, that is, made up of parallel bands of different colors that produce a shadowing or highlighting effect. The color composition of the distal band is always a mirror image of that of the proximal band, so that an axis of symmetry for coloration is created in the field between the two bands. Furthermore, the background coloration of the field between the two bands is usually slightly different in color (darker, as a rule) than that of the surrounding wing surface, so that the bands of the central symmetry system could be interpreted as the boundaries of a broad central color field. The two bands of the central symmetry system can be partially fused (as in *Charaxes analava*, Fig. 2.5A). With more extensive fusion the central symmetry system is reduced to a system of circular patterns (Fig. 2.5B,C), as in *Mantaria* (Satyrinae), *Smyrna* (Limenitinae), and *Mesosemia* (Riodininae).

3. The *wing root band* is the most basal pattern element. It appears to be fairly rare in butterflies but is quite common in moths. In some cases, such as the Geometridae, the wing root band may actually consist of a pair of bands, forming a small symmetry system.

4. The *basal symmetry system* is a pair of bands just distal to the wing root band. The bands of the basal symmetry system are often connected near the anterior and posterior margin of the wing, and occasionally the pair of bands is broken up into a set of circular spots (Nijhout, 1978).

5. The *border ocelli system* (or *ocellar symmetry system*) consists of a series of eyespots in the distal half of the wing. They are particularly common in the Satyrinae and related subfamilies. In the most general case, there is one ocellus per wing cell. The center of an ocellus is usually clearly marked by a white spot (the focus) and always lies on the midline of the wing cell. As can be seen in Figure 2.1, the most posterior wing cell in both forewing and hind wing between veins Cu_2 and 2A appears to bear two ocelli, rather than the singular one that is characteristic of the other wing cells. This variation occurs because early in the development of the wing an additional wing vein (1A) exists between Cu_2 and 2A, but it atrophies during the early pupal stage (see Chapter 1). The tract of this vein is often discernible on a cleared wing and passes between the two ocelli. When ocelli are complex—that is, when

they are composed of several concentric rings—the outer rings of adjacent ocelli may fuse to form a single very large ocellus (as on the hind wings of many species of *Precis*), or if there is a row of ocelli, they may form a set of scalloped bands that flank the row of ocelli proximally and distally (Fig. 2.18A). Süffert (1927) referred to the border ocelli and the concentric rings of pigment around them as the ocellar symmetry system. In most taxa outside the Satyrinae the border ocelli are not perfectly round, and in many taxa their shape is so distorted as to be barely recognizable as homologs to these neat circular patterns. The developmental and theoretical analysis of these deviations from circularity will occupy much of Chapter 7.

6. The *marginal and submarginal bands* are a pair of narrow bands, one usually at or very near the wing margin and the other a short distance removed from it. These bands are often reduced to dots or short dashes on the wing cell midline or the wing veins.

7. The *parafocal element* is a pattern element that occurs between the border ocelli and the submarginal band (Nijhout, 1985b). Neither Schwanwitsch nor Süffert recognized this as a pattern element distinct and independent from the others. Schwanwitsch (1924) considered it to be the innermost band of the submarginal band system and named it Externa III, whereas Süffert (1927, 1929b) considered it to be the distal member of the ocellar symmetry system. It is, in fact, the only element of the color pattern on whose character and affinities Schwanwitsch and Süffert disagree. More recent studies (Nijhout, 1984; Nijhout and Grunert, 1988) have shown that the parafocal element develops independently from the border ocelli and the submarginal system, and comparative morphological studies presented below have shown that this pattern element varies and has evolved indepen-

dently from the other elements of the color pattern. The parafocal element stands out as being, without question, the most structurally diverse and complex of the pattern elements. Theoretical studies of its development and of the origin of its diversity have been instrumental in the development of a cohesive model for color pattern formation (see Chapter 7).

The nymphalid ground plan represents the maximal pattern that is found in the family Nymphalidae. However, no species is known that has all the elements of the nymphalid ground plan in its wing pattern. Usually border ocelli are not expressed in all wing cells, and the two most basal pattern elements (the wing root band and the basal symmetry system) occur only rarely. The distinctive color patterns of each species arise from the selective expression and distortion of the various elements of the ground plan. Each element appears to be able to change in evolution in-

dependently from the others, and it is the permutation of size, shape, color, and presence or absence of this relatively small set of pattern elements that is responsible for much of the observed diversity of wing patterns in the butterflies and moths (Plates 3 and 4A). Some examples of the diversity that can be generated by selective expression of different elements of the ground plan are shown in Figure 2.2. Over the years, the nymphalid ground plan has been shown to be an excellent representative of the organization of color patterns in almost all major families of moths and butterflies (Figs. 2.3 and 2.4). Detailed analyses of patterns in the Noctuidae, Geometridae, Arctiidae, Nymphalidae, Papilionidae, Pieridae, and Lycaenidae are given by Schwanwitsch (1926, 1929a,b, 1930, 1943, 1949, 1956a,b), Süffert (1927, 1929b), Henke (1928, 1936), Sokolow (1936), and Henke and Kruse (1941). Each taxon, even within the Nymphalidae, tends to deviate from the ground plan in stereo-

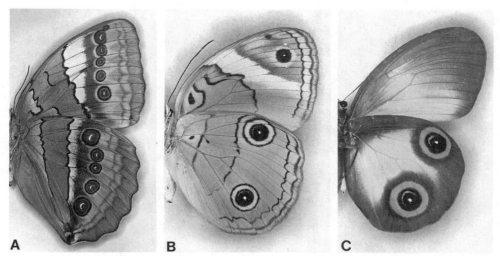

A   B   C

Figure 2.2. Three patterns derived from the nymphalid ground plan by progressive loss of pattern elements. *A, Stichophthalma camadeva* (Nymphalidae: Morphinae) has nearly the entire complement of pattern elements, except for those at the base of the wings. *B, Faunis menado* (Nymphalidae: Morphinae) has lost all but three of its border ocelli. *C, Taenaris macrops* (Nymphalidae: Morphinae) has lost all elements of the ground plan except for two border ocelli on the hind wing.

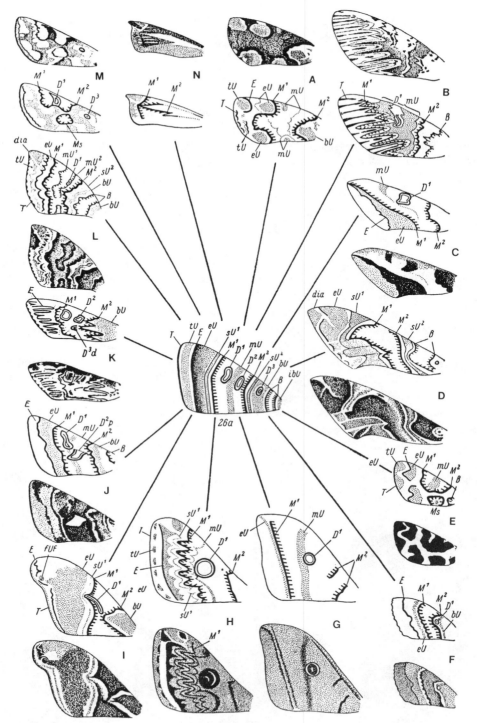

Figure 2.3. Derivation of the wing patterns of various moths from the nymphalid ground plan. For nomenclature of the pattern elements, see Figure 2.1 and Table 2.1. *Clockwise from 12 o'clock: A, Thyatira batis* (Thyatiridae); *B, Cerura vinula* (Notodontidae); *C, Deilephila euphorbiae* (Sphingidae); *D, Daphnis nerii* (Sphingidae); *E, Arctia caja* (Arctiidae); *F, Dendrolimus pini* (Lasiocampidae); *G, Antheraea* sp. (Saturniidae); *H, Caligula boisduvali* (Saturniidae); *I, Philosamia cynthia* (Saturniidae); *J, Phytometra aemula* (Noctuidae); *K, Cucullia asteris* (Noctuidae); *L, Lygris prunata* (Geometridae); *M, Nymphula nymphaeta* (Pyralidae); *N, Crambus hamellus* (Pyralidae). (From Schwanwitsch, 1956a)

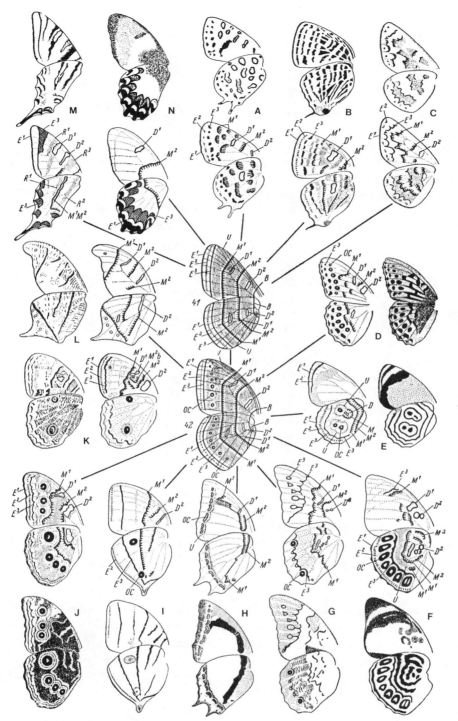

Figure 2.4. Derivation of the wing patterns of various butterflies from the nymphalid ground plan. For nomenclature of the pattern elements, see Figure 2.1 and Table 2.1. *Clockwise from 12 o'clock: A, Aphari-tis cilissa* (Lycaenidae); *B, Japonica saepestriata* (Lycaenidae); *C, Hesperia sidae* (Hesperiidae); *D, Argynnis paphia* (Nymphalidae); *E, Diaethria meridionalis* (Nymphalidae); *F, Agrias claudina* (Nymphalidae); *G, Prepona eugenes* (Nymphalidae); *H, Polyura jalysus* (Nymphalidae); *I, Amathuxidia dilucida* (Nymphalidae); *J, Morpho* sp. (Nymphalidae); *K, Opoptera acadina* (Nymphalidae); *L, Caerois chorinaeus* (Nymphalidae); *M, Iphiclides podalirius* (Papilionidae); *N, Papilio* sp. (Papilionidae). (From Schwanwitsch, 1956a)

typed ways, and for most taxa it is possible to devise a subordinate ground plan that captures the major pattern theme on which variations have evolved in that particular taxon. Such subordinate ground plans always have fewer components than the standard nymphalid ground plan. Some examples of subordinate ground plans are given in Chapter 3.

Although the structural schemes that Schwanwitsch (1924, 1926, 1929c) and Süffert (1927, 1929b) recognized as the nymphalid ground plan were nearly identical, the nomenclature they used reveals different perspectives in their analyses. Schwanwitsch (1924), for instance, placed greater emphasis on the shading of various bands and of various areas of the wing. This emphasis is revealed by his use of names like "granulata" and "umbra" for areas that are no longer recognized as real pattern elements (they refer to the central field of the central symmetry system and to the surrounding field of the border ocelli, respectively). Schwanwitsch's nomenclature also grouped the pattern elements into families suggesting that the elements were somehow related to each other (like Externa I, II, and III, or Media I and II), without commenting further on the nature of that relationship or the rationale behind the choice of names. Süffert (1927, 1929b), by contrast, recognized that the relationships between these families of pattern elements resided in the fact that several of them constituted symmetry systems. A symmetry system consists of a pair of bands whose pigmentation and fine structure are mirror images of one another. The term "symmetry" refers only to the coloration and not to the shape or size of the two bands, which can be different. Süffert recognized two symmetry systems explicitly, the central symmetry system and the ocellar symmetry system, and one implicitly, the basal symmetry system, which he called the *Hohlbinde,* or hollow band. The "hollow-

ness" of this band arises from its being made up of two bands that have fused near the anterior and posterior wing margins, so that the symmetry system now forms a closed elliptical figure.

Similar closures or partial fusions of the proximal and distal bands are common in the central symmetry system (e.g., in *Charaxes analava,* Fig. 2.5A), and in the ocellar symmetry system, where the bands are homologous with the outer concentric rings of complex eyespots. The most extreme form of this closure in the central symmetry system is found in some Geometridae and in butterflies of the genera *Mantaria* (Satyrinae) and *Mesosemia* (Riodininae), in which the bands become fused at nearly every wing vein, so that the central symmetry system in effect becomes broken up into a series of circles or eyespotlike patterns. A particularly nice illustration of such a fragmented symmetry system is seen in *Mantaria maculata* (Fig. 2.5B). When distal and proximal bands of the central symmetry system fuse, they become smoothly continuous. In those cases in which the bands are made up of a gradation of colors, each of the colors fuses smoothly with its homolog in the opposing band (Figs. 2.6 and 5.6). The picture that emerges is an obvious one—namely that, in their most fundamental form, symmetry systems are large concentric patterns of pigmentation. These concentric systems are always elongated along an anterior-posterior axis, and when they are expanded or stretched to a very large size, their continuity near the anterior and posterior wing margins is lost. It is this expansion of an essentially concentric system of pigment bands that gives rise to the symmetrical distribution of pigments in opposing bands. In Chapter 5 we will further refine this view of symmetry systems in the light of experimental studies that have revealed their developmental origin and have established

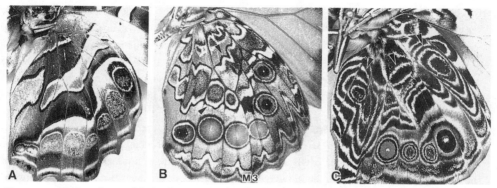

Figure 2.5. Various forms of fusion between the bands of the central symmetry system. Anterior is in the upper right corner of each picture; all specimens are ventral hind wings. *A*, *Charaxes analava* (Nymphalidae: Charaxinae), with a single point of fusion in the posterior region of the wing. *B*, *Mantaria maculata* (Nymphalidae: Satyrinae), with multiple fusions so that the central symmetry system posterior to vein M_3 is reduced to a series of ring-shaped patterns. *C*, *Smyrna blomfildia* (Nymphalidae: Limenitinae), in which the central symmetry system bands in both the posterior and the anterior portions of the wing are fused to form circular or elliptical rings.

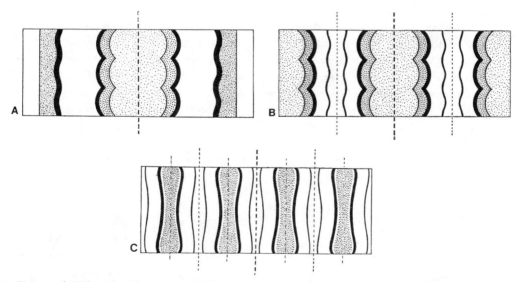

Figure 2.6. Elaboration of a simple central symmetry system (*A*) into several higher-order symmetry systems. In *B*, two subordinate symmetry axes are added on either side of the central axis of symmetry to form two secondary symmetry systems. In *C*, four more intervening axes of symmetry are added to produce tertiary symmetry. Such high-level self-symmetrical pigment patterns are found on the wings of some butterflies and many moths. In the butterflies each of the subordinate symmetry systems has usually evolved unique specializations, so that the self-symmetrical character of the basic pattern is obscured. The distal systems have evolved into the border ocelli symmetry system, and the proximal ones into the basal symmetry system. (From Süffert, 1929b)

the close developmental similarity between symmetry systems and eyespot patterns.

Symmetry Systems and Pattern Evolution

The butterfly wing color pattern has at least three and perhaps four symmetry systems lying side by side: the border ocelli system, the central symmetry system, the basal symmetry system, and the wing root band, which appears to be the distal member of an even more basal symmetry system. Süffert (1929b) and Henke (1933b, 1948) have suggested that wing patterns among the Lepidoptera as a whole evolved by a successive addition of symmetry systems. They proposed that the surface of the wing has become subdivided into a series of developmentally equivalent fields that are arrayed in parallel on the wing (Figs. 2.6 and 2.7). Each of these fields is composed of a paired set of parallel bands with a line of symmetry running between them. Süffert and Henke's idea about this fundamental property of color pattern organization presaged contemporary developmental theory about compartmentalization of developmental fields by standing wave functions such as those developed by Kauffman

(1977) for the imaginal disks of *Drosophila*, by Bard (1977, 1981) and Murray (1981a,b) for striping patterns in zebras, and by Meinhardt (1982) and Murray (1989) for a broad variety of developmental systems. It appears that species of Lepidoptera differ in the number of wavelengths they can fit on their wing surface and that this characteristic is responsible for differences in the "fineness" of the banding pattern among species (Fig. 2.7). The number of waves need not be an integral number, and half symmetry systems are common, particularly near the base of the wing. It could be, of course, that there is an integral number of waves on the wing disk at the time the pattern is first determined and that peripheral portions of these systems are subsequently lost by cell death during the sculpturing of the wing shape (see Chapter 1).

Süffert (1929b) has shown that in certain moths the symmetry systems are themselves further subdivided into lower-order symmetry systems (Fig. 2.8). Such hierarchical ranks of symmetry systems are reminiscent of (and perhaps developmentally homologous to) the complex banding patterns seen in lionfishes (*Pterois* spp.) and many other fishes. One can think of such subordinate symmetry systems as arising in a growing developmental field that, as it expands, can support pro-

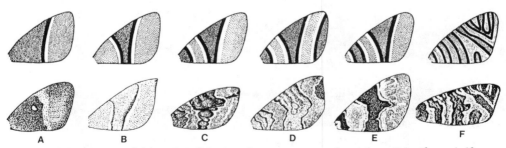

Figure 2.7. Major types of elaboration of the central symmetry system in moths, ranging from a half symmetry system (A) to a distorted tertiary symmetry (F). A, *Lasiocampa quercus*; B, *Metrocampa margaritata*; C, *Larentia juniperata*; D, *Lygris testata*; E, *Larentia sagittata*; F, *Argyroploce rivulana*. (From Henke, 1933a)

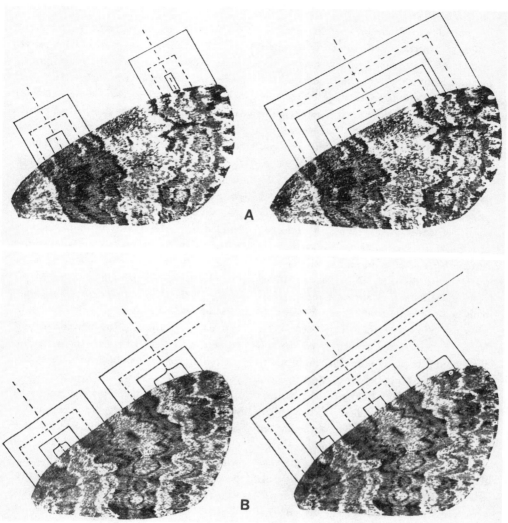

Figure 2.8. High-order symmetry systems in the Geometridae. On the left, the secondary symmetry system is isolated by brackets, and tertiary symmetry within each is indicated by dashed lines. On the right, the primary axis of symmetry is indicated, and the homologous bands on either side are connected by brackets; lower levels of symmetry within the primary symmetry system are indicated by dashed lines. *A, Larentia ruberata; B, Larentia miata.* (From Süffert, 1929b)

gressively more wavelengths. At some intermediate size, crests and troughs of low amplitude and short wavelength can become superimposed on an existing long-wavelength pattern to generate shadow stripes roughly halfway between existing bands. There are many analogs to such space-filling systems of pattern formation. For instance, some races of zebras have shadow stripes in the white portions of their pattern where the black stripes are widely spaced, and the widest spaces between the spots in a cheetah's coat are usually filled with smaller spots (Bard, 1977, 1981; Murray, 1981a,b).

Patterns with little or no differentiation among symmetry systems are fairly common among the moths, particularly in the Geometridae. But butterflies almost always have a distinctive and often extreme differentiation and specialization of each symmetry system, to the extent that in most species the underlying homology *among* symmetry systems is totally obscured. Only in a few species, particularly in the Lycaenidae, is it easy to deduce that the pattern consists of a proximo-distal repetition of fundamentally similar elements. In other cases, as in the genus *Charaxes*, homology among symmetry systems can also be deduced but with considerably more difficulty (Nijhout and Wray, 1986). The homology among symmetry systems helps to explain why many species have identical colors and color sequences in the bands of their central symmetry system and their border ocelli system. The multiplicity of mirror-image pigment sequences that can arise from homologies among these systems occasionally makes it difficult to unambiguously assign a particular piece of the pattern to one symmetry system or another. Later, when we look in detail at the comparative morphology of patterns, and in particular when we develop a theoretical model for pattern formation, it will become evident that even in well-differentiated systems it is not always easy, and sometimes is impossible and inappropriate, to assign a given element of a color pattern to a specific system, even though, paradoxically, its homology relations may be quite obvious.

Dependent Patterns

Schwanwitsch (1924, 1926) recognized two pattern elements that do not feature in the ground plan presented by Süffert (1927), namely the venosas, or venous stripes, and the intervenosas, or intervenous stripes (Fig. 2.1). The venous stripes are a color pattern that is a precise outline of the wing veins. In some cases venous stripes are merely a darkening of the scales that overlie the wing veins, but in others they consist of a pattern that contrasts sharply with the coloration of the remainder of the wing (Fig. 2.9). The intervenous stripes, by contrast, consist of pigmented stripes that lie on the midline of each wing cell (Fig. 2.10). Schwanwitsch's ground plan shows these lines as extending from the wing margin about halfway into each wing cell as a kind of shadowy background pattern upon which the remaining elements of the nymphalid ground plan can be expressed. Although Süffert did not recognize venous and intervenous stripes as pattern elements in his first paper on the analysis of color patterns (Süffert, 1927), he later added a category of pattern elements that he named dependent patterns (Süffert, 1929b). This category included venous and intervenous stripes as members of a special category of pattern elements whose expression was closely tied to (dependent on) structural features of the wing such as the wing veins and wing margin. The pigmentation of venous stripes can sometimes be broad and quite complex, as well as symmetrical about the wing vein (Fig. 2.9C). This suggests that the wing vein lacunae serve as the source of determination for this pattern element. In many species with an extensive dependent color pattern, the positions of veins that atrophy during development (vein 2A and the basal branches of the media) are clearly detectable in the color pattern (Fig. 2.10A,B). Determination of these patterns must thus have occurred before atrophy of the veins.

Intervenous stripes are somewhat less common than venous stripes, but they are far more interesting because they appear to form a mutually exclusive category of patterns

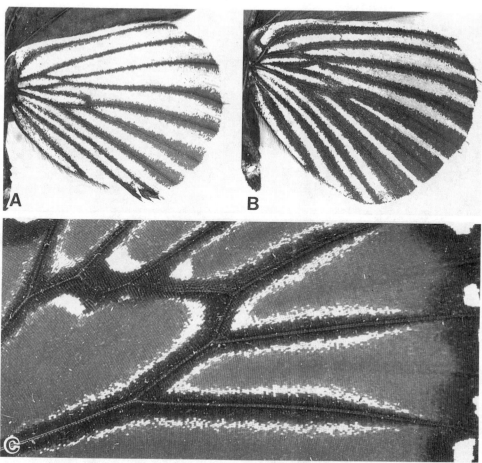

Figure 2.9. Venous patterns. *A* and *B*, *Chlosyne ehrenbergii* (Nymphalidae: Nymphalinae), showing different degrees of expansion of the dark pattern. In the more extreme case (*B*), the pale background gives the impression of intervenous stripes. *C*, *Danaus gilippus* (Nymphalidae: Danainae), showing stepwise graded colors in its venous pattern. In *Chlosyne* the venous pattern tends to expand near the wing margin, whereas in *Danaus* the venous stripes become narrow and vanish toward the margin.

with the elements of the nymphalid ground plan. To my knowledge, only a few species, primarily in the genus *Adelpha,* have a clear intervenous stripe superimposed on a well-defined element of the nymphalid ground plan. Cases in which a minuscule border ocellus (actually, only the central focus of an ocellus) appears within an intervenous stripe are slightly more common, though still rare. In a few taxa, the intervenous stripes have one or two bulges or local condensations of pigment, or they have become fragmented into short dashes or spindle-shaped patterns whose positions may correspond to those of elements of the nymphalid ground plan. As we will see in Chapter 7, intervenous stripes and the classical elements of the nymphalid ground plan appear to be opposite extremes of a single developmental series and differ from each other only in the relative timing of

Figure 2.10. Intervenous patterns. *A, Atrophaneura varuna* (Papilionidae), with narrow stripes, and *B, Atrophaneura nox,* with wide stripes. Note that these species also have a narrow venous pattern. *C, Acraea encedon* (Nymphalidae: Heliconiinae); *D, Acraea oberthuri.* In the last species, the distal portion of the intervenous stripe is split. This represents the first step in a transition to a scalloped submarginal pattern common in other butterflies (see Chapter 7).

certain developmental events. The mutual exclusivity of these two categories of pattern is thus readily explainable on developmental grounds. A number of patterns appear to be developmental intermediates between intervenous stripes and ground plan elements, and these are discussed below in the section "Themes and Variations."

Ripple Patterns

The final category of pattern elements is the ripple patterns. These finely striated patterns are common on the ventral wing surface of Satyrinae, Brassolinae, and a few nymphalid genera like *Prepona* and *Anaea*. They are relatively rare elsewhere in the butterflies, though very similar ripple patterns occur in many moths. Ripple patterns derive their name from their resemblance to the ripples in windblown sand (Fig. 2.11). The ripples are evenly spaced but otherwise apparently random in length and position. Although they are clearly continuous across each wing cell, they are often abruptly discontinuous at the wing veins. In keeping with their random character, ripple patterns on homologous areas of the left and right wings are always different.

In their relation to the nymphalid ground plan, ripple patterns are neither pattern nor background. In *Caligo* (Fig. 2.11A), as in other species, different elements of the nymphalid ground plan interact differently with the ripples. Border ocelli are always superimposed on ripple patterns. The parafocal elements by contrast, clearly interact with the ripples and acquire a fragmented or shredded appearance when they occur in combination with a ripple pattern. The whole pattern gives one the impression that the ripples and parafocal elements are part of a surface sheet of color that has become stretched and stress-fractured, with a pale background showing through the shreds or ripples of the overlying sheet. The ripples can

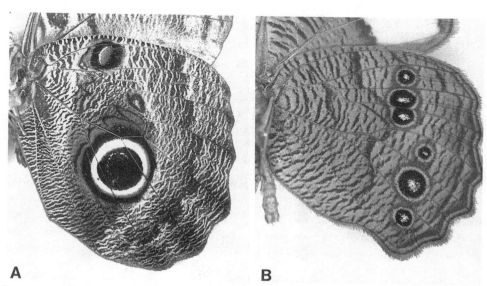

A   **B**

Figure 2.11. Ripple pattern in (*A*) *Caligo eurilochus* (Nymphalidae: Brassolinae) and (*B*) *Cercyonis pegala* (Nymphalidae: Satyrinae).

thus be thought of as being superimposed on the parafocal elements. Parafocal elements are therefore part of the background as far as ripples are concerned, whereas ripple patterns are background for the border ocelli. This suggests that these patterns are determined sequentially: first the parafocal elements, then the ripples, and finally the border ocelli. The manner in which ripples interact with the other elements of the ground plan is somewhat variable. Pieces of the central symmetry system are usually superimposed on the ripples, but in a few satyrines the bands of the central symmetry system appear to be rent in much the same way as the parafocal elements of *Caligo*. Ripples are usually superimposed on submarginal bands. It would seem that if the details and variability of the relations between ripples and other pattern elements could be analyzed rigorously, we might be able to use that information as evidence for the relative order of pattern determination on the wing, but this has not yet been done.

Significance of the Nymphalid Ground Plan

It is important to recognize that the nymphalid ground plan does not represent the ancestral pattern or the primitive condition of butterfly color patterns. The nymphalid ground plan is merely a diagrammatic list of homologies among the elements of the color pattern. It represents the maximum pattern that could be present in a generalized butterfly. But no species actually possesses the entire array of elements shown in Figure 2.17. On the other hand, few species have pattern elements that are not either part of the nymphalid ground plan or a dependent pattern (see Chapter 3).

Perhaps the best way to appreciate the sig-

nificance of the nymphalid ground plan is in terms of archetypes and morphotypes, as defined by Patterson (1982). An archetype is an idealization with which the features of an organism may be homologized by abstract transformation. A morphotype, on the other hand, is a description of the shared derived characters (synapomorphies) of a group. An archetype therefore entails no hypothesis about, nor does it make predictions about, ancestry or hierarchical grouping. A morphotype describes the character states that characterize a monophyletic group and are thus primitive for the group as a whole. According to these definitions, the nymphalid ground plan is an archetype. The actual primitive character states for the elements of the color pattern of the butterflies as a whole is not known. Homologies among pattern elements can be traced across the entire range of butterfly diversity, however, which gives us hope that reconstruction of the primitive patterns for various taxa among the butterflies will eventually be possible.

Pattern and Background

The elements of the nymphalid ground plan together with the venous and intervenous stripes constitute the "pattern" of butterfly wing patterns. This pattern generally occupies only part of the wing surface; the remainder is considered background. The background is sometimes homogeneously colored, but most often it has distinctive patches or gradients of color. The color of the background, particularly when it contains patches of contrasting colors, contributes a great deal to the overall appearance of the wing pattern. When the background color is "patterned," it always has a bold pattern of a few large areas of color without sharp boundaries. The boundaries between different areas of background color do not correspond to any of the landmarks on the wing, such as the venation pattern, or to any of the elements of the ground plan. The color of the background can be under independent genetic control in different parts of the same wing (see Chapter 6).

Dislocation of Pattern Elements and Wing Cell Autonomy of the Pattern

It was stated above that species-specific differences in the expression and pigmentation of the individual elements of the nymphalid ground plan account for much of the diversity of color patterns. There is, however, another developmental feature of color patterns that increases the diversity of possible patterns. The bands of the various symmetry systems seldom if ever run smoothly from the anterior to the posterior margin of the wing as suggested by the nymphalid ground plan of Figure 2.1. In real color patterns these bands are offset to a greater or lesser degree wherever they cross a wing vein (Fig. 2.12). Schwanwitsch (1924) referred to this phenomenon as dislocation, by analogy to visually similar discontinuities in sedimentary strata along geologic faults. Species differ in the degree and direction of this dislocation in various wing cells. As is true of the other modes of variation of the color pattern, different taxa among the butterflies tend to have characteristic dislocation themes (Fig. 2.13).

Dislocation arises because pattern development in each wing cell is largely independent of that in other wing cells. In later chapters we will see that this wing cell autonomy of the pattern can be demonstrated experimentally and that it can also be deduced from a statistical analysis of individual variation of color patterns. Here we will deal only with

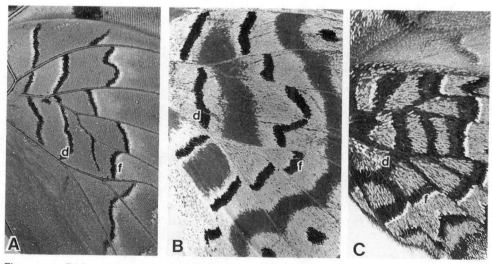

Figure 2.12. Dislocation in the bands of the central symmetry system. *d*, proximal band; *f*, distal band. *A*, *Charaxes bipunctatus* (Nymphalidae: Charaxinae); *B*, *Asterope pechueli* (Nymphalidae: Limenitinae); *C*, *Incisalia niphon* (Lycaenidae).

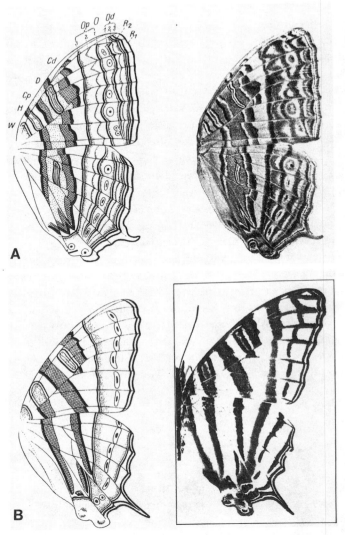

Figure 2.13. Analysis of the patterns of (*A*) *Cyrestis cocles* (Nymphalidae: Limenitinae) and (*B*) *Cyrestis camillus*, showing different degrees of dislocation and realignment of the pattern elements. (From Süffert, 1927; courtesy Georg Thieme Verlag, Leipzig)

the morphological evidence for wing-cell autonomy in pattern formation, and with the consequences for pattern diversity.

The idea that the wing cell is the unit of color pattern formation emerges from the following observations: (1) In the majority of species, the color pattern consists of a serial repetition of identical or nearly identical patterns from wing cell to wing cell (Fig. 2.14). (2) In some of these species, the pattern in one, two, or three wing cells differs from that of all the others (Figs. 2.15 and 2.16), or the background color of a single wing cell differs from that of the other wing cells. (3) In many of the species that have identical serially repeated patterns in their wing cells, one pattern element is missing in one or two wing cells only (Fig. 2.15). (4) When there are discontinuities in bands or blotches that range across a large area of the wing, they often oc-

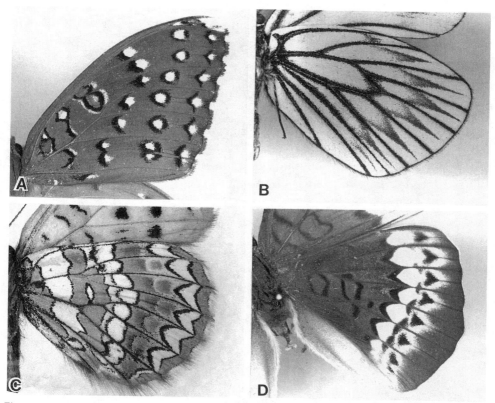

Figure 2.14. Serial repetition of nearly identical patterns in each wing cell. *A*, *Hamanumida daedalus* (Nymphalidae: Limenitinae); *B*, *Aporia goutellei* (Pieridae); *C*, *Melitaea bellona* (Nymphalidae: Nymphalinae); *D*, *Tanaecia trigerta* (Nymphalidae: Limenitinae).

cur at a wing vein. (5) Ripple patterns are often discontinuous at wing veins (Fig. 2.11).

These observations establish the importance of wing veins as boundaries in pattern formation. The serially repetitive nature of the majority of patterns, and the evident ability of single wing cells to deviate from the general pattern theme, suggest that individual wing cells behave as fields or compartments in color pattern determination and that the overall wing pattern consists of serially homologous arrays of pattern elements (Nijhout, 1985b, 1986). Although serial ho-

mology of the wing pattern is most easily detected in the border ocelli and parafocal elements, it is now clear that the other elements of the nymphalid ground plan are likewise built up of serial repetitions of nearly identical patterns from wing cell to wing cell (Nijhout, 1985b). Therefore, a correct representation of the nymphalid ground plan should illustrate the serial repetition of pattern elements. This is done in Figure 2.17, which presents the form of the nymphalid ground plan that is used throughout this book. Figure 2.17 also presents the convention for letter coding of the elements of the

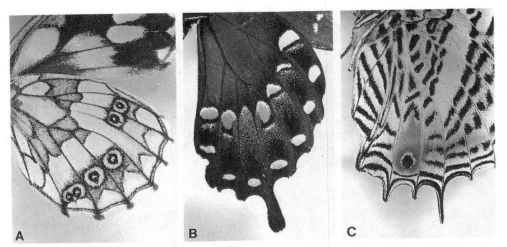

Figure 2.15. Species in which the pattern in a single wing cell is distinctly different from that in the other wing cells. A, *Melanargia galathea* (Nymphalidae: Satyrinae); B, *Papilio troilus* (Papilionidae); C, *Baeotus baeotus* (Nymphalidae: Limenitinae).

Figure 2.16. Two examples of species that have distinctively different pattern themes in pairs and triplets of wing cells on forewings and hind wings. A, *Euselasia ferrugo* (Lycaenidae: Riodininae); B, *Abisara cudaea* (Lycaenidae: Riodininae).

ground plan: a simple alphabetical sequence from base to distal margin. The equivalent names, as noted before, are given in Table 2.1.

If the central symmetry system and basal symmetry system develop autonomously in each wing cell, that would explain why the bands of these symmetry systems usually appear to be dislocated when they cross a wing vein. Because the pattern in each wing cell is largely independent of that in other wing cells, a failure of serially homologous elements in adjoining wing cells to line up perfectly is less noteworthy than is a precise match that produces the appearance of an un-broken band running across the wing. Perhaps the strongest evidence for the wing cell autonomy of the bands of the central symmetry system comes from the observation that, just as with the border ocelli, the bands of the central symmetry system are often missing in one or more wing cells. This ability to express each band of the central symmetry system—and any other pattern element, for that matter—selectively in each wing cell enables butterflies to achieve the diversity of pattern and flexibility of design that culminate in accurate mimicry and in the remarkable camouflage effects epitomized by the dead-leaf butterflies of the genus *Kallima* (Fig. 3.7).

The M_3 Boundary

Wing vein M_3 has a special function in pattern development. In a number of species the pattern elements in the wing cells anterior to M_3 are different from those posterior to it. The differences that normally exist between wing cells are exaggerated on either side of M_3, so that a distinctively different pattern theme can occur on each side (e.g., forewing in Fig. 2.16A). Figure 2.18 illustrates two cases of this phenomenon. In *Morpho achilleana* the eyespots anterior to M_3 form a cluster that shares its outer rings, whereas those posterior to M_3 are separated. The wing veins anterior to M_3 do not appear to have as strong a compartmentalizing function as those posterior to M_3. In *Polyura pyrrhus,* pattern element **g** anterior to M_3 is neatly lined up and forms an almost continuous arc across four wing cells, whereas element **g** posterior to M_3 is considerably dislocated between wing cells. Thus, as in *M. achilleana,* the pattern shift in *P. pyrrhus* suggests that there is more

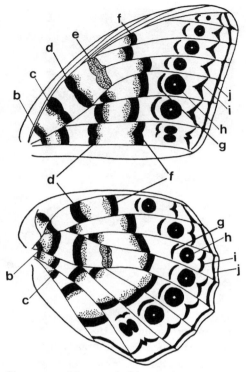

Figure 2.17. The nymphalid ground plan. For nomenclature, see Table 2.1.

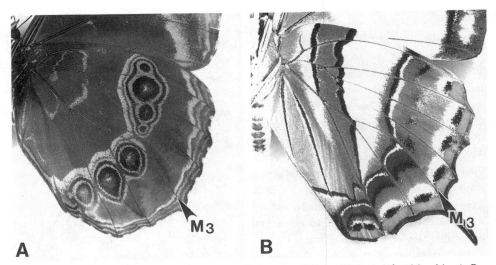

Figure 2.18. Pattern boundary at vein M₃ (*arrow*). A, *Morpho achilleana* (Nymphalidae: Morphinae); B, *Polyura pyrrhus* (Nymphalidae: Charaxinae).

continuity among the wing cells anterior to M_3 than among those posterior to it. A difference in the character of the central symmetry system on either side of M_3 is seen in *Mantaria maculata* (Fig. 2.5B). Posterior to M_3 the central symmetry system is contracted into a series of circular patterns, one per wing cell, whereas anterior to M_3 the system is expressed as a more conventional pattern of roughly parallel bands. The position of the M_3 boundary corresponds approximately to the anterior-posterior compartment boundary in *Drosophila*, which is of considerable importance in the specification of segmental identity (Garcia-Bellido et al., 1979).

The Border Ocelli System

The greatest difficulty in resolving homologies resides in the relationships among the pattern elements within what Süffert referred to as the ocellar symmetry system (pattern elements **g**, **h**, and **i**). We noted above that the distal band of the ocellar symmetry sys-

tem (element **i**) can be regarded as a pattern element in its own right, the parafocal element, whose shape appears to vary and evolve independently from that of the ocellus proper. The proximal band of the ocellar symmetry system is often absent and, when present, frequently forms an arc that is perfectly concentric with the ocellus, giving the impression that it is merely part of an outer ring for that ocellus. Yet, as the analyses in Chapters 3 and 7 will show, this proximal band varies independently of the ocellus and should thus also be regarded as a discrete pattern element. Because of extreme variability in the details of the shape and in the presence or absence of elements **g** and **h**, the analyses of Süffert and Schwanwitsch and those of Nijhout and Wray (1986, 1988) have not been entirely consistent in the nomenclature and homology assignments among the elements of the border ocelli system. The ground plan illustrated in Figure 2.17 presents what I believe to be the correct and complete account of the nymphalid ground plan, and the nomenclature and relations shown in Figure 2.17 will be the standard reference throughout this book.

Themes and Variations

When we restrict our attention to the morphology of patterns in single wing cells, we lose some of the principles of organization discussed above and we gain some new ones. The dominance of the central symmetry system is lost completely. Because of the development of crossveins and severe dislocation of portions of the central symmetry system, each of the wing cells that reach the margin of the wing usually bears only one band of the central symmetry system, and sometimes none. The distribution of border ocelli is one of the most obvious features of the overall wing pattern, but when we focus on a single wing cell, their shape—and in particular the variation in their shape—becomes their most obvious property. The parafocal elements, which looked like little more than a fringe of pattern near the wing margin, now become one of the most prominent features of the pattern, and here, as with the border ocelli, an exuberant diversity of shapes catches the eye.

These variations in shape are not boundless. Upon closer examination, it becomes clear that most of the diversity of shapes among the border ocelli and parafocal elements can be grouped into a relatively small number of themes. Border ocelli are usually round but can be elliptical, triangular, or cardioid. Parafocal elements are usually V-shaped but can be stretched and indented into W and M shapes or inverted Vs. Intervenous stripes are usually straight and narrow but can have bulges or be flared out at their base.

In order to document and analyze this diversity, I undertook an extensive survey of pattern shapes for the border ocelli and para-

focal elements in 321 genera and 2,207 species of Nymphalidae, excluding only the Danainae and Libytheinae (Figs. 2.19 and 2.21). To ensure that as much of nymphalid pattern diversity as possible was surveyed, an attempt was made to include all monotypic genera and all species whose patterns seemed to deviate from the norm for their genus. The genera and the number of species in each genus included in this survey are listed in Appendix C. The diversity shown in Figures 2.19 and 2.21 is estimated to cover nearly 90% of the entire pattern diversity in the Nymphalidae (Nijhout, 1990). The patterns in Figures 2.19 and 2.21 are arranged in groups of similar shapes. The numbers below each figure indicate the numbers of genera and species in which each pattern was found, so that the relative prevalence or rarity of a given pattern can be determined at a glance.

It is clear that in the border ocelli, circular patterns are the most common, though such patterns occur in fewer than half the species surveyed. Triangular and kidney-shaped patterns are also quite common. Many border ocelli do not look like eyespots at all, and that is why a letter code for pattern elements (h in this case; see Table 2.1) is often more appropriate than a general descriptive name. The most extreme form of a border ocellus is a spindle-shaped dash (pattern 31 in Fig. 2.19). This pattern would normally be classified as an intervenous stripe, but it is quite obvious from species like *Perisama plistia* (Limenitinae) and *Argyrophorus argenteus* (Satyrinae) that these line-shaped patterns are homologous to border ocelli (Figure 2.20). Many patterns in Figure 2.19 have peaklike extensions along the wing cell midline as if they were formed by a combination of these short midline stripes and some of the other pattern shapes among the border ocelli. As we will see in Chapter 7, such close developmental relationships exist between the inter-

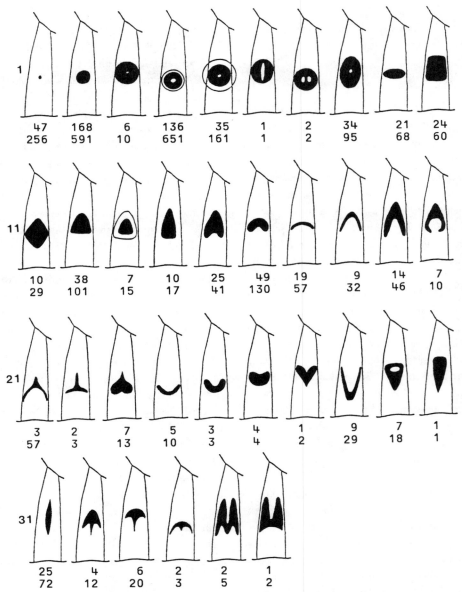

Figure 2.19. Diversity of border ocelli in the Nymphalidae. The numbers under each pattern give the number of genera (*top*) and the number of species (*bottom*) in which each pattern type was found. The genera surveyed are listed in Appendix C. (From Nijhout, 1990)

venous patterns and the border ocelli and parafocal elements that it is in fact inappropriate and technically impossible to draw an unambiguously clear distinction between the two types of patterns.

The diversity of parafocal elements among the 2,207 species surveyed is much greater than that of the border ocelli and is illustrated in Figure 2.21. The 110 different shapes are arranged according to their overall

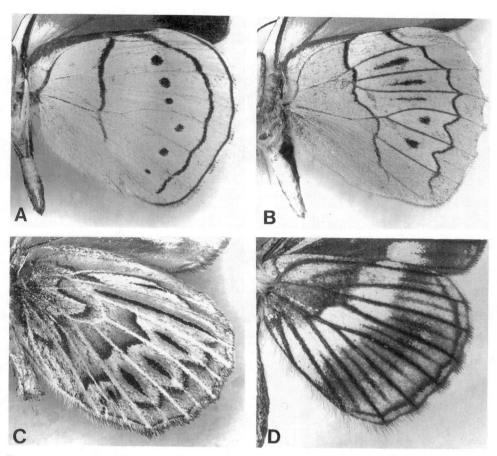

Figure 2.20. Diversity of form in the border ocelli, illustrating transitions between dot-shaped ocellar patterns and intervenous midline patterns. *A, Perisama euriclea* (Nymphalidae: Limenitinae); *B, Perisama plistia; C, Argyrophorus argenteus* (Nymphalidae: Satyrinae); *D, Gnathotriche exclamationes* (Nymphalidae: Nymphalinae).

similarity. Each pattern in this series actually represents a small category of similar-looking patterns. For instance, patterns 3 and 4 are both arcs but with different aspect ratios. Among the species surveyed, there occurs a nearly complete array of intermediate aspect ratios, which for purposes of classification were assigned to the discrete pattern class (3 or 4) they most closely resembled. In fact, morphologically intermediate patterns exist between almost every adjacent pair in this figure.

Several pattern themes can be seen in this arrangement. Simple curves can be concave relative to the wing edge (patterns 1–6) or convex (patterns 60–64). Many patterns are dominated by peaks along the wing cell midline that point toward the apex of the wing cell (15–46 and 72–80) or toward the wing margin (49–59). Some parafocal elements are compact and rounded and closely resemble shapes previously encountered in the border ocelli (2, 23–31, 100–103, and others). Intervenous stripes are present among

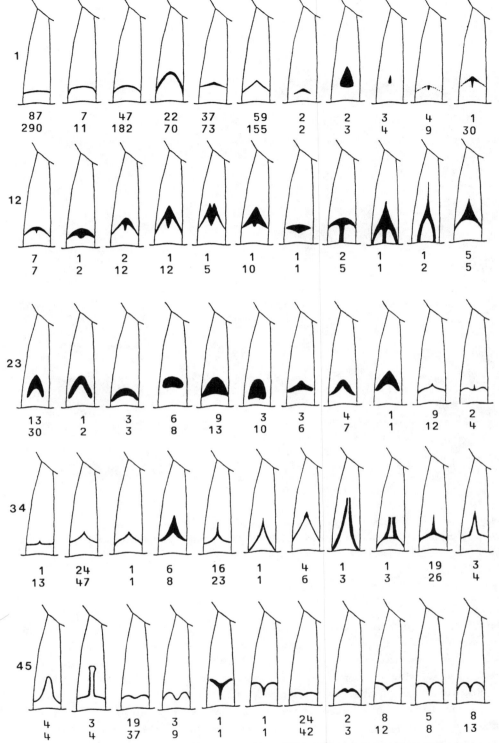

Figure 2.21. Diversity of parafocal elements in the Nymphalidae. The numbers under each pattern give the number of genera (*top*) and the number of species (*bottom*) in which each pattern type was found. The genera surveyed are listed in Appendix C. (From Nijhout, 1990)

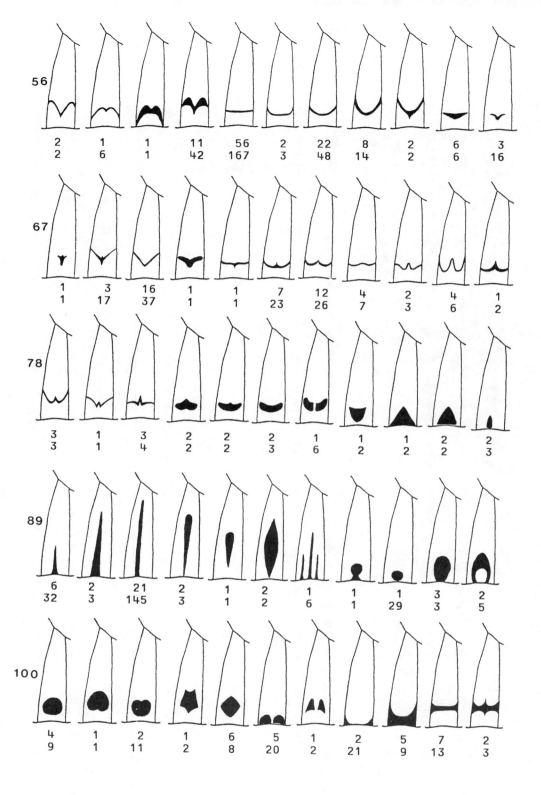

56

2   1   1   11   56   2   22   8   2   6   3
2   6   1   42   167   3   48   14   2   6   16

67

1   3   16   1   1   7   12   4   2   4   1
1   17   37   1   1   23   26   7   3   6   2

78

3   1   3   2   2   2   1   1   1   2   2
3   1   4   2   2   3   6   2   2   2   3

89

6   2   21   2   1   2   1   1   1   3   2
32   3   145   3   1   2   6   1   29   3   5

100

4   1   2   1   6   5   1   2   5   7   2
9   1   11   2   8   20   2   21   9   13   3

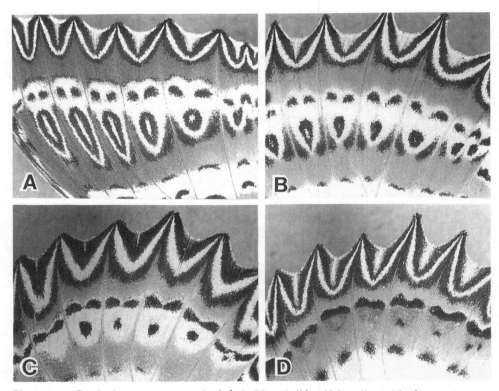

Figure 2.22. Continuity among patterns in *Cethosia* (Nymphalidae: Heliconiinae). The four specimens in this morphological series show a gradual transformation of the border ocellus (*h*) from a teardrop shape to a vanishing dot, and of the parafocal element (*i*) from a pattern of double dots to a scalloped line. A, forewing of *C. biblis; B,* hind wing of *C. biblis; C,* hind wing of *C. hypsea; D,* hind wing of *C. gabina.* (From Nijhout, 1985b)

the parafocal elements as well (89–95). The close relation between intervenous stripes and the more classical parafocal elements is revealed by patterns that form transitions between the two (e.g., 8–17; 21 and 22; 43 and 44; 87 and 88).

Individual variation and diversity of species patterns provide continuity between many of the patterns of shown in Figure 2.21. Examples of this continuity are shown in Figure 2.22, for the ocelli and parafocal elements of *Cethosia,* and in Figure 6.8, for intervenous and ocellar patterns in *Papilio memnon.*

In spite of the continuity between many groups of adjacent patterns in Figure 2.21, a linear system of classification is clearly impossible because too many aspects of the shape can vary independently. In the M-shaped themes of patterns 47–59 and the W-shaped themes of patterns 71–80, for instance, there is independent variation in the length and curvature of the lateral arms and in the length and shape of the central peak. A method for creating order in systems such as these will be presented in Chapter 4, and in Chapter 7 we will see how each of these features can be controlled independently.

Few studies have systematically analyzed the taxonomic and evolutionary significance of butterfly color patterns. The majority of contemporary studies merely use the coloration of particular regions of the wing as a means of distinguishing between related species. Few if any have attempted to utilize the system of homologies revealed by the Schwanwitsch-Süffert nymphalid ground plan in studying the systematics of butterflies, and none have yet used the pattern elements of the nymphalid ground plan as characters in phylogenetic reconstruction. Until the recent work of Nijhout and Wray (1986, 1988), the classical studies of Schwanwitsch (1924, 1925, 1926, 1929a,b, 1930, 1931, 1935a,b, 1948, 1956a,b; Schwanwitsch and Sokolov, 1934) constituted the only major atempts at a systematic analysis of the morphology and evolution of butterfly color patterns.

In this chapter, I will present a series of analyses of the wing patterns of various taxa of butterflies. These analyses are intended to explore the scope and usefulness of the nymphalid ground plan (Fig. 2.17) in identifying the elements of the color pattern in specific taxa. Each taxon has been chosen to illustrate particular points about the structure and evolution of wing patterns. Some taxa chosen for analysis are exemplars of common patterns for their family or subfamily, and others have been chosen to illustrate various important themes in color pattern evolution, such as dislocation, selective expression of pattern elements, and pattern integration. Some taxa have been chosen because they represent "worst-case" scenarios in having either extraordinarily diverse and complicated patterns (*Charaxes*) or deceptively simple ones (*Heliconius*). With each taxon, one or two new concepts or phenomena are introduced that are important in understanding the developmental morphology and evolution of butter-

Chapter 3

The Analysis of Wing Patterns

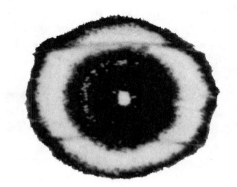

fly color patterns. This chapter, therefore, constitutes a primer on pattern identification and analysis.

Finding the Central Symmetry System

Before proceeding with an explication of the ways in which a broad diversity of color patterns are derived from the nymphalid ground plan, it is necessary to look a bit more closely at the distribution of the elements of the central symmetry system. Identification of the central symmetry system is almost always the first step in the analysis of an unknown pattern. As can be seen from Figure 2.17, the proximal and distal members of the central symmetry system (elements **d** and **f**) sometimes occur within the same wing cell, as they do in the longer, more posterior wing cells of the forewing. In the anterior half of the forewing, however, each wing cell bears only one of these elements, either the distal one in the long wing cells that reach the wing margin, or the proximal one in the discal cell. This disjunct distribution of the central symmetry system bands among wing cells and the discal cell needs to be reconciled with the bands' obvious paired nature, revealed by the pigmentation and developmental physiology of the bands. We need a method for identifying which pattern element in each wing cell is part of the central symmetry system in the absence of clues that are given by pigmentation, because in many cases all pattern elements, including those of the central symmetry system, are uniformly colored and nearly identical in pigmentation.

To understand the derived structure of the central symmetry system, it is necessary to realize that during development the venation pattern of the wing changes considerably after the color pattern has been determined.

At the time that color pattern determination begins, the venation pattern is much simpler than the final adult venation. The earlier venation pattern is basically the primary venation pattern (see Chapter 1) and consists exclusively of longitudinal veins. Figure 3.1 traces the hypothetical steps in the development and evolution of the central symmetry system on the forewing from the simplest possible pattern superimposed on the primary venation system (Fig. 3.1A) to the normally dislocated pattern seen in the nymphalid ground plan (Fig. 3.1C). The disjunction of the central symmetry system comes about by virtue of a progressively greater dislocation of the pattern in the more anterior wing cells, followed by the atrophy of veins within the discal cell and the development of crossveins that close off that cell. As a consequence, most species in the Nymphalidae have three pigment bands in the discal cell (Fig. 2.17). The most distal of these is located on the crossveins that close off the discal cell and is thus identified as the discal spot (pattern element **e**), which marks the central field of the central symmetry system. The band immediately proximal to it is the local representative of the proximal band of the central symmetry system. The most proximal of the three is pattern element **c**, or occasionally a fusion of pattern elements **c** and **b** (the basal symmetry system).

The term "local representative" used above points out an ambiguity in the nomenclature of the elements of the color pattern. This ambiguity arises because the terminology is derived from a ground plan in which bands, coursing across the entire expanse of the wing, were considered to be the fundamental units of the color pattern. We now know, however, that wing cells are semiautonomous compartments in color pattern formation and that the pattern elements within each wing cell are, in fact, the fundamental units of the

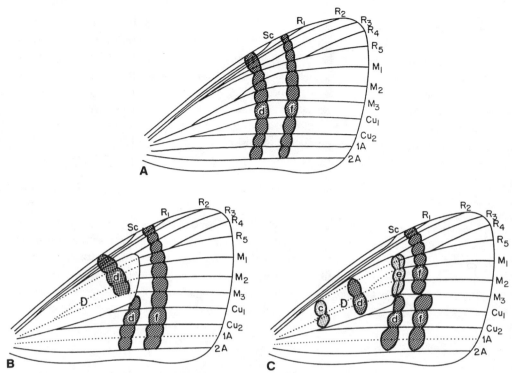

Figure 3.1. Derivation of the typical morphology of the central symmetry system. *A*, a simple symmetry system superimposed on the primary venation pattern of the larval imaginal disk. *B*, slightly dislocated intermediate stage superimposed on the secondary venation system of a pupal and adult wing. *C*, typical morphology of the central symmetry system, with the portion of the proximal band (*d*) within the discal cell (*D*) severely dislocated.

color pattern. The bands of color that make up the nymphalid ground plan are chains of serially repeated homologous pattern elements. Thus a term such as "proximal band of the central symmetry system" can refer to the entire band or to any one of the serially homologous elements that make up the band. In the sections that follow in this chapter, pattern elements will always be assumed to be the smallest units of pattern, and when the name of a particular pattern element is used, that name will always refer to the element's presence or form in a specific or generic wing cell. When a particular statement is intended to apply simultaneously to all members of a serially homologous series, they will be referred to as ranks or arrays of serial homologs.

Stichophthalma camadeva (Morphinae: Amathusiini)

The ventral wing pattern of *Stichophthalma camadeva* (Fig. 2.2A) probably comes closer than any other to exhibiting the complete nymphalid ground plan. It is, therefore, an excellent choice for familiarizing oneself with the general features of wing patterns. The discal spot is well developed as a line along the veins that close off the discal cell. The proximal and distal bands of the central symmetry system are easily recognizable and illustrate the idea of color symmetry, because both are made up of a dark (medial) and light (lateral) pigment. Border ocelli are expressed in almost every wing cell. However, they vary considerably in size from wing cell to wing cell and are not developed into well-defined and perfectly circular eyespots in this species. The parafocal elements and submarginal bands form flattened arches or M-shaped patterns in each wing cell. The basal symmetry system and marginal bands are absent. The other amathusiines shown in Figure 2.2 illustrate one of the most common modes of pattern evolution, the loss of elements in specific wing cells and the loss of entire ranks of serially homologous elements.

Pierella (Satyrinae)

Schwanwitsch (1925) described what he called "a remarkable dislocation" in the banding pattern of members of the Neotropical genus *Pierella* (Satyrinae). The phenomenon he discovered was subsequently found to be common among the Satyrinae and appears to constitute a major patterning theme in the butterflies as a whole. What is involved is a massive dislocation of the central symmetry system along the vein that borders the discal

cell posteriorly and continues as M_3 to the wing margin (Fig. 3.2). The distal band of the central symmetry posterior to these veins has undergone a substantial proximal shift. The dislocation is so extreme that in several species (e.g., *P. luna* and *P. pallida*) the posterior portion of the distal band of the central symmetry system has come to lie in the exact position of the anterior portion of the proximal band (Fig. 3.2C). The posterior portion of the proximal band is shifted so that it lines up with pattern element **c** in the discal cell (Fig. 3.2C). Thus what appears to be a pattern of simple thin lines that traverse the wing is, upon analysis, a remarkably complex and precisely integrated system of compound lines whose anterior and posterior portions are not homologous and have different developmental origins.

Schwanwitsch (1925) demonstrated that this dislocation is not imaginary, because various intermediate stages are present as individual variability in the color pattern of a single species (*P. luna*) and as species-specific patterns in several other members of the genus (e.g., *P. rhea*, Fig. 3.2B). The clearest indication of the nature of the dislocation is shown in Figure 3.2B, in which the distal band of the central symmetry system can be seen to take an abrupt turn along the posterior margin of the discal cell before continu-

Figure 3.2. Pierellization: dislocation of bands of the central symmetry system along vein M_3. *d*, proximal band; *f*, distal band. *A*, *Pierella incanescens* (Nymphalidae: Satyrinae); *B*, *P. rhea*; *C*, *P. luna*.

ing on its normal anterior-posterior course. Furthermore, species of Satyrinae differ in whether one or both bands of the central symmetry system are involved in this basipetal shift. Because of the genus in which it was discovered, and the apparent general nature of this particular dislocation, Schwanwitsch (1925) proposed "pierellization" as a convenient term for what would otherwise require rather an elaborate description.

Two points about pierellization are worthy of note here because they reappear in various contexts throughout this book. First is the extreme importance of comparative studies, and in particular the study of individual variability, in elucidating the nature of homologies among pattern elements in any species. Were it not for intermediate cases such as the one shown in Figure 3.2B, it would be difficult to correctly deduce the structure of the central symmetry system of *Pierella*. Indeed, it is difficult to interpret the homologies of the color pattern in many species unless one has either a long series of individual variants or a number of closely related species from which to reconstruct transitions to previously known patterns. The second noteworthy feature is the remarkable precision with which the dislocated bands become aligned with nonhomologous bands of the color pattern. In the majority of specimens the match is so precise that little or no discontinuity is present to reveal the compound nature of the pigment bands. This means that the developmental events responsible for generating the central symmetry system in any one wing cell are separable from those responsible for lining up the pattern elements in adjoining wing cells to form an overall pattern. This must be the case, because in *Pierella,* as in a great many other species (e.g., *Kallima* and *Iphiclides,* below), pattern elements with different developmental origins can become lined up to form visually continuous lines

across the wing. Perhaps the most dramatic demonstration that developmentally independent pattern elements can form a well-coordinated overall pattern is seen in species that have striped wing patterns. In such animals the lines on forewing and hind wing often line up perfectly when the wings are held in the resting position (Fig. 8.2).

Pierellization thus makes us realize that the perfect alignment of pattern elements in adjoining wing cells should be considered a special or derived condition, not a general one. Perfect alignment must be the result of specific selection, and this implies the operation of special, or at least separable, processes in the development of the wing pattern to accomplish the effect.

Satyrinae

A comprehensive account of the comparative morphology and modes of evolution of the color patterns of Palearctic Satyrinae has been presented by Schwanwitsch (1929a, 1931, 1935a,b, 1948; Schwanwitsch and Sokolov, 1934). Among the species he studied, Schwanwitsch recognized several themes of divergence from the ground plan, and among these divergent themes, he found many cases of parallelism. Figure 3.3 shows a summary figure from Schwanwitsch (1935b) that focuses only on the radiation that has occurred in the shape of the distal band of the central symmetry system. In this diagram, it is assumed that the linear pattern of the nymphalid ground plan is the ancestral, or primitive, condition, and the diagram is intended to illustrate an evolutionary radiation of patterns from the primitive ground plan. For the reasons discussed above, however, a dislocated pattern such as that shown as specimens 8 and 9 in Figure 3.3 is likelier to be the more primitive condition, because the pattern elements in each wing cell develop

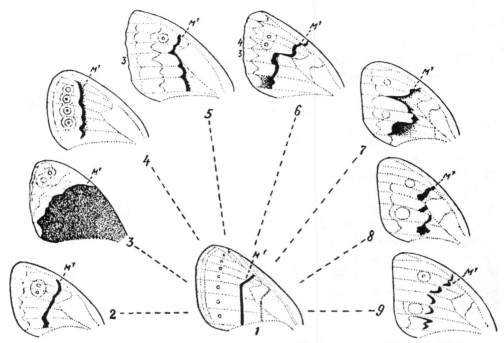

Figure 3.3. Diversification of the distal band of the central symmetry system in the Satyrinae. Although Schwanwitsch interpreted this diversification as a radiation from the ground plan, it is more likely that a pattern such as pattern 9 is primitive and that the divergence pattern is much more complicated than that suggested here. (From Schwanwitsch, 1935b)

largely independently from those in adjoining wing cells (see Chapter 5). Independent variation in adjoining wing cells is the norm in pattern formation. As in the case of pierellization, the perfect alignment of pattern elements in adjoining wing cells is due to evolutionary convergence in the position of those patterns. To make sensible deductions about the actual pattern of divergence in this unresolved radiation, it is, of course, necessary to know whether the group under consideration is monophyletic, the phylogenetic interrelationships within the group, and the aspects of the color pattern that are primitive for the group. None of these items of information is available for this particular group of Satyrinae, but Schwanwitsch's studies strongly

suggest that a proper phylogenetic analysis of color patterns ought to be feasible.

Lycaenidae

Schwanwitsch (1949) has produced a comprehensive comparative morphology of the color patterns of the Lycaenidae. In this work he explicitly elucidated the homologies of the patterns of more than 100 species worldwide in this family. Many lycaenid patterns are severely dislocated; pierellization is common, and Schwanwitsch (1949) noted that a breakup of the pattern into wing-cell-specific segments is the norm. Although he marveled at the pervasiveness and diversity of pattern dislocations, Schwanwitsch failed to recognize that the dislocated pattern is the primitive condition. The Lycaenidae adhere to the generalized nymphalid ground plan except that, according to Schwanwitsch, the border ocelli are absent in the entire family. The marginal ocelli on the hind wing, so common in this family (Fig. 3.4), appear to be elaborations of the submarginal band (Externa II, Table 2.1) and are therefore not homologous to the ocelli of the Nymphalidae. The distal band of the central symmetry system (Schwanwitsch's Media I) is the most diverse pattern element in this group (Fig. 3.5), but even this element is always easy to recognize across the family. Diversification of the element that Schwanwitsch identified as Externa III (homologous to the parafocal element) is shown in Figure 3.6. The reason for inferring

the absence of border ocelli in the Lycaenidae is never fully justified in Schwanwitsch's work. In my view, it is in fact most likely that the marginal ocelli (Fig. 3.4) represent true homologs of the nymphalid border ocelli and that the element that Schwanwitsch identified as Externa III (Fig. 3.6) is homologous to element **g**, the proximal member of the ocellar symmetry system. This means that the marginal and submarginal elements, and in most cases the parafocal elements, are absent from the ground plan of the Lycaenidae.

Kallima inachus (Nymphalinae)

Both the wing shape and ventral color pattern of the dead-leaf butterfly *Kallima inachus* serve to present an extraordinarily effective

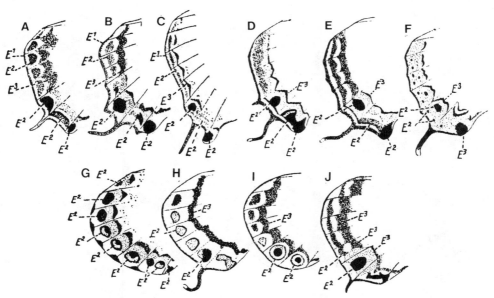

Figure 3.4. Diversity of marginal ocelli in the Lycaenidae. Schwanwitsch suggested that these patterns are homologous to the submarginal bands, but they are probably homologs of the border ocelli of Nymphalidae. A, *Lycaena arota*; B, *Jalmenus ictinus*; C, *Deudorix* sp.; D, *Strymonidia w-album*; E, *Zephyrus orientalis*; F, *Zephyrus lutea*; G, *Plebejus argus*; H, *Tarucus balkanicus*; I, *Tarucus* sp.; J, *Jamides elpis*. (From Schwanwitsch, 1949)

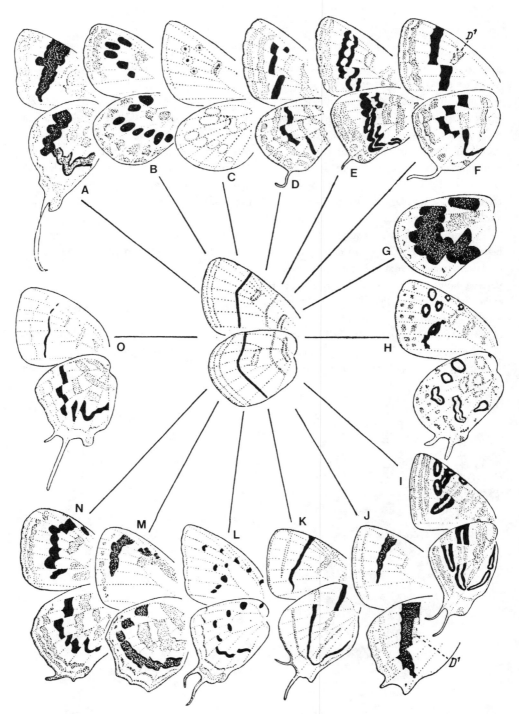

Figure 3.5. Diversity of the distal band of the central symmetry system in the Lycaenidae. *A*, *Bindahara phocides*; *B*, *Phengaris atroguttata*; *C*, *Lycaena pheretes*; *D*, *Tarucus balkanicus*; *E*, *Lampides boeticus*; *F*, *Jamides celeno*; *G*, *Tomares mauritanicus*; *H*, *Apharitis acamas*; *I*, *Spindasis lohita*; *J*, *Thecla betulae*; *K*, *Arawacus sito*; *L*, *Pseudolycaena marsyas*; *M*, *Hypochrysops polycletus*; *N*, *Amblypodia anthore*; *O*, *Marmessus ravindrina*. (From Schwanwitsch, 1949)

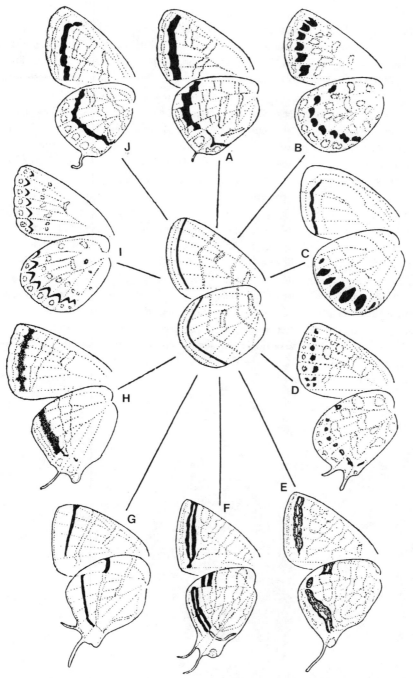

Figure 3.6. Diversity of pattern element **g** in the Lycaenidae. Schwanwitsch suggested that this element is one of the submarginal bands (his Externa III, homologous to the parafocal element as used in this work), but it is most likely homologous to **g** in Figure 2.17. *A, Jamides elpis; B, Lycaena lanti; C, Thysonotis danis; D, Apharitis cilissa; E, Apharitis acamas; F, Spindasis lohita; G, Arawacus sito; H, Jamides aratus; I, Lycaena semiargus; J, Tarucus balkanicus.* (From Schwanwitsch, 1949)

camouflage mimicry. The color pattern is an excellent representation of leaf venation, and the dead-leaf mimicry is further enhanced by the presence of fungus-spot mimics at various locations (Fig. 3.7). Süffert (1927) deduced the homologies of the pattern elements of this species and showed that they could be readily derived from the nymphalid ground plan; his analysis is shown in Figure 3.7. Two modifications of the ground plan account for the main characteristics of this pattern: (1) the alignment of both homologous and non-homologous pattern elements to form single lines of pigment across the wing, and (2) the loss of certain pattern elements in some wing cells. The large "midrib" pattern on the forewing is a composite. The posterior portion consists of the distal band of the central symmetry system, but this pattern element takes a proximal turn and becomes one of the "side veins," with the continuation of the midrib being made up of the outer portions of a row

of border ocelli. Just as in the case of *Pierella*, this composite nature of an apparently continuous pigment band becomes obvious only upon comparative study of pattern variation in which slight dislocations and discontinuities at the transition point become apparent.

The short pieces of "side veins" that contribute to the overall leaf mimicry on the proximal half of the wing are composed of bands of the central symmetry system that have become selectively expressed or suppressed in various wing cells. It is in the expression of these pattern elements that the wing cell independence of pattern is best illustrated. Figure 3.7 shows that the central symmetry system is actually expressed in few wing cells, and at the locations where it is expressed, the bands are angled in such a way as to give the visual impression of a radial leaf vein. No correlation or interdependence of position or angle exists among these portions of the central symmetry system. Thus in *Kal-*

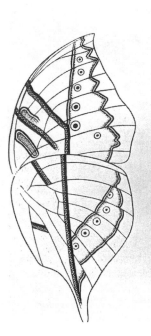

Figure 3.7. Derivation of the color pattern of *Kallima inachus* (Nymphalidae: Nymphalinae). Homologies of the pattern elements of this species can be found by comparison with Figures 2.1 and 2.13. (From Süffert, 1927)

lima we have on a single wing surface an illustration of a simultaneous selection for precise alignment of elements of one band of the central symmetry system and for a high degree of dislocation and selective suppression of elements in the other band. The evolution of this kind of a pattern is possible only because the development and variation of each pattern element within each wing cell can be controlled independently.

Iphiclides podalirius (Papilionidae)

The derivation of the color pattern of Papilionidae meets with some difficulties because no species in the family have a color pattern sufficiently close to the nymphalid ground plan to allow for the reconstruction of a continuous series of intermediates. Thus the homologies of the elements of papilionid color patterns must be deduced from their positional relations to the wing veins and to each other. Many species have striped patterns that are sufficiently similar to those of certain Nymphalidae (e.g., compare *Graphium* spp. with *Cyrestis camillus* and several *Marpesia* spp.) to allow connections to the nymphalid ground plan to be drawn with confidence. Süffert (1927) was able to reconstruct the pattern homologies in *Iphiclides podalirius*, one of the kite swallowtails (Fig. 3.8), because in this species the striped pattern on the ventral wing surface is much more differentiated than the stripes on the dorsal surface. The ventral pattern shows that some of the stripes are composite (Fig. 3.8B). Just as in *Pierella,* one major dislocation occurs in the pattern along vein M_3. The pattern elements posterior to this vein are dislocated in the proximal direction and come to line up with nonhomologous pattern elements in the anterior portion of the wing.

The dislocation is not quite as extreme as in *Pierella.* In *I. podalirius* the posterior portion of the distal band of the central symmetry system comes to be lined up with the discal spot (Fig. 3.8C) instead of with the anterior portion of the proximal band, as in *Pierella.*

The correct identification of pattern elements in *I. podalirius* would not have been possible without knowledge of a relatively rare form (*I. podalirius undecimlineata,* also called *I. p. podalirinus*). In the discal cell of this form, an additional band is present, which Süffert (1927) identified as the proximal band of the central symmetry system (element **d**). The pattern of *undecimlineata* is much less dislocated than that of the normal form, allowing an unambiguous identification of the bands on either side of vein M_3 (Fig. 3.8D). A comparison with other related species would not have been very useful in this case, because other kite swallowtails (*Graphium* spp. and *Eurytides* spp.) mostly lack the central symmetry system in the wing cells posterior to M_3. Many species of *Graphium* do, however, have an additional stripe in their discal cell as in *undecimlineata,* and we can thus use Süffert's analysis to identify this "extra" dash as element **d** in these species and go on from there to identify all the other pattern elements in these and other kites with abbreviated patterns.

The analysis presented above applies to a number of species in the genus *Papilio* as well, in particular to those of the tiger swallowtail group (subgenus *Pterourus*). Transitional forms that allow one to sort out the homologies are common in *Papilio eurymedon* and *P. pilumnus.*

Pieridae

Although most Pieridae have a simple color pattern, the pattern elements are clearly de-

Figure 3.8. Derivation of the color pattern of *Iphiclides podalirius* (Papilionidae). *A*, normal dorsal forewing pattern; *B*, normal ventral forewing pattern; *C*, interpretation of the homologies and dislocation of the bands (also compare with Fig. 2.13); *D*, dorsal forewing of the form *undecimlineata,* which shows the composite nature of the bands. (*C*, from Süffert, 1927)

rivable from the nymphalid ground plan as shown by Schwanwitsch (1956b). Discal spots and parafocal elements are virtually always present, even in genera like *Colias* and *Pieris* (ventrally in most species), which have extraordinarily abbreviated patterns (Fig. 3.9). Schwanwitsch (1956b) noted that the well-patterned genera such as *Catasticta* show many resemblances to the patterns of primi-

tive Papilionidae. *Catasticta* species, however, also have in the structure of their color pattern a most unusual feature, which, if it proves to be widespread among the Pieridae, would set this family well apart from the other butterflies. The unusual feature is that the entire central symmetry system in these species appears to lie proximal to the discal spot. In all other butterflies (and moths)

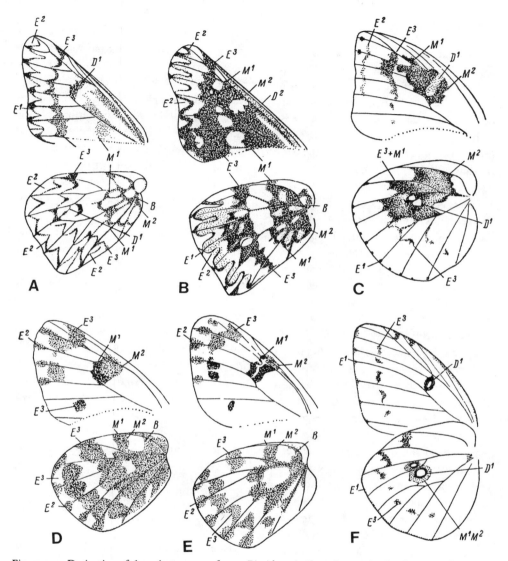

Figure 3.9. Derivation of the color pattern of some Pieridae. *A, Catasticta masica; B, Catasticta chrysolopha; C, Catopsilia catilla; D, Pontia daplidice; E, Pontia daplidice raphani; F, Colias hyale.* For nomenclature and homologies of the labeled pattern elements, see Figure 2.1 and Table 2.1. (From Schwanwitsch, 1956b)

studied, the discal spot lies in the central field of the central symmetry system. This basipetal shift of the central symmetry system is best seen in the hind wing pattern (because of the lack of pattern clarity on the forewing, there is some question of whether the shift also occurs on the forewing). Schwanwitsch showed the distal band of the central symmetry system on the forewing to be proximal to the discal spot in *Catasticta masica* (Fig. 3.9A) and distal to the discal spot in *Catasticta chrysolopha* (Fig. 3.9B). The basipetal shift on the hind wing may be fairly widespread, as Schwanwitsch also shows it on the hind wing of several species of *Dismorphia* and *Pontia,* though it is much less clear here than in *Catasticta.* In *Catopsilia* and *Colias,* however, the distal band of the central symmetry system lies distal to the discal spot, from which we can conclude that the basipetal shift of the central symmetry system may be common but not universal among the Pieridae. A similar shift of the distal band of the central symmetry system has been described in *Heliconius* (Nijhout and Wray, 1988; also see the section *"Heliconius"* below).

The explanation of this apparent breakdown in the otherwise constant relation between the central symmetry system and the discal spot probably resides in the fact that the crossveins that mark off the discal cell develop well after the color pattern has been determined. Therefore no necessary relation exists between the relative positions of the central symmetry system and the discal spot. This was already clear from the experiments on *Ephestia* by Kühn and von Engelhardt (1933), who showed that the bands of the central symmetry system could be shifted together without affecting the position or size of the discal spot. In addition, studies of the comparative morphology of patterns show that the discal spot usually lies well off the line of symmetry and varies independently of the bands of the central symmetry system. We can conclude that the processes responsible for positioning the sources of the central symmetry system and the discal spot are separable and probably quite independent from one another. That the discal spot nearly always lies within the central symmetry system is probably a secondary consequence of both being determined by processes that seek the approximate center of the wing.

The hind wings of Pieridae are then unusual in that the processes that determine the position of the sources for the central symmetry system cause them to be placed much more proximal than is the norm in other butterflies. Only in two other taxa, the Brassolinae and the Morphinae, do we find a similarly severe proximal dislocation of the central symmetry system. In the members of these two groups, the border ocelli lie on a line just distal to the apex of the discal cell. The central symmetry system is difficult to identify in most brassolines and many morphines.

Many Pieridae have a pattern of venous stripes in addition to the elements of the ground plan. In some species the venous stripes are the dominant feature of the wing pattern, particularly on the ventral hind wing. The venous stripes may be identical in color and fuse smoothly with some of the elements of the nymphalid ground plan, as in *Belenois,* or they may be distinctively colored and remain well separated from the ground plan elements, as in some species of *Colotis* and *Tatochila.* Similar diversity in the interaction between venous stripes and elements of the ground plan is found throughout the butterflies.

Charaxes (Charaxinae)

Charaxes, a genus of tropical butterflies from Africa and Asia, has a greater diversity of color patterns than is seen in any other taxon of similar rank. Among its members are species with the most complex color patterns found in the butterflies. Nijhout and Wray (1986) studied the comparative morphology of the color patterns of this genus as a means of testing the sufficiency of the nymphalid ground plan in accounting for truly difficult and divergent patterns. We found that the identity of every dot and bar of the *Charaxes* color patterns could be homologized to elements of the nymphalid ground plan. The peculiar structure of the pigment bands of some species allowed us to develop deeper insights into the relations that exist among the elements of the nymphalid ground plan. We can now recognize at least five different levels of homology within the nymphalid ground plan; these will be explained later in this section, after we learn how to read the patterns in this genus.

As in most species of butterflies, the color pattern on the ventral hind wings of *Charaxes* is the most conservative and is therefore the pattern most easily derived from the nymphalid ground plan (Nijhout, 1985b; Nijhout and Wray, 1986). By "most conserv-

ative," we mean that this pattern usually has most of the elements of the ground plan present, whereas the patterns on the other wing surfaces tend to be more or less modified from the ground plan by the absence of elements or by fusion among them. Thus the identities of pattern elements and their homologies to the pattern elements of other species are most easily discerned on the ventral hind wing surface. In many species it is, in fact, difficult if not impossible to understand the structure of the dorsal wing pattern without first having developed an understanding of the ventral hind wing pattern.

The *Charaxes* ground plan is illustrated in Figure 3.10. It differs from the general nym-

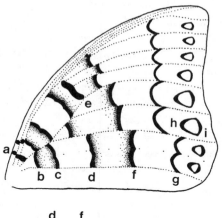

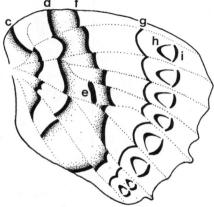

Figure 3.10. The ground plan of the genus *Charaxes*.

phalid ground plan only in the absence of the submarginal bands (element j) and in the emphasis of the distinction between elements g and h in the border ocelli system. Element g is an arc-shaped element that represents the outer boundary of the border ocelli system; it is usually expressed as a distinctly pigmented band, but occasionally only as a more or less sharp color transition (e.g., *C. analava*, Fig. 2.5A). Pattern element h is homologous to the border ocelli. In *Charaxes* this pattern element is complex, much more so than in any other taxon of butterflies. Variation in the morphology of elements h and i on the hind wing among species of the *brutus* subgroup is illustrated in Figure 3.11 (see also Plate 4B). The proximal black portion of this pattern corresponds to the border ocellus (element h) of other species, and the distal black portion is homologous to the parafocal element (element i). Because of the complexity of patterns in the border ocellus region, Nijhout and Wray (1986) did not recognize the parafocal element as a distinct pattern element in *Charaxes* but as a portion of element h. That interpretation is probably incorrect, because element i seems to vary independently from element h (Fig. 3.11). The *Charaxes* ground plan shown in Figure 3.10 reflects this interpretation.

The manner in which a broad diversity of real *Charaxes* patterns is derived from this ground plan is illustrated in Figure 3.12 for the ventral hind wing and in Figure 3.13 for the ventral forewing. In each of these figures a pair of panels to the right of the photograph presents a semidiagrammatic dissection of the color pattern. To avoid crowding and to facilitate the comparison between species, each panel shows alternate elements of the ground plan so that the complete pattern is reconstructed by overlaying the two diagrams. Thus the diversity of form of any

given pattern element can be readily seen by scanning up and down a column of diagrams.

In addition to illustrating the diversification of form within a closely knit evolutionary lineage, the analyses shown in Figures 3.12 and 3.13 reveal two features that are of interest for pattern evolution in general. First is the evolution of complex bands in the *etesipe* and *brutus* species groups. Different species in these groups have evolved various degrees of elaboration in the coloration of the bands of their central symmetry system. These elaborations consist of the development of a core of contrasting color, so that the band itself takes on the appearance of a miniature symmetry system (Fig. 3.14). In addition, small circular spots, made up of concentric circles of pigment identical to those of the complex band, often accompany such bands. These small satellite spots, as they are called, may be separate from the band (Fig. 7.18), though more often than not they are partially fused with it, as shown in Figure 3.14. The concentric structure of complex bands is remarkably similar to that of small symmetry systems. The satellite spots, too, resemble small symmetry systems or ocelli. The concentric character of both elements strongly suggests that the bands are structured around sources of pattern induction, just as symmetry systems and ocelli are (Chapter 5). The implications of this observation will be examined in Chapter 7.

The second feature of interest is the finding that many bands and all broad areas of solid black are made by the fusion of several adjacent pattern elements. On the ventral wing this is best illustrated by the patterns of *C. nobilis* (Figs. 3.12 and 3.13). In this species almost every one of the broad black bands on the forewing is composite. The fusion of pattern elements is of a very different type from that which we saw earlier in *Pi-*

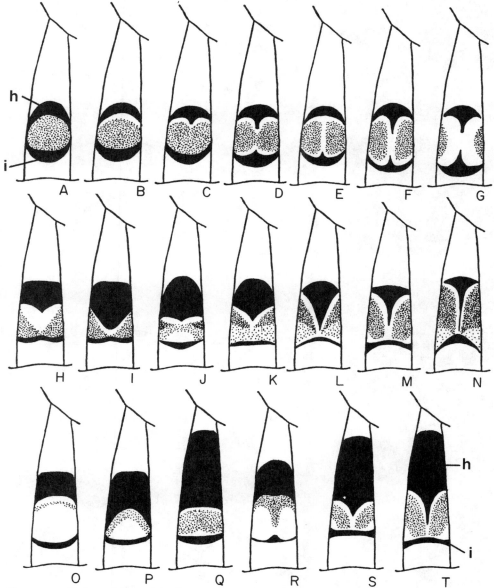

Figure 3.11. Unity and diversity of form in pattern elements h and i in the *brutus* species group of *Charaxes*. A, F, H, and I, *Charaxes ansorgei*; B–D, *C. pollux*; E and G, *C. brutus*; J–N, *C. jasius;* O and P, *C. andranodorus*; Q, S, and T, *C. castor*; R, *C. andara*. (From Nijhout and Wray, 1986)

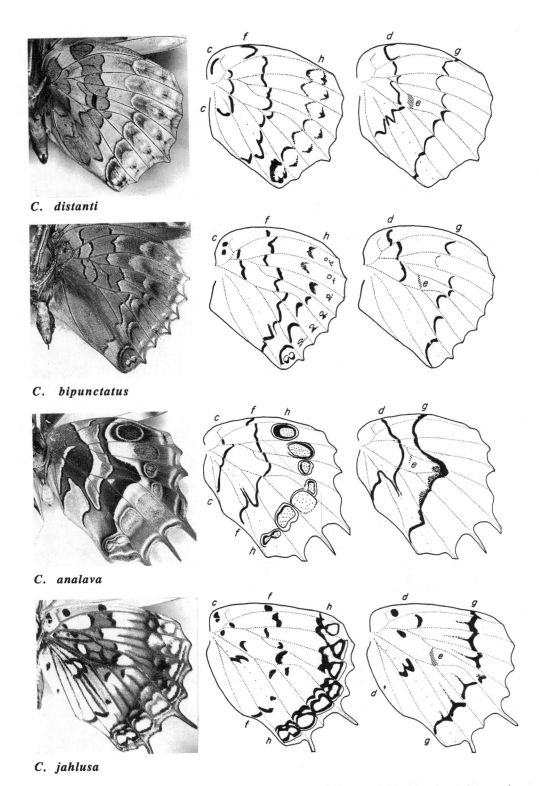

C. distanti

C. bipunctatus

C. analava

C. jahlusa

Figure 3.12. Homologies in the ventral hind wing patterns of *Charaxes*. For each species, a photograph is accompanied by two diagrams, each of which isolates alternate elements of the ground plan. (From Nijhout and Wray, 1986)

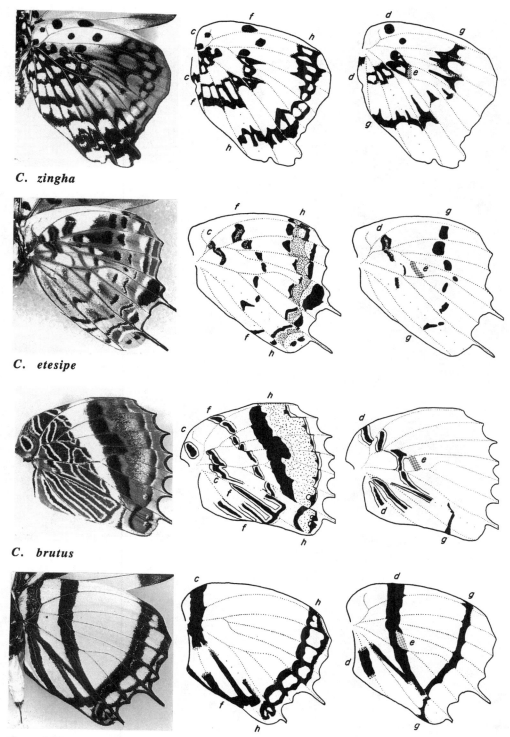

C. zingha

C. etesipe

C. brutus

C. nobilis

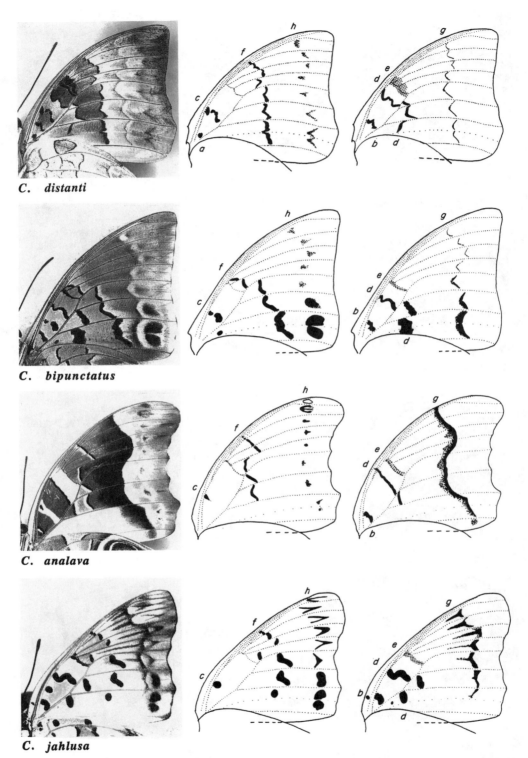

C. distanti

C. bipunctatus

C. analava

C. jahlusa

Figure 3.13. Homologies in the ventral forewing patterns of *Charaxes*. For each species, a photograph is accompanied by two diagrams, each of which isolates alternate elements of the ground plan. (From Nijhout and Wray, 1986)

C. zingha

C. etesipe

C. brutus

C. nobilis

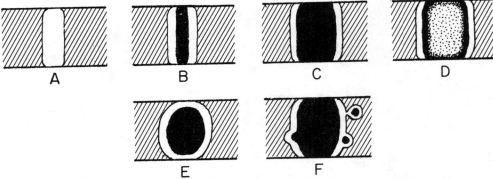

Figure 3.14. Complex bands in *Charaxes*. In species of the *brutus* group, each of the bands of the central symmetry system exhibits various degrees of elaboration from normal single bands (*A*) to bands with complex internal differentiation that form a secondary symmetry system (*D*). In many cases, such a complex band becomes circular (*E*) and resembles an ocellus. Occasionally, small ocelluslike satellite spots develop near these complex bands (*F*). (From Nijhout and Wray, 1986)

erella and *Kallima*. The fusion in the latter two species is a serial fusion in which homologous (and occasionally nonhomologous) pattern elements in adjoining wing cells line up to make a continuous band. *Charaxes nobilis* also has this type of fusion, but in addition its pattern elements exhibit lateral fusion, in which two nonhomologous elements within the same wing cell fuse to form a broader-than-usual pattern.

As with many of the preceding pattern analyses, this lateral fusion could not have been deduced from the study of a single specimen. All that one can generally detect in a single specimen is that there are too few bands in the color pattern to account for all the elements of the nymphalid ground plan. We have already seen that the loss of pattern elements is quite a common phenomenon in pattern evolution, and this should be the first hypothesis upon finding that a given species' pattern appears to be lacking one or more elements. In this case, however, when one has several specimens at one's disposal, it is possible to find individuals that actually show a thin line of white scales (white is the background color in *C. nobilis*) running down the

middle of some of their black bands, which one can take as initial evidence of the composite nature of the bands. In addition, and particularly in fresh specimens, it is possible to detect the composite nature of many bands by slight local difference in the texture, intensity, and hue of their black pigmentation. Such subtle differences in texture and coloration are usually evident only under particular angles of illumination and do not photograph well. But because the boundaries between such areas vary little among individuals and also correspond to the locations of the sparse white scaling just mentioned, the combined observations provide strong evidence for the composite nature of these bands. The main fusion on the ventral forewing of *C. nobilis* is between the two bands of the central symmetry system (elements **d** and **f**).

The patterns on the dorsal wing surfaces of *Charaxes* appear to be much simpler than those on the ventral surfaces. Most species have a dorsal pattern of broad black areas against a brown or off-white background, and that pattern appears on first sight to bear no resemblance to the more finely detailed patterns on the ventral wing surfaces or to the nymphalid ground plan. However, a few species have fairly fragmented dorsal wing patterns, and it is possible to recognize in them the elements of the nymphalid ground plan. By studying pattern variability in such species, it is possible to find transitional stages to patterns in which the black pattern elements are much enlarged and become fused so that the ground plan is not readily detectable. Figure 3.15 illustrates the analysis of two very different dorsal wing patterns in *Charaxes,* one in a species with a sparse black pattern (*C. protoclea*) and the other in a species with a more typical wing pattern that is mostly black (*C. castor*). It is clear that the mostly black patterns are far from simple. All elements of the nymphalid ground plan are present but are much enlarged and fused, so that the black surface is in reality a complex composite pattern. Enlargement and fusion of pattern elements to produce large fields of

a solid dark color appear to be a common evolutionary theme among the butterflies. All patterns that have substantial areas of black are probably built up in this way.

In addition to the specific features of pattern morphology we have discussed so far, the analysis of *Charaxes* color patterns revealed a hierarchical system of organization in which several levels of iterative homology can be recognized (Nijhout and Wray, 1986). Figure 3.16 summarizes the various forms of homology that can be recognized in *Charaxes* and that apply by extension to all butterflies. In deducing these homologies, we adopted Roth's (1984) operational definition of homology, which holds that structures are homologous if they share a developmental pathway. This definition applies to the classically recognized cases of phylogenetic homology and to the various forms of serial or iterative homology as well. Thus, in butterfly wings,

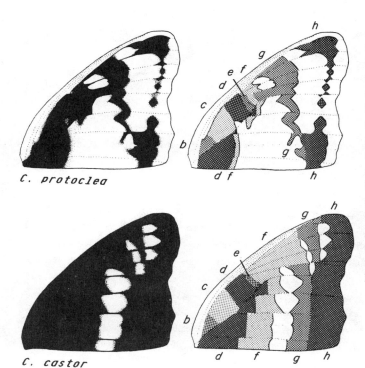

C. protoclea

C. castor

Figure 3.15. Derivation of the forewing pattern of *Charaxes.* Black is pattern, and other areas are background. Large areas of black are made by enlargement and fusion of the elements of the nymphalid ground plan. The makeup of the pattern is most easily recognized from species with relatively little black, like *C. protoclea*. From such species, it is possible to trace gradual transitions to more solidly black patterns, like those of *C. castor*. (From Nijhout and Wray, 1986)

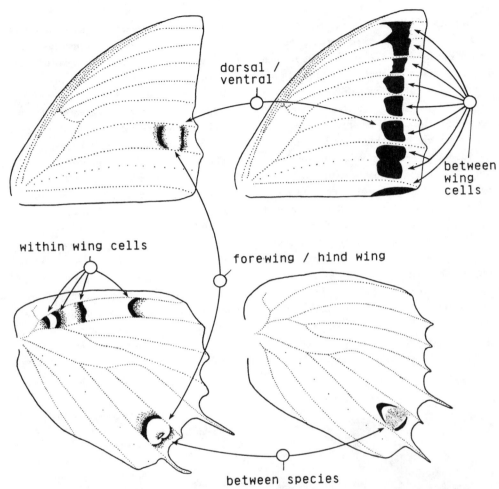

Figure 3.16. The five levels of homology that exist among the elements of the nymphalid ground plan as exemplified by the patterns of *Charaxes*. Classical phylogenetic homology is represented by between-species homology; the other four cases are different kinds of iterative homology. (After Nijhout and Wray, 1986)

pattern systems that are induced by a gradient of determination that emanates from discrete inductive sources are homologous in that respect. This means that border ocelli, central symmetry systems, and basal symmetry systems are all homologous as symmetry systems. In *Charaxes* species of the *brutus* group, the bands of the basal and central symmetry systems are identical in structure and pigmentation, and the border ocelli of these species, though different in morphology, have patterns with the same sequences of pigments that are found in the two symmetry systems. Thus the processes that are responsible for generating these three pattern elements produce different morphologies but make them out of the same array of pigments. As we will see in later chapters, this similarity in coloration probably results from these three pattern elements' being induced by gradients of inductive signals that differ in shape but not in kind. If this interpretation is correct, then all three pattern elements are perfect iterative homologs of one another. In other species of butterflies, in which the pigmentation as well as the morphology of these three elements differ, it will be necessary to hypothesize the existence of additional factors that account for the diversification in pigmentation.

Heliconius (Heliconiinae)

Heliconius butterflies are of major interest in evolutionary biology because of the extensive systems of Müllerian mimicry that have evolved among many of the species in this ge-

nus. Studies on the genetics of mimicry systems in *Heliconius* by J. R. G. Turner, L. E. Gilbert, and their co-workers (see Chapter 6) have gone a long way toward developing an understanding of the evolution of polymorphism, mimicry, and the genetic control of pattern formation in this group. The explication of the genetics of color patterns in *Heliconius* will be a major concern in Chapter 6; here we will restrict ourselves to developing the necessary preliminary understanding of the homologies among wing pattern elements in this genus and of the manner in which the rather unusual patterns of these butterflies (Fig. 3.17) are derived from the nymphalid ground plan.

The color patterns of most species and

Figure 3.17. Color patterns of *Heliconius,* illustrating transitions from patterns in which the elements of the nymphalid ground plan are readily recognized (*A*) to patterns that are largely black (*F*) because of enlargement and fusion among the elements of the pattern. *A* and *C, Heliconius ethilla; B, H. numata; D, H. ismenius; E, H. wallacei; F, H. burneyi.*

races of *Heliconius* are made up of large and usually irregular areas of black and various bright colors with no obvious relation to any of the elements of the nymphalid ground plan. The key to understanding the structure of these color patterns lies in the patterns of the primitive genera of Heliconiinae such as *Dione* and *Agraulis* (Nijhout and Wray, 1988). The patterns of these primitive heliconines resemble those of many fritillaries and are readily derived from the nymphalid ground plan. The derivation of the primitive heliconine ground plan from the general nymphalid ground plan is shown in Figure 3.18B. The ground plan in Figure 3.18B is based on the combined patterns of *Dione* and *Agraulis*. It differs from the generalized nymphalid ground plan by a loss of elements in the border ocelli system and by a proximal dislocation of pattern elements f and h in the wing cells anterior to vein M_3. This dislocation is along the same boundary as the pierellization we dealt with earlier, but in the opposite direction. None of the heliconines possess clearly developed border ocelli or parafocal elements. Nijhout and Wray (1988) interpreted the pattern in the border ocellar region of the heliconines to be homologous to element g. It is probably more correct to assume either that this element is a fusion of g and h or that it is element h alone and that element g is in fact missing from the heliconine ground plan. The M-shaped element at the wing margin is then either a submarginal band or a parafocal element (i), but because there are no transitional forms to a more "standard" nymphalid pattern within the tribe, we may never be entirely confident about the homology of this element. The shape of this element most resembles a common shape theme found among the parafocal elements across the butterflies (see "Themes and Variations" in Chapter 2), so here it is

tentatively identified as homologous to the parafocal element (i).

To derive the patterns of the advanced heliconines (the genera *Heliconius, Laparus,* and *Neruda*), it proved essential to start the analysis with those species that have the most broken up and detailed patterns. The species composing the *H. silvana* group (Emsley, 1965; Brown, 1972, 1981) have clearly discernible pattern elements (Fig. 3.17A–D), most of which repeat from wing cell to wing cell and can be shown to be homologous to the elements of the nymphalid ground plan (Nijhout and Wray, 1988). Figure 3.18C shows the ground plan that applies to the entire genus *Heliconius* and its sister genera *Laparus* and *Neruda*. The ground plan of the forewing differs from that of the primitive heliconines only in the loss of pattern elements f posterior to vein M_3 and an even more severe proximal dislocation of pattern element h anterior to M_3. Differences in the relative degree of expansion of these pattern elements, and in the preferred direction of that expansion, determine the overall wing pattern of *Heliconius* species as shown in Figure 3.19.

The ground plan of the *Heliconius* hind wing (Fig. 3.18C) is considerably simplified and somewhat different from that of the nymphalid ground plan (Fig. 2.17). The anterior quarter of the hind wing usually has a broad dark band that is made up of the fused elements c and d, and possibly element b as well. Pattern elements f, h, and i make up most of the wing pattern, and, unlike the forewing, there is a general tendency in the hind wing for all wing cells to bear all of the elements of the ground plan (Nijhout and Wray, 1988). In addition to the normal elements of the ground plan, many *Heliconius* hind wings bear venous and intervenous stripes. These elements give the hind wing

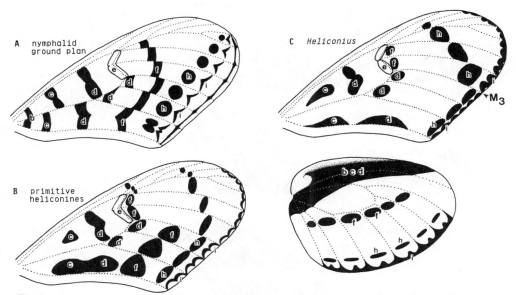

Figure 3.18. Derivation of the *Heliconius* ground plan from the generalized nymphalid ground plan. *A*, the nymphalid ground plan projected onto the wing shape of *Heliconius*. *B*, the ground plan of the primitive heliconines, *Dione* and *Agraulis*, illustrating the proximal dislocation of the elements anterior to M_3 that is characteristic of this group. *C*, the ground plan of the genus *Heliconius*, which is derived from the primitive heliconines by a more severe dislocation and the loss of element f posterior to M_3 on the forewing. (Modified from Nijhout and Wray, 1988)

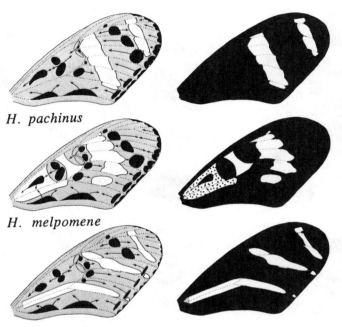

H. pachinus

H. melpomene

H. charitonia

Figure 3.19. Derivation of the patterns of three species of *Heliconius*. Species differ in the degree of expansion and in the preferred direction of expansion of the pattern elements. (After Nijhout and Wray, 1988)

pattern a rayed appearance. Some of the more common ways in which hind wing patterns are built up are illustrated in Figure 3.20. As noted earlier, it is extremely rare to have the simultaneous occurrence of intervenous stripes and nymphalid ground plan elements. Yet on the *Heliconius* hind wing the superposition of these two types of elements occurs quite commonly and is responsible for a characteristic feature of *Heliconius* hind wing patterns, namely the presence of a single or double row of white spots parallel to the wing margin. Figure 3.21 illustrates the manner in which this spotting pattern is believed to originate. Similar submarginal spotting patterns occur in the genus *Hypolimnas* and are a nearly universal feature of the wing pattern in the Danainae, where they also arise by the superposition of pattern elements, as will be seen in the treatment of *Idea* (below) and in the analysis of the pattern of several danaines in Chapter 4.

To understand the morphology of the *Hel-*

iconius color pattern, it is essential to recognize that the black and red areas constitute the pattern, and the yellow and white portions are background (Nijhout and Wray, 1988; L. E. Gilbert, pers. com.). Like the dorsal wing patterns of *Charaxes,* the large black portions of the *Heliconius* wing pattern are composite, resulting from the fusion of many pattern elements. In members of the *silvana* group the composite nature of black areas is clearly evident (Fig. 3.17A–D), but in most other species in the genus the boundaries between pattern elements are not always easily visible. The only complication that is added to what we have learned about composite patterns from *Charaxes* is that in *Heliconius,* pattern elements can also be red. As we will see in Chapter 6, different alleles at certain genetic loci are involved in switching some pattern elements between red and black. This feature has proved invaluable in confirming deductions about the shape and boundaries of particular pattern elements and the composite nature of the black areas of the wing, because particular pattern elements can be switched from black to red and thus be made to stand out within an otherwise black area (L. E. Gilbert, pers. com.; also see color illustration in Gilbert, 1984).

In *Heliconius* the elements of the ground plan are themselves centers of origin of pattern, suggesting that they are functionally homologous to secondary symmetry systems. In secondary symmetry systems (Figs. 2.6 and 2.8), each of the bands of the central symmetry system becomes a symmetry system itself and must therefore have become a source of pattern induction (Chapter 5).

Idea and *Ideopsis* (Danainae)

The large Indo-Australian danaids *Idea* and *Ideopsis* have relatively simple patterns of

black spots on a white background (Ackery and Vane-Wright, 1984; Fig. 3.22). Yet their patterns are surprisingly diverse for such closely knit groups of species. They are of particular interest because the spotting pattern near the wing margin presents some unusual difficulties when one attempts to homologize the pattern to the nymphalid ground plan. However, the substantial diversity of patterns available in these two genera makes it possible to dissect the nature of these submarginal patterns, and this in turn has proved useful in understanding the makeup of patterns in the Danainae in general. In Chapter 4 we will look at pattern diversity of these two genera in some detail; here we will merely attempt to reconstruct the homology relations among various pattern elements.

It is convenient to use the pattern of *Idea*

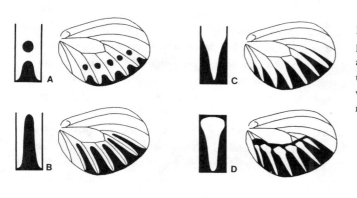

Figure 3.20. The hind wing patterns of *Heliconius* are usually made up of a serial repetition of midline (*A* and *B*) or venous (*C* and *D*) pattern elements.

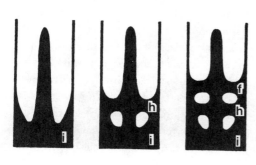

Figure 3.21. Origin of the "double dot" pattern along the wing margin of many species of *Heliconius*. The white dots are background that remains when an intervenous stripe and pattern elements f and h do not fully overlap. (From Nijhout and Wray, 1988)

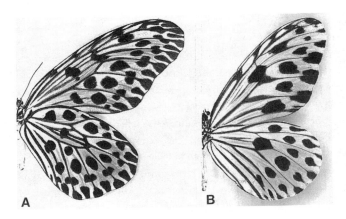

Figure 3.22. The color patterns of (A) *Idea lynceus* and (B) *Ideopsis gaura* (Nymphalidae: Danainae).

A   B

lynceus (Fig. 3.23) as our standard or ground plan for these taxa, because this species most fully expresses all the pattern elements typical to these two genera. The ground plan for *Idea* consists of pattern elements **d** (proximal band of the central symmetry system), **e** (discal spot), **h** (border ocelli), and a complex submarginal pattern, possibly homologous to the parafocal elements (element **i**), but with a highly derived morphology that arises from its interaction with the wing veins. The ground plan of *Ideopsis* differs from the *Idea* ground plan in that element **d** is lacking in almost all species. No pattern elements other than those illustrated are present in *Idea* or *Ideopsis,* and it is probable that the ground plan shown in Figure 3.23 will be found to apply with relatively few modifications to all the Danainae. Pattern element **f**, the distal band of the central symmetry system, is absent in both genera. The rationale for identifying the large spots on the wing cell midline as element **h**, and not **f**, is as follows. The much simplified patterns of *Idea blanchardii* and *I. durvillei* show that the two main pattern elements in the middle of each wing cell are derived from an intervenous stripe (Fig. 3.24A–C). This intervenous stripe breaks up into two short segments that can become tri-

angular or rounded. This joint origin of the two pattern elements means that the more proximal of the two cannot be part of the central symmetry system (element **f**) and must therefore be the homolog of the border ocelli (element **h**). The marginal spot on the wing cell midline is difficult to homologize to any element in the nymphalid ground plan. From its position, we may assume that it is derived from the parafocal element. Developmentally, it is the product of the distal portion of the broken intervenous stripe of *Idea durvillei*. As we will see in Chapter 7, a developmental continuity exists between intervenous stripes and the elements of the nymphalid ground plan. The marginal patterns of several species of *Idea* and *Ideopsis* are intermediates in this transition; therefore they are not unambiguously one or the other.

An evolutionary continuity also exists between intervenous stripes and the elements of the nymphalid ground plan (Fig. 3.24). We just saw that across the species of *Idea*, one can construct a morphological series of patterns that shows a transition from strict intervenous stripes (*Idea blanchardii*) to patterns in which a long intervenous stripe is broken up into two shorter dashes (*I. durvillei*) to patterns in which each of these dashes becomes

Figure 3.23. Homologies of the pattern elements of *Idea lynceus*.

rounded into one of the ground plan elements (*I. iasonia, I. lynceus*). Similar transitions between intervenous stripes and discrete ground plan elements are seen in many other taxa (e.g., *Acraea, Papilio* [see Fig. 6.8], *Precis, Cepora, Catasticta, Cethosia, Pseudacraea, Neptis, Epitola,* and *Euselasia*).

The species of *Idea* and *Ideopsis,* and the Danainae in general, have a tendency to develop well-defined venous stripes. These venous stripes tend to fuse smoothly with any nearby element of the ground plan (Fig. 3.24). This can give the overall color pattern a very irregular appearance, which occasionally makes it difficult or impossible to distinguish the boundaries of each pattern element precisely. The smooth fusion of venous stripes and ground plan elements suggests that both patterns arise by identical developmental processes or morphogenetic gradients that control the same determinative switch from white to black (see Chapter 7). A particularly interesting feature of this inductive compatibility is that it can reveal otherwise invisible (presumably subthreshold) pattern potential within a wing cell. For example, it is common to find that a venous stripe slightly bulges out exactly next to a pattern element in the middle of the wing cell (Fig. 3.24D,E). Sometimes this bulge is so large as to fuse smoothly with that element. In cases where the element is small or not expressed at all, a bulge is often observed in the venous stripe, which I interpret as indicating that the factors that normally induce the central pattern element are present at a subthreshold level that is nevertheless high enough to interact additively with the venous signal.

The complexity of the submarginal patterns of *Idea* and *Ideopsis* can be explained with the help of this idea. The basic element of the submarginal pattern is a short inter-

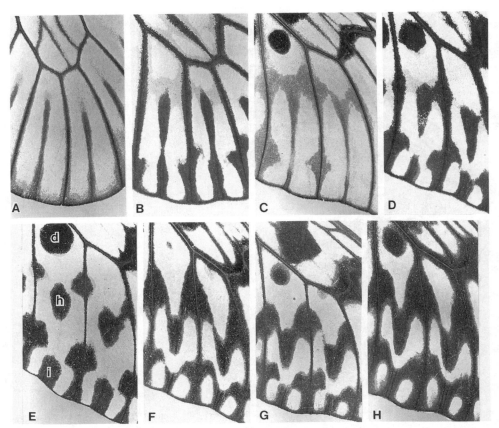

Figure 3.24. Morphological series of marginal patterns in *Idea*. Arranged as a progression from simple midline patterns (*A*), the series illustrates a breakup of the midline (*B* and *D*), followed by the broadening of each of the two remaining pieces and their fusion with pattern lobes that appear to emerge from the wing vein (*E–G*), and resulting patterns with a zigzag or scalloped appearance (*H*). *A* and *C*, *Idea blanchardii*; *B*, *I. durvillei*; *D*, *I. idea*; *E*, *I. iasonia*; *F* and *G*, *I. hypermnestra*; *H*, *I. leuconoe*. The identities of the pattern elements are shown in *E*.

venous stripe that may bulge slightly at its apex. Added to this are two lobes that emerge from the venous stripes, one near the wing margin and one just proximal of the apex of the intervenous stripe (Fig. 3.24). According to the hypothesis presented above, the venous bulges reveal that the local threshold for the white-to-black transition is influenced by the pattern element on the wing cell midline. The shape of the inhomogeneity in the pattern threshold is revealed by individuals on the extreme end of the range of variability. In

such individuals the venous bulge and the central pattern element connect to form a pair of arches or loops than enclose two circular white spots (Fig. 3.25). Differences in the degree of dominance of the intervenous stripe and of the shape of the bulge that develops at its apex, together with differences in the degree of fusion of these pattern elements with venous lobes, account for most of the pattern diversity in the submarginal area of *Idea* and *Ideopsis*. In *Idea* there is a nearly perfect parallel between pattern diversity in

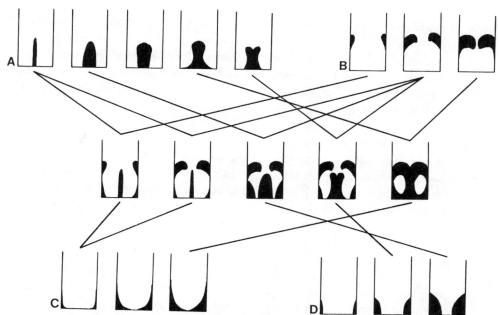

Figure 3.25. A model for the component parts of the submarginal patterns (element i) of *Idea* and *Ideopsis*. Three independently varying elements make up this pattern: a short midline stripe (*A*), a pair of venous lobes (*B*), and a corner element, which occurs in two forms (*C* and *D*). Real patterns (*middle row*) are made up of the superposition and interaction among these three components. The morphology and diversity of pattern element **h** arise through similar interactions (see Fig. 3.24).

the submarginal area and pattern diversity of element **h**. In both cases the fundamental element is a circular spot on the wing cell midline that becomes connected to the lateral bulges emerging from the wing veins. The method of superposition of pattern elements shown in Figure 3.25, will form the basis of a new way of analyzing the derivation and diversification of wing patterns, which we will explore in Chapter 4.

Envoy

The extraordinary orderliness of butterfly color patterns expressed in the Schwanwitsch-Süffert ground plan, exemplified by Figures 2.3 and 2.4, and summarized in this chapter stands in contrast to the general usage of pattern nomenclature among lepi-

dopteran systematists. Systematists today use an informal nomenclature developed for each of the species or genera in which they specialize. Terms such as "medial line," "mesial line," "submedial line," "postmedial line," "inner postmedial line," "subapical spot," "submarginal spot," and "apical bar" are used without regard to their homology in other taxa and without regard to the enormous amount of information on homology that is available for nearly all major families of butterflies and moths (see references to Schwanwitsch's work at the back of this book). By analogy, it is as if we used terms like "apical lobes" in cats, "appendicular protrusions" in dogs, and "claspers" in monkeys for what we call toes in humans. Such nomenclature may be fine if all we want to know is whether there are three, four, or five of these things in a given species, but ignoring that these structures are homologous and share developmental and evolutionary origins discourages the possibility of discovering unity among taxa and degrees of relationship among species. It is not that the concept of homology is shunned. Homologies of wing veins, leg segments, and body parts are universally used in the classification and systematics of butterflies, just as homologies among wing pattern elements are universally ignored. Knowledge of homologies among the elements of wing patterns is old and well documented, and a significant portion of the literature in this field is in English, so inaccessibility of the concept is not the explanation for the general failure by systematists to exploit the information inherent in the commonality of pattern among Lepidoptera.

Although I am not proposing uncritical adherence to Schwanwitsch's schemes—Schwanwitsch oversimplified many things and misinterpreted others—I would like to suggest that he has produced a treasury of information that ought to be examined carefully by anyone who wishes to use color pattern characters in lepidopteran taxonomy and systematics. Wing patterns provide a potentially rich source of taxonomic and phylogenetic information. Wing cell autonomy in the presence, size, shape, and position of each of the pattern elements provides a large number of characters (6–8 wing cells $\times$ 6–8 pattern elements $\times$ 4 wing surfaces) with various degrees of developmental and evolutionary association among them. All these characters are two-dimensional and observable without dissection. They are therefore readily amenable to morphometric analysis.

In the previous two chapters, we looked at pattern structure and diversity from the perspective of the nymphalid ground plan. We have seen that pattern elements are serially repeated from wing cell to wing cell and that much of color pattern evolution has proceeded by the independent modification of the pattern elements in each of the wing cells. Within a wing cell, the diversity of pattern elements exhibits distinct morphological themes, and many of these themes have a large number of variations that bridge the gaps between them. This observation enables us to look at pattern diversity in a new and completely different way. If we focus on the wing cell as the unit of pattern evolution, then the commonalities and differences in the shapes of pattern elements, rather than their distribution and relative locations, become the features of interest.

In this chapter, therefore, instead of looking at the way in which the arrangement and morphology of the elements of the nymphalid ground plan form a particular overall pattern, we will analyze the morphology of the patterns in a single wing cell and explore the nature of the diversity of these patterns. In doing this analysis we will restrict our attention to taxa that possess wing cell patterns that are clearly serially repeated. Because these constitute the majority of pattern types among the butterflies, the analyses below will be of significance for understanding much of pattern evolution. The analyses will be performed at the level of the genus, and we will take our examples from small to medium-sized genera of several families of butterflies.

Each analysis will start by documenting the diversity of patterns in the genus. This pattern diversity will be based on the pattern in one of the central wing cells of the hind wing. When the pattern of an individual wing cell is isolated in this way, a great deal

Chapter 4

Exploring Pattern Morphospace

of order and regularity is revealed. Although no two patterns in a genus are quite alike, most patterns will be found to have, in addition to their unique features, certain features in common with one or more of the others. When patterns are compared in this manner, we see that several features of their morphology appear to vary independently from other features. Thus, we see within each wing cell the repetition, on a small scale and in a slightly different manifestation, of the characteristic of the overall wing pattern: a structure composed of a discrete number of semi-independent units of variation.

As we did for the patterns on the whole wing surface, we will now attempt to break the diversity of wing cell pattern shapes into those parts or aspects that vary independently from one another. Each independent variable that can be identified will be described as an axis of variation, or morphocline, that varies continuously between the extremes of the character. Each morphocline will be represented by a panel of four patterns that illustrate the endpoints and two intermediate stages in the value of the character in question; each morphocline is, however, to be regarded as an axis of continuous variation with an indefinitely large number of intermediates between the extremes. If these morphoclines are imagined to be arranged orthogonally to each other, they would form a multidimensional space of potential morphologies: a morphospace. (For an analogy, see the morphospace developed by Raup, 1966, for the diversity of shell shapes in mollusks.) The pattern of each species occupies a point in this morphospace, and this point can be projected onto each of the axes of variation (Fig. 4.1). The morphology of a species' pattern can therefore be summarized by its coordinates in morphospace.

This method of analysis provides us with several kinds of insight into pattern morphology that are useful in developing a fuller understanding of pattern evolution. First, it makes us focus on a mode of variation that is not dealt with by the nymphalid ground plan but is fundamental to the fine tuning of the structure of the pattern. Second, it gives us a method of summarizing pattern morphology by identifying the independent variables in the pattern. These variables, in turn, may be produced by independent developmental processes and may thus represent the potential axes of evolution. Third, it allows us to determine how species are clustered in pattern morphospace. This may enable us to define useful taxonomic characters in the color pattern.

Below are the pattern morphospace analyses for seven genera, beginning with simple patterns with few variable components and progressing to fairly complicated patterns with more variables. Before we start, it will be useful to outline the conventions that will be used to define and interpret the axes of morphological variation. Assume we are dealing with an arc-shaped pattern. The modes of variation in such an arc are its position, its width, and its aspect ratio (Fig. 4.1). All three of these modes of variation can be expressed as a morphocline, and each can be projected onto the others. It is, therefore, not necessary to plot an axis of variation in the width of the arc for a proximal and a distal location independently. This may seem trivial, because the three axes in this example define a three-dimensional space that is easy to visualize. But as we will see, as many as ten axes of variation are needed for some genera, and the ten-dimensional space they define cannot be visualized. Thus, in examining the morphoclines in the figures below, it will be necessary to continually remind oneself that they all interact.

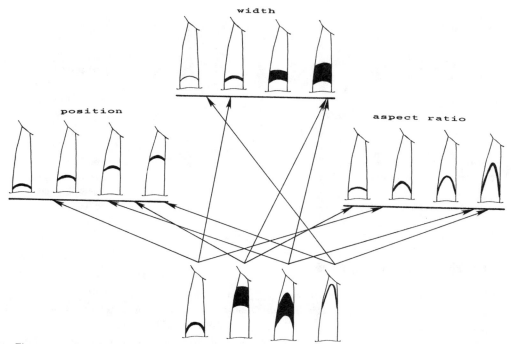

Figure 4.1. Three independent modes of variation in the morphology of an arc-shaped pattern element. Species may differ in the position of this element, in its width, and in its aspect ratio. Independent variation along these three axes can lead to the diversity of form shown in the lower row.

The overall pattern of the wing cell is achieved by a superposition of the patterns from each axis on which the pattern is represented. In doing this superposition, it is necessary to "round off" any sharp corners that are made where two patterns overlap (Fig. 4.2), because this is what happens at the intersection of patterns in the actual specimens. Venous stripes, for instance, almost always interact with pattern elements centered on the wing cell midline and fuse smoothly with them when they get close enough. Such smooth fusions between portions of the pattern that are otherwise apparently independent, indicates that the two are produced by identical processes.

The choice of independent variables has an admittedly arbitrary component, particularly when the patterns are complicated. As with any morphological analysis, it is necessary to pick the level of detail, or resolution, at which to do the analysis. Details of shape and variation below this level of resolution are ignored, because they are assumed to have no effect on the principles being investigated. That is not to say, however, that such details are not important. Small-scale individual variability, if heritable, is the basis of all change in evolution, and the analysis of individual variability can reveal a great deal about pattern development and evolution, as we will see in later chapters. But for the present, we will be interested only in the large-scale components of the wing cell pattern for each species. With these ground rules in mind, we will now examine the pattern morphospace of several genera.

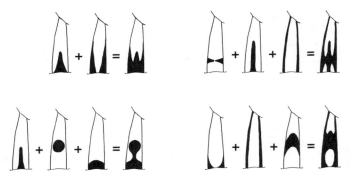

Figure 4.2. Complex patterns arise from the superposition of simpler component parts. This figure illustrates several hypothetical examples to show that when two elements fuse, the intersection between them becomes rounded. This phenomenon occurs because all the elements arise from identical or compatible developmental processes and morphogenetic gradients.

Ornithoptera (Papilionidae)

The dominant feature of the pattern of *Ornithoptera* is a large ocelluslike pattern, probably homologous to element **h**, the border ocellus (Figs. 4.3 and 4.4). Species differ in the size and shape of this element. Its size ranges from a small dot to a black field that covers nearly half the area of the wing cell. Its shape can be circular, elliptical, or cardioid. In addition, a pattern emerges from the wing margin. This pattern also appears in three different forms, depending on the species. It can be a simple black border that remains roughly parallel to the margin, it can be triangular with an apex along the wing cell midline, or it can be deeply convex, creeping up the wing veins. Also, an apical pattern element can extend various distances into the wing cell and projects distally (downward in Figs. 4.3 and 4.4) more strongly at the wing veins than at the midline of the wing cell. Finally, as in almost all taxa of butterflies, the potential exists for the expression of venous stripes.

The independent variables that make up the 25 patterns in Fig. 4.3 and form the basis of the pattern diversity in this genus are shown in Fig. 4.4. All three axes of variation in the central ocellus (Fig. 4.4A–C) are size axes, and they may therefore represent variation in the same size parameter working on three different shape-determining mechanisms. The three axes of variation of the marginal element (Fig. 4.4F–H) are size axes as well, as is the axis for the apical element (Fig. 4.4D). Thus, species patterns are constructed by picking a point on one of the ocellar shape morphoclines, on one of the marginal morphoclines, and on the apical morphocline. Of course, the nonexpression of one or more of these pattern classes is also an option. Whether or not the venous pattern (Fig. 4.4E) is expressed is also a species-specific character.

In species with highly enlarged pattern elements such as those on the bottom row in Figure 4.3, the light-colored background becomes the most noticeable feature of the pattern. Such specimens look as if they have a pattern consisting of light-colored triangles on a black background. This is a common feature of the color pattern in the Papilionidae, and we must remember throughout that in almost all butterfly wing patterns, the true pattern is dark, usually a black or brown melanin, and the more brightly colored features on the wings are background.

Several species, such as *Ornithoptera victoriae,* are represented by several patterns in

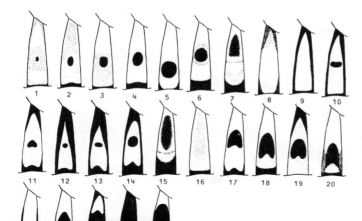

Figure 4.3. Diversity of wing cell patterns in *Ornithoptera* (Papilionidae). *1* and *17*, *O. croesus; 2, 4,* and *16*, *O. goliath; 3, O. chimaera; 5, O. urvilliana; 6, 7,* and *15*, *O. rothschildi; 8–10* and *12–14, O. alexandrae; 11* and *19*, *O. priamus; 18, 20–25*, *O. victoriae.*

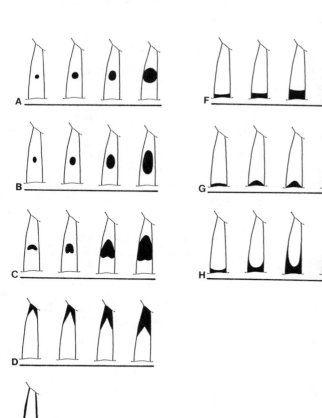

Figure 4.4. The component elements of the wing cell patterns of *Ornithoptera*.

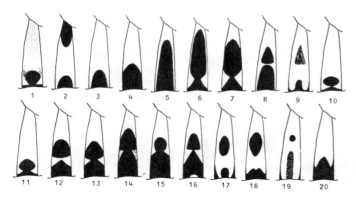

Figure 4.5. Diversity of wing cell patterns in *Troides* (Papilionidae). *1, 2, 10, 11*, and *17, Troides hypolitus; 3, T. aeacus; 4, T. vandepolli; 5, T. staudingeri; 6, T. flavicollis; 7, T. andromache; 8* and *16, T. helena; 9, T. cuneifer; 12–14, T. oblongomaculatus; 15, T. plato; 18* and *19, T. criton; 20, T. mirandus.*

Figure 4.3 because those species are both geographically diverse and individually variable. Each of those species' patterns will thus occupy many points in morphospace. A species is probably best thought of as occupying a small volume of morphospace that represents its individual and geographic variations. Many other species discussed in this chapter are quite variable and will therefore be represented by two or more pattern diagrams.

Troides (Papilionidae)

The hind wing patterns of *Troides* (Fig. 4.5) are based on the same designs as those of *Ornithoptera*. The dominant pattern element is a central ocellus, but in *Troides* this element tends to become much larger than in *Ornithoptera*. The enlargement occurs in two independent modes (Fig. 4.6A–C): an overall, nearly isometric increase in diameter, and an elongation along the long axis of the wing cell that can greatly enlarge the ocellus without allowing it to fuse with the wing veins. The marginal element is developed much more strongly than in *Ornithoptera*; it is mostly arc-shaped or triangular and can be greatly elongated along the long axis of the wing cell as well (Fig. 4.6E–G). The ocellus

can be either elliptical or triangular, and the marginal element can be either rounded or pointed. The ocellus and the marginal element can fuse to form a single figure. In many specimens the enlargement and coalescence of these two pattern elements cause the hind wing to be almost entirely black. *Troides hypolitus* is sometimes separated into the monotypic genus *Ripponia* (Haugum and Low, 1979) and differs from the other species considered here in that the ocellus in males is placed very near the margin of the wing cell (e.g., specimens 1, 10, and 11 in Fig. 4.5).

Troides has no strong venous stripes. The wing veins are almost always marked in black (against a generally yellow background) but are never enlarged as in *Ornithoptera*. Some kind of interaction of the pattern with the wing veins is evident, though, from the fact that greatly enlarged pattern elements often do not meet the wing veins but stay a small distance from them (e.g., specimens 5, 6, and 8 in Fig. 4.5).

Atrophaneura (Papilionidae)

The members of *Atrophaneura* are sometimes also classified as the Asian members of the genus *Parides*. The ocellus in this group can be circular, triangular, or arc-shaped, with a

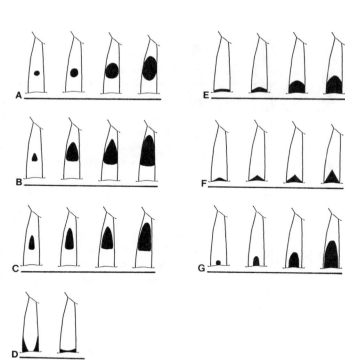

Figure 4.6. The component elements of the wing cell patterns of *Troides*.

great degree of size variation in all three shapes (Figs. 4.7 and 4.8A–D). The marginal element in most cases resembles that of the previous two genera in being either an arc (as in *Troides*) or a band parallel to the margin (as in *Ornithoptera*). However, in some specimens (e.g., 6 and 7 in Fig. 4.7), the marginal element is free and placed at some distance from the margin. This does not appear to be a species characteristic, because in *A. semperi* both types of positions occur on the same wing in different wing cells. The inhibition of dark pattern near the margin is a common feature in other papilionids; it is widespread in *Battus*. In *Papilio* it is mostly expressed as a small white or yellow arc on the margin at the wing cell midline, and its presence gives the wing a scalloped appearance. In *Parnassius* both elements can be developed into eyespotlike patterns.

The broad intervenous stripe that extends through the entire wing cell of *A. nox* could be an elongated marginal element or the fusion of a small marginal element with an elongated triangular ocellus. The separation of these two elements in the otherwise fairly similar pattern of *A. alcinous* (specimen 14 in Fig. 4.7) suggests that the latter is most likely the correct interpretation. As in the case of *Troides,* the fact that these intervenous patterns remain parallel to the wing veins suggests some kind of interaction. In later chapters, we will see that wing veins can have both an activating and an inhibiting effect (depending on the species) on pattern development within the wing cell.

Ideopsis (Danainae)

The wing cell patterns of *Ideopsis* and of the other two genera of Danainae we will discuss in this chapter, *Idea* and *Euploea,* have a very different appearance from that of the papilionids we looked at above. Yet we will see

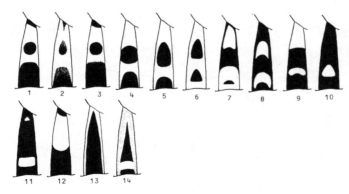

Figure 4.7. Diversity of wing cell patterns in *Atrophaneura* (Papilionidae). *1–3*, *Atrophaneura priapus*; *4*, *A. hageni*; *5*, *A. sycorax*; *6*, *A. horishanus*; *7* and *8*, *A. semperi*; *9*, *A. latreillei*; *10*, *A. coon*; *11*, *A. polyeuctes*; *12*, *A. neptunus*; *13*, *A. nox*; *14*, *A. alcinous*.

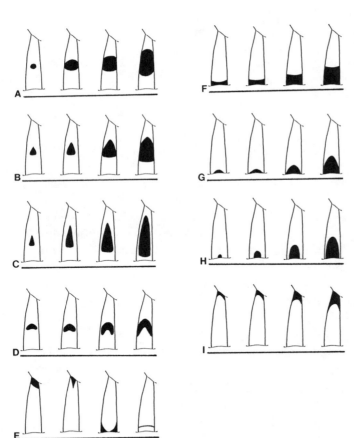

Figure 4.8. The component elements of the wing cell patterns of *Atrophaneura*.

that the danaines have several important features in common with the papilionids. The danaines also have several new features of pattern construction that they share with each other. The diversity of patterns of *Ideopsis* is shown in Figure 4.9, and the analysis of their component parts in Figure 4.10.

An ocellus (element **h**) is evident in the patterns of many species, but the dominant feature of the patterns in this genus lies in the

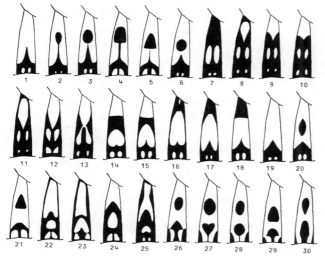

Figure 4.9. Diversity of wing-cell patterns in *Ideopsis* (Nymphalidae: Danainae). *1, 16, 17,* and *19, Ideopsis vitrea; 2–6, 15, 18, 20–22,* and *24–30, I. gaura; 7, I. klassika; 8* and *10, I. juventa; 9, I. oberthurii; 11, I. similis; 12* and *13, I. vulgaris; 14* and *23, I. hewitsonii.*

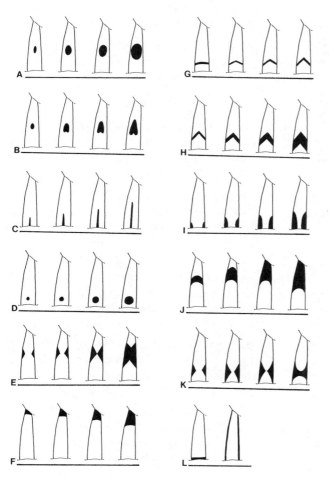

Figure 4.10. The component elements of the wing cell patterns of *Ideopsis.*

complex elaboration of the marginal elements in the distal half of the wing cell. The marginal pattern is built up from several independent components, each fairly simple in morphology, whose combinations and variations produce a family of related patterns. Almost all patterns have a midline (intervenous) stripe that can be short or long (Fig. 4.10C). This stripe interacts with a submarginal pattern that can be either circular or chevron-shaped (Fig. 4.10D,G,H). Interaction with the circular pattern leads to the marginal "lollipop" patterns, and interaction with a chevron forms the basis for the double white dot patterns, one of the most common features of the danaine wing pattern. The double dots are essentially background that is not occluded by the overlap of the midline stripe, the submarginal pattern, and two additional elements (Fig. 4.10I,L) along the margin and at the base of the veins. These marginal and vein patterns are similar to those we have already seen in the papilionids. The rounding that occurs at the intersection of patterns accounts for the roundness of the double dots.

Ideopsis has a new pattern feature that it shares with other danaines, namely the localized enlargement of a venous pattern into a lobe that extends into the wing cell. Venous lobes can occur in two locations: one near the midpoint of the wing cell, the other near the margin (Fig. 4.10E,K). Enlargement of these lobes can cause them to meet and thus produce a dark band across the wing cell. When these venous lobes fuse, they leave an arched background (white) pattern. When a midline stripe is added, this background pattern becomes broken up into a pair of oval spots so that the wing pattern now seems to have a double pair of white dots in each wing cell.

The venous lobes do not correspond to a known element of the nymphalid ground plan. They are common in the Danainae, and we see them occasionally in other taxa. Where they occur, they are associated with bands that run across the wing cell and flare out at the wing veins. When such bands diminish in width, often all that is left are two peaklike bulges on the wing veins. The venous lobes could thus be due to the presence of a subthreshold pattern element (the interpretation given in the discussion of *Idea* patterns in Chapter 3). It is probably no coincidence that the venous lobes occur at the locations of both the large ocellus and the small submarginal ocellus. In fact, one gets the impression that the lobes and ocelli are functionally related, because they are often connected by thin bridges of pigmented scales, just as in *Idea*. It is possible that the chevron pattern and the more proximal arc-shaped pattern are homologous to the ocellus-plus-lobe combinations.

Idea (Danainae)

The wing cell patterns of *Idea* (Fig. 4.11) are built from parts (Fig. 4.12) that are very similar to those of *Ideopsis*. But the proportions and shapes of those parts are slightly different, and that gives the patterns of *Idea* a completely different appearance. The central ocelli can be circular, but most often they are triangular, sometimes very thin and elongated. A continuity exists between thin midline (intervenous) stripes and triangular ocelli, similar to the continuity between intervenous stripes and ocelli in other taxa (Fig. 2.20). A marginal element, also centered on the wing cell midline, is similar to the one in *Ideopsis*. The two midline elements can form a single continuous intervenous stripe (specimens 1, 2, and 27 in Fig. 4.11), which is, however, always slightly indented about halfway to the margin, revealing its composite nature. Two sets of venous lobes are possible,

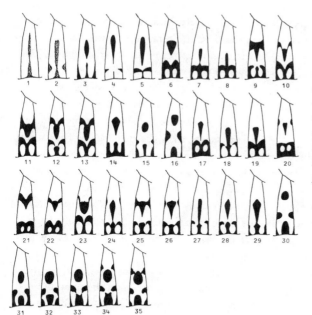

Figure 4.11. Diversity of wing cell patterns in *Idea* (Nymphalidae: Danainae). *1* and *2*, *Idea blanchardii*; *3–6*, *9*, *14*, *17*, *20*, *21*, *24*, and *25–29*, *I. idea*; *7*, *8*, *10*, *18*, and *19*, *I. durvillei*; *11–13*, *22*, and *23*, *I. leuconoe*; *15* and *16*, *I. iasonia*; *30*, *I. lynceus*; *31–35*, *I. hypermnestra*.

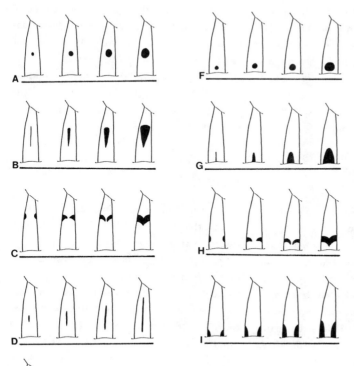

Figure 4.12. The component elements of the wing cell patterns of *Idea*.

as in *Ideopsis,* but their shape is more delicate (Fig. 4.12C,H). The venous lobes of *Idea* are almost always curved, and when they meet on the wing cell midline, they form a double-arched figure. The origin of the white dots from the area enclosed by these arches is more clearly visualized here than in *Ideopsis* (see also Fig. 3.22).

An interesting parallelism exists between the ocellar (**h**) and marginal (**i**) patterns in *Idea.* Both are built from nearly identical components, and as specimens 12 and 34 and their neighbors in Figure 4.11 illustrate, it is possible to get nearly identical morphologies in both systems. The two pattern systems seem to be iterative homologs of each other. But because of the proximity of the margin, and the additional pattern elements that occur at the margin and the tips of the veins, the marginal pattern usually develops a different morphology from that of the ocellar pattern, as most of the specimens in Figure 4.11 illustrate. In contrast to *Ideopsis,* no fusion occurs in *Idea* between the marginal and ocellar systems to form a broad dark area on the wing, and that gives the patterns of *Idea* a more open and lacy appearance.

a clear affinity with the *Ideopsis* and *Idea* patterns, but the remaining patterns elaborate the common theme in a way unique to *Euploea.* The proximal and distal set of venous lobes can fuse with the intervenous stripe to produce one or two pairs of white dots, the proximal pair almost always being substantially larger than the distal pair. The larger size of the proximal dots appears to be due to the tendency of the proximal venous lobes in *Euploea* to lie farther from the margin than in the previous two genera. The intervenous stripe is continuous along most of the length of the wing cell in most species. Only the distal (bottom) half of the intervenous stripe is expressed in specimens 3, 4, and 20, but in specimens 21 through 24 the small projection from the proximal (top) black pattern suggests the presence of a proximal portion of the intervenous stripe, as in *Idea.* Although the proximal segment is in no case clearly isolated and identifiable, we will assume that patterns such as those of specimen 21 indicate that it is expressed but is "hidden" through fusion with other elements in the proximal half of the wing cell (Fig. 4.14B,G,I).

Euploea (Danainae)

The patterns of *Euploea* (Fig. 4.13) are built with the same components (Fig. 4.14) as those of the previous two danaines, but the shapes and sizes of the building blocks are again slightly different. The main difference is that the central ocellus (element **h**) is missing in *Euploea,* and a fairly broad intervenous stripe dominates the pattern along the wing cell midline. All elements of the pattern tend to be fairly broad, giving the overall wing pattern of most species a dark appearance.

The first four patterns in Figure 4.13 show

Belenois (Pieridae)

The patterns of *Belenois* (Fig. 4.15) are more diverse than those of the other six genera discussed in this chapter. *Belenois* has three major pattern themes. The top row of specimens in Figure 4.15 illustrates patterns in which an ocelluslike element dominates. The second row shows a transition to a wishbone-shaped pattern that can develop a strong proximally oriented peak. Patterns with various elaborations of this wishbone are shown in the third row. The third theme is shown in the fourth row. It is also a wishbone-shaped

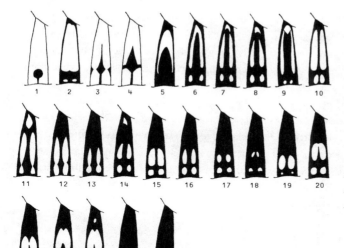

Figure 4.13. Diversity of wing cell patterns in *Euploea* (Nymphalidae: Danainae). *1, 3, 4,* and *10, Euploea tobleri; 2, E. sylvester; 5, 22,* and *23, E. dentiplaga; 6–9, 11,* and *14, E. mulciber; 12, E. doubledayi; 13* and *16, E. algea; 15, E. core; 17* and *21, E. euploea; 18, E. phaenareta; 19* and *25, E. batesii; 20, E. andamanensis; 24, E. leucostictos.*

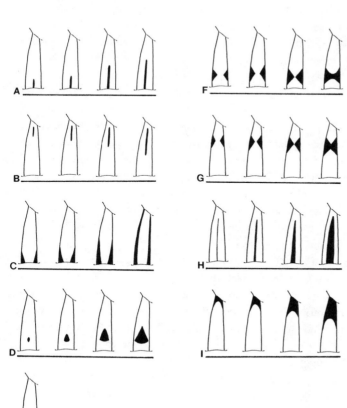

Figure 4.14. The component elements of the wing cell patterns of *Euploea.*

pattern but with its peak pointing distally. The patterns in Figure 4.15 can be constructed out of the components shown in Figure 4.16.

A small pair of venous lobes emerges near the margin in some species (Fig. 4.16G). These lobes can fuse to form an arc and produce a set of marginal patterns reminiscent of those of *Idea* and *Ideopsis* (specimens 36–40 in Fig. 4.15). These patterns differ from those of the danaines, however, in that fusion between the lobes and the midline pattern is not common.

The principal challenge in *Belenois* is to explain the origin of the two wishbone-shaped patterns (Fig. 4.16E,F). It is possible that these are different expressions of the same pattern element. Transitions between them seem to occur in specimens like 21 through 25 in Figure 4.15. The long peak of the wishbone could be due to the elongation of an intervenous stripe (e.g., Fig. 4.16B) in the proximal or distal direction, depending on the species. The two morphoclines that lead from arc to wishbone (Fig. 4.16E,F) could therefore be continuous (the connection between them can be made visually, if the sequence of either E or F is reversed). If this interpretation is correct, then chevron-shaped patterns, like those in the second and third rows in Figure 4.15, all arise as interactions between simple arcs and more or less well-developed intervenous stripes. The theoretical model for pattern formation that we will develop in Chapter 7 suggests that this is indeed the correct interpretation.

It is not difficult to visualize smooth continuous transitions between almost any two species in Figure 4.15, and this suggests that a continuity may exist between many of the patterns that were drawn as separate morphoclines in Figure 4.16. It is therefore possible that a continuity in the developmental processes that produce these patterns is not

captured by the morphoclines in Figure 4.16. Thus, although the morphoclines allow us to dissect the pattern into its component parts, we begin to suspect that these components are not entirely independent from each other. Some information about the developmental origin of the components themselves is likely to be essential in order to interpret the relations among the component parts and the mechanism of their interaction.

Conclusions

A large amount of within-species pattern variability exists in many of the species we just analyzed. Some of this variation is due to geographic variability, but most of it is normal continuous individual variation. It is possible to find a whole range of intermediate patterns between the ones that were chosen for illustration, and thus the patterns of each species occupy a cloud of points in morphospace. It is clear from examination of the figures that different species may overlap on one or more morphoclines but be clearly separated on others. These morphoclines may therefore have some value in taxonomy, as exploratory devices for identifying characters by which species may be reliably summarized, even in the face of much individual variability.

This simple and logical method of first identifying those parts of the pattern that vary independently from other parts, and then analyzing the range of variability of each of the components independently, allows one to see a great deal of order in what is otherwise a wildly confusing array of visual information. Within each family are strong pattern themes that all members have in common. But perhaps the most interesting finding to emerge from this analysis is that even among these seven genera, taken from

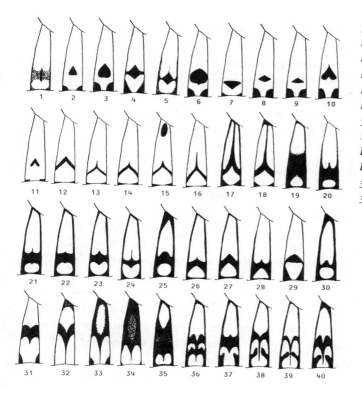

Figure 4.15. Diversity of wing cell patterns in *Belenois* (Pieridae). *1*, *19–21*, and *28*, *Belenois creona*; *2*, *8*, and *9*, *B. thysa*; *3* and *10*, *B. rubrosignata*; *4*, *B. calypso*; *5*, *7*, *23*, and *24*, *B. subeida*; *11*, *B. ogygia*; *12–16*, *B. zachalia*; *17* and *18*, *B. victoria*; *22*, *25–27*, and *30*, *B. aurota*; *29*, *B. hedyle*; *31–40*, *B. gidica*.

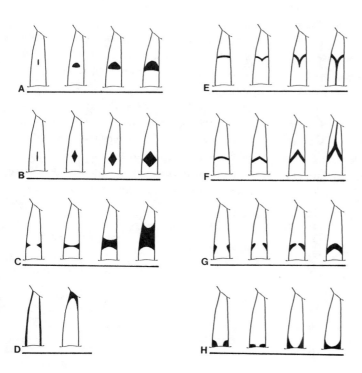

Figure 4.16. The component elements of the wing cell patterns of *Belenois*.

three taxonomically distant families and with each genus having distinctive patterns, there is little diversity in the building blocks out of which their patterns are constructed. To begin to realize how simple the underlying morphology really is, it is instructive to compare all the figures of the species patterns among themselves first and then to compare the morphocline figures of all seven genera. Almost all species use either a central ocellus or an intervenous stripe. The ocellus is almost always either elliptical or triangular (actually, shaped like an arrowhead in most cases), and the intervenous stripe can be short, long, or broken into two short segments. Differences in the shapes, and particularly in the sizes, of the components of the pattern give rise to much of the within-genus pattern diversity. The differences between genera, by contrast, are largely due to differences in the relative positions and proportions of the components.

Most of the species surveyed in this analysis have readily identifiable border ocelli (element h in Fig. 2.17). The homologs of the parafocal elements (element i) are more difficult to visualize, however, because they are generally fused with the wing margin and their shape is modified by interaction with several of the vein-dependent patterns. When the marginal patterns are displayed as separate morphoclines, they are seen to be made up of patterns that are arcs, triangles, or chevrons. These are standard morphologies for the parafocal elements (Fig. 2.21).

Many fine details of the shapes of several parts of the real wing cell patterns cannot be produced with the tools provided by these relatively simple morphoclines. To produce the exact shape of the curvatures, and the small indentations and protuberances that are characteristic of some patterns, clearly requires more information than is presented by the morphoclines that define the large-scale features of the pattern. It is possible to keep adding additional variables to account for progressively smaller features of variation and diversity, but that is probably not the optimal way to proceed in developing a deeper understanding of patterns and the origin of their diversity. The similarities in the shapes of the components that build up the pattern in most genera, and in particular the continuity that appears to exist between many of the patterns of *Belenois* (Fig. 4.15), suggest that hidden connections exist among many of the morphoclines. Understanding what these connections are is likely to be important for understanding the relations among patterns. The important things missing from this analysis are not the variables that account for the fine details of the patterns but the hidden variables that are responsible for specifying the elements' shape (why are some ocelli round, and others triangular?), size (why are some bands narrow, and others wide?), and position (why are some elements restricted to the margin, and others to the center of the wing cell?). To understand the nature of these hidden variables, we will need to learn a great deal more about the developmental processes that give rise to the color pattern. This we will do in the next three chapters.

The elucidation of the Schwanwitsch-Süffert ground plan, and in particular the recognition of the paired nature of the bands of the central symmetry system, provoked a flurry of interest in exploring the developmental and physiological basis of these color patterns. The experimental approach to the problem of pattern formation has consisted primarily of perturbation studies, which are easy to perform and are almost always the best initial approach to a system about which one has little prior knowledge. Such studies consist of inflicting a minor injury at various places in a developmental field and at various times of development and then observing how the pattern responds. If a consistent alteration of pattern can be obtained after a specific type of injury, the investigator has a toehold from which the causal pathway of the perturbed pattern can be analyzed. Focused perturbation experiments can then be used to build a map of how the response varies in time and space. From such a mapping it is sometimes possible to deduce the processes that are likely to be at work in the normal development of the pattern. The formulation of specific hypotheses about these processes can then serve as the basis for refining the experimental approach to pattern formation. Three types of perturbations have traditionally been used to study color pattern formation in the Lepidoptera: (1) microcautery with a fine hot needle, which kills only a small patch of cells on the wing; (2) surgical ablation of a large or small portion of the wing disk during the larval stage, which results in a small and deformed wing, often with a severely altered pattern; and (3) temperature shock, consisting of a short exposure to nearly lethal high or low temperatures.

These methods have been used to study development of most of the elements of the nymphalid ground plan. Each particular pattern element has, however, usually been

Chapter 5

Experimental Studies on Color Pattern Formation

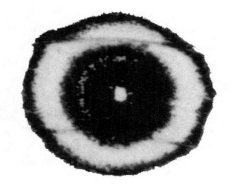

studied in only one species or in a few related species. Furthermore, because of the greater ease with which moths can be reared, most of the experiments on pattern formation have been done on moths. We are able to generalize the findings from those studies to the Lepidoptera as a whole because of the high level of confidence about the homologies of pattern elements across the Lepidoptera. The existence of homologous structures implies continuity of developmental information, so we may assume that homologous structures develop by similar developmental rules and processes (Roth, 1984, 1988). As noted above, the system of pattern homologies that we know as the nymphalid ground plan pertains to all major families of butterflies and moths in which pattern morphology has been studied. We can therefore be reasonably confident that developmental phenomena discovered in one or a few species may be generalized to other species with equivalent patterns. In the sections that follow, we will first discuss the experimental studies of pattern formation as they pertain to each pattern element, and then we will conclude from these findings some general features of color pattern formation.

Cautery and Surgery

The Central Symmetry System

The first and still the most detailed experiments on pattern formation were done by Kühn and von Engelhardt (1933) on the determination of the central symmetry system of *Ephestia kuehniella*. Because these studies provide the philosophical basis for all subsequent work on pattern determination, they will be discussed in some detail. The normal color pattern of *E. kuehniella* consists of the central symmetry system, composed of a pair of pale bands flanked by dark bands, and the discal spot, composed of two dark spots. This pattern is illustrated in Figure 5.1 (and diagrammatically in Fig. 5.5). It can be dramatically altered by small cauteries done early in the pupal stage at various places on the wing surface. The characteristics of the pattern alteration that is obtained depend on both the timing and the location of the cautery. A priori, we may assume that cautery of small groups of cells can interfere with development in two different ways. If pattern determination is accomplished by classical embryological induction mechanisms (Gilbert, 1985), then cautery can be used to kill cells that serve as sources for the organizing or inducing signals. The result would be the loss of the pattern that would normally have been induced. Cautery could also locally inhibit or prevent normal cell-to-cell communication mechanisms required for pattern induction. In that case, the result would be a pattern that is distorted from its normal form. The extent of the distortion and the detailed morphology of the altered pattern can be used to deduce things about the direction of cell-to-cell signals and the rate and mechanism by which those signals propagate.

When cautery is done between 0 and 24 hours after pupation in the field between the two bands of the central symmetry system, the shape of the band nearest the site of cautery is severely distorted. This distorted band, observed in all experimental animals, forms an arch around the site of cautery and has the effect of excluding the damaged area from the central field of the symmetry system (Fig. 5.1). After placing cauteries of various sizes at different places in the central symmetry system field, Kühn and von Engelhardt (1933) concluded that the bands of the central symmetry system mark the front of a progression zone of pattern determination. This progressive determination originates from two locations at the leading and trailing edges of the forewing and spreads across the

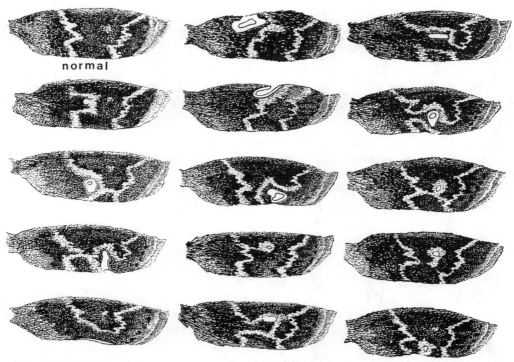

normal

Figure 5.1. Class I response to cautery in *Ephestia kuehniella* (Pyralidae). The wing in the upper left is normal. The other wings illustrate the distortions of the central symmetry system that follow cautery at various locations of the wing. (From Kühn and von Engelhardt, 1933)

wing during the first day of the pupal stage (Fig. 5.2). The type of response to cautery illustrated in Figure 5.1 is referred to as the Class I response. Kühn and von Engelhardt (1933) suggested that Class I effects probably come about through the cautery's production of a patch of dead cells that presents a simple obstacle to the progression of the determination front.

When cautery is done 36 to 60 hours after pupation, a different result, referred to as the Class II response, is obtained. During this period, cautery anywhere on the wing surface causes a distortion of both bands of the central symmetry system. It appears to be immaterial whether the cautery is within the central symmetry field or not, and the effect on the two bands appears to be symmetrical or nearly so. The observed distortions consist of displacement of the two bands toward the axis of symmetry in the middle of the wing and, in the most severe cases, fusion between the two bands in the middle of the wing (Fig. 5.3). These effects were interpreted by Kühn and von Engelhardt (1933) as resulting from an inhibition or freezing of the determination front at different points in its progression across the surface of the wing. Although Figure 5.2 gives the impression of a progressive sequence of effect, it is noteworthy that the degree of the response is not correlated with the timing of the cautery within the sensitive period for the Class II response. However, there appears to be a slight increase in the frequency of severe contractions of the central symmetry system when cauteries are particularly large. Thus, if Kühn and von Engelhardt's suggestion that cautery during the

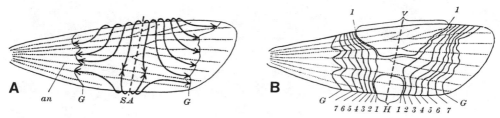

Figure 5.2. Interpretation of the results from cautery experiments illustrated in Figure 5.1 as demonstrating the progression of a front of a determination stream that travels across the wing. The determination stream arises from areas along the anterior and posterior margins (*A*). The final position of this front (*B*) determines the position of the bands of the central symmetry system. Wounds made by cautery are presumed to present an obstacle to this flow, resulting in a distortion of the final position of the bands. (From Kühn and von Engelhardt, 1933)

Figure 5.3. Class II response to cautery in *Ephestia kuehniella*. (From Kühn and von Engelhardt, 1933)

Class II sensitive period simply freezes the determination front at the point it had reached is correct, then their results imply that the progress of determination is poorly synchronized among individuals of the same age.

Cauteries done 72 or more hours after pupation have no effect on the morphology of the pattern and constitute the Class III response. Figure 5.4 shows the temporal pattern of the effects described above. Kühn and von Engelhardt (1933) reported finding individuals with intermediate responses between Classes I and II—that is, a local deviation of one band, and a slight approximation of the two bands—when cauteries were done between 24 and 48 hours after pupation. Such intermediates were not found (or were rare) in similar studies performed on the

closely related species *Plodia interpunctella* (Wehrmaker, 1959).

A number of studies on several other species of moths have confirmed the findings of Kühn and von Engelhardt (1933). Class I responses have been documented in *Plodia interpunctella* (Pyralidae—Wehrmaker, 1959; Schwartz, 1962; Brändle, 1965), *Abraxas grossulariata* (Geometridae—Kühn and von Engelhardt, 1936), *Samia (Philosamia) cynthia* (Saturniidae—Henke, 1933a,b, 1948), *Antheraea pernyi* (Saturniidae—Henke, 1944), *Hyalophora cecropia* (Saturniidae), and *Malacosoma americana* (Lasiocampidae—Nijhout, unpub. data). In all these cases the sensitive period for the response lies within the first 24 to 48 hours after pupation (except for *Abraxas grossulariata,* in which the critical period extends for nearly 10 days, and *Hyalo-*

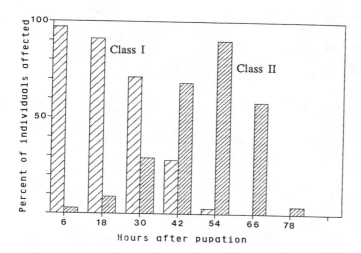

Figure 5.4. Sensitive periods for the Class I and Class II responses to cautery in *Ephestia kuehniella*. (After Kühn and von Engelhardt, 1933)

phora cecropia, in which the critical period occurs during the first days of adult development after breaking a several-month-long pupal diapause). Class II responses were confirmed in *E. kuehniella* by Toussaint and French (1988) and have been documented in the related species *Plodia interpunctella* by Wehrmaker (1959). Wehrmaker found that the degree of response in *P. interpunctella* did not depend on the location or size of the injury. He was also unable to demonstrate a time dependence within the 10-hour sensitive period for the Class II response. In other species, Class II responses are documented poorly if at all. Henke (1933a) reported a slight displacement of the bands of the central symmetry system after damage to the wing on day 2 after pupation, but the effect was largely restricted to the distal band. Class II responses do not occur in any other of the aforementioned species. Class II responses do, however, occur in some species as phenocopies (see "The Response to Temperature," below) when, as early pupae, they are subjected to temperature shocks. Such responses have been documented for *Ephestia kuehniella* (Feldotto, 1933) and *Ptychopoda seriata* (Kühn and von Engelhardt, 1944). These and other

pattern phenocopies will be discussed in more detail below.

The experiments of Kühn and von Engelhardt (1933), Wehrmaker (1959), and Schwartz (1962) produced sufficient spatial and temporal resolution to allow them to conclude that both *E. kuehniella* and *Plodia interpunctella* have at least two sources of pattern determination for the central symmetry system, one at the anterior margin of the wing and one at or near the posterior margin. In Figure 5.2 these sources are suggested to be on the underside of the wing, but no direct evidence exists for such a ventral source. In fact, the summary diagrams of Kühn and von Engelhardt (1933) clearly show the pattern to be contracting on three points (Fig. 5.5) and suggest not two but three sources on the dorsal wing surface (Nijhout, 1985b). The diagrams of Wehrmaker (1959) also suggest multiple sources (perhaps as many as four or five) for the central symmetry system along the central axis of the forewing. Because of the small size of these two species of Pyralidae, it is difficult to obtain sufficient spatial resolution by cautery to determine the exact position of the sources or to determine whether there is one source per wing cell.

normal

Figure 5.5. Summary diagram of the various forms of pattern reduction obtained from cautery and temperature-shock experiments in *Ephestia kuehniella*. The results reveal a progression of pattern emerging from three source areas that lie on the axis of the central symmetry system. (From Kühn and von Engelhardt, 1933)

Ephestia kuehniella certainly appears not to have a source in every wing cell, which means that one source must determine pattern across several wing cells. Thus, for this species there is no evidence for wing cell autonomy of the pattern or for compartmentalization by wing veins. Kühn and von Engelhardt (1933) noted, however, that in *E. kuehniella* the progress appears to be faster along the wing veins than elsewhere on the wing surface, as shown by the peaks formed by the outermost dark portion of each band in Figure 5.5.

The surgical ablation experiments on *Samia cynthia* (Henke, 1933a) and the cautery experiments on *Antheraea pernyi* (Henke, 1944) were not sufficiently detailed, nor did they cause a sufficient degree of fusion of the proximal and distal bands, to allow us to deduce the number of sources these species have for the development of their central symmetry system. Some of the pattern aberrations described for *S. cynthia* (Henke, 1933a),

however, suggest the presence of at least three or four sources on the forewing and provide circumstantial evidence that two of these sources are centered in specific wing cells and that two occur near the anterior and posterior margins, respectively.

Taken together, the experimental work described above demonstrates that the central symmetry system arises from sources of determination that occur along the central midline of the wing. In the species studied, there appear to be sources near the middle of the anterior and posterior wing margins, plus one or more additional sources along the line connecting the marginal sources. Kühn and von Engelhardt (1933) envisioned each band of the central symmetry system as the front of a progressing wave of determination. An alternative interpretation is that each band represents a threshold (or contour) on the gradient of a diffusible morphogen that arises from the alleged sources (Nijhout, 1978, 1985b). The location of the sources on an anterior-to-posterior axis in the center of the wing explains why the coloration of the prox- imal and distal bands is symmetrical on that axis. More recently, Toussaint and French (1988) have repeated the classical experi- ments of Kühn and von Engelhardt (1933) and have shown that the results of cautery ex- periments in *E. kuehniella* are inconsistent with the original determination wave hy- pothesis but are more readily explained by a diffusion gradient model. They arrived at this conclusion by noting that the determi- nation wave hypothesis predicts a strict cor- respondence between time of cautery and the position of the bands of the central symmetry system in the Class II response, and this is clearly not the case. They noted, however, that the gradient model by itself does not ex- plain the Class II response either (one would have to assume that the Class II response in- volves a haphazard alteration of thresholds). Nor does the gradient model, in their opin- ion, explain certain ring-shaped patterns that occasionally form around cauteries within the central field of the central symmetry system. Both of these features, however, may become explicable after we examine a few additional aspects of pattern determination in Lepidop- tera.

The Discal Spot

Although the cautery experiments in *Ephestia kuehniella* and *Plodia interpunctella* described above had dramatic effects on the position and shape of the bands of the central symme- try system, they had no effect on the discal spot, suggesting that these two pattern ele- ments are determined independently (Kühn and von Engelhardt, 1933; Wehrmaker, 1959).

Henke's (1933a) ablation experiments on *Samia cynthia* likewise show that the discal spot and the central symmetry system are de- termined independently and that the discal spot is determined several days prior to the central symmetry system. In *S. cynthia* the discal spot is elaborated into an elongated ocellus made up of concentric rings of several colors, which, as it happens, are identical to those found in the bands of the central sym- metry system, but in reversed order (Fig. 5.6). When the banding pattern is severely distorted by surgical ablation of a portion of the wing so that the bands of the central sym- metry system become contracted toward the midline, it is found that the correspondingly colored bands of the discal ocellus and the central symmetry system fuse smoothly to- gether (e.g., Fig. 5.7). Thus, the discal ocel- lus and the bands of the central symmetry system must be manifestations of similar, or perhaps identical, physiological or biochem- ical processes.

Such a fusion shows that these pigment bands are developmentally homologous (non- homologous patterns would be expected to intersect or be superimposed on one another, and this is not the case). The reversed se- quence of colors of the discal ocellus relative to that of the central symmetry system indi- cates that the central field of the discal ocellus has the same developmental value as (or is topologically identical to) the surface of the wing outside the field of the central symme- try system. Thus, if the bands of the central symmetry system are specified by a determi- nation signal that spreads across the wing as in Figure 5.2, then the central field of the discal ocellus must be a relative sink or an obstacle for that signal. In fact, in *Antheraea pernyi* (Henke, 1944) and *Hyalophora cecropia* (Nijhout, unpub. data), a small cautery within the central symmetry field can be- come completely enclosed by a loop of the nearest band, producing a rounded pattern with a reversed sequence of colors, otherwise identical to that found in the discal spot. Pat- terns in which the sequence of colors in the

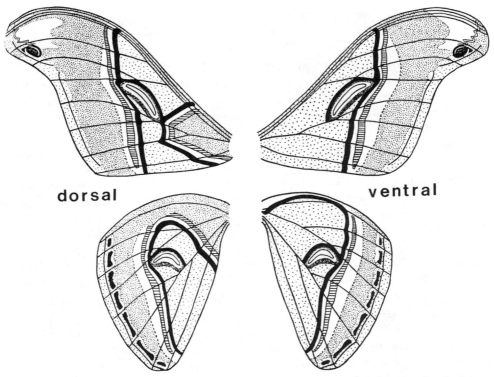

dorsal   **ventral**

Figure 5.6. The wing pattern of *Philosamia cynthia* (Saturniidae), illustrating fusion of the bands of the central symmetry system on the hind wing, and fusion of the bands of the central symmetry system with the outer ring of the discal spot on all wings. (From Henke, 1933a)

discal spot are the reverse of those seen in the central symmetry system are common in the moths and also occur in many butterflies.

The Border Ocelli

Border ocelli are simple circular patterns of pigment, usually with a distinctive white central focus. The development of a perfectly radially symmetrical pattern such as an eyespot is most likely to be controlled by a source of determination at its center. The developmental control of eyespot formation has been studied in *Precis coenia* (Nijhout, 1980b, 1985a). When a small group of cells at the focus of the large forewing eyespot of *P. coenia* is killed by microcautery within a few hours after pupation, the large eyespot fails to de-

velop (Fig. 5.8B). When cauteries are done progressively later in the pupal stage, small but ever larger eyespots develop (Fig. 5.8C) until, at 2 to 5 days after pupation (depending on the temperature), cautery has no further effect on development of the eyespot (Fig. 5.9). If the cells of the focus are instead excised and transplanted to a different location on the wing, a small eyespot is induced in the host tissue around the transplant (Fig. 5.8D). The only restriction on the success of such a transplant experiment is that the focus must be transplanted to a location in the white area of the wing. Foci transplanted to one of the brown areas of the wing fail to induce an eyespot in the surrounding wing. Presumably the brown areas had already been

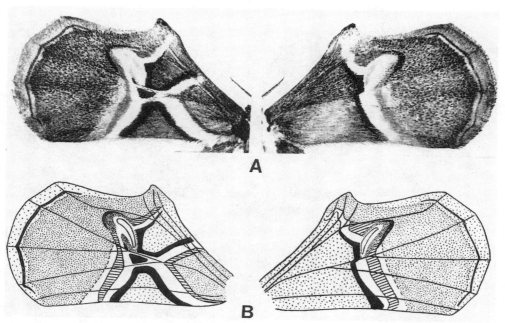

Figure 5.7. Pattern distortion following surgery on the forewing imaginal disk of *Philosamia cynthia*. The central symmetry system fuses into two arcs, much as is seen after cautery in *Ephestia*. *Left,* dorsal side; *right,* ventral side of same wing. (From Henke, 1933a)

determined and were not subject to further pattern induction by a transplanted focus. These cautery and transplant experiments show that the cells at the focus are inducers of pattern formation; that is, these cells produce an inductive signal that spreads radially from cell to cell.

When reciprocal transplants of foci are made between *Precis coenia* and the closely related *P. lavinia,* normal eyespots are induced in each, showing that the inductive signal is not species-specific at this level. The eyespots of these two species differ in that the one of *P. lavinia* lacks the outermost dark ring found in *P. coenia.* In each of the reciprocal transplants, the induced eyespot has a morphology similar to a normal eyespot of the host, not to that of the focus donor, showing that the species specificity of eyespot morphology resides in the response, not in the inductive signal (Nijhout, unpub. data). As

of this writing, it has been impossible to transplant foci successfully between more distantly related species. I have tried transplanting *P. coenia* foci to *Vanessa cardui, V. atalanta,* and *Limenitis archippus,* but in each case the transplant failed to become established, and no pattern induction occurred.

The border ocelli on the hind wing of *P. coenia* present a special and interesting case because the mechanism of their determination is in a sense the inverse of what we just discussed for the eyespot on the forewing. I discovered this while trying to disrupt development of the large hind wing eyespot by means of cautery experiments of the kind that work so well on the forewing. In the pupal stage the hind wing is completely covered by the forewing, and cauteries must be done by blindly inserting a fine cautery needle through the forewing to the hind wing be-

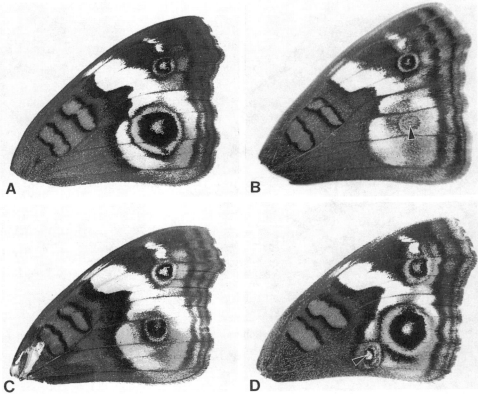

Figure 5.8. Development of a border ocellus in *Precis coenia* (Nymphalidae: Nymphalinae). *A*, normal wing. *B*, when a small region at the center of the presumptive ocellus (*arrow*) is killed by cautery a few hours after pupation, the ocellus fails to develop. *C*, if cautery is done at 24 hours after pupation, an ocellus about half normal size develops. *D*, when a small clump of cells from the center of an ocellus (the white focus) is transplanted to the wing of another individual (white region indicated by arrow), a small ocellus is induced to develop in the host around this transplant. (After Nijhout, 1980b)

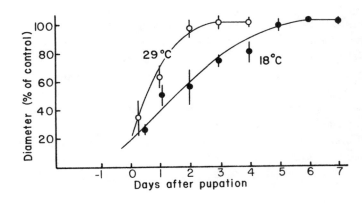

Figure 5.9. Time course of the response to cautery in *Precis coenia*. Horizontal axis is the time at which cautery is done. Vertical axis is the size of the ocellus that develops, measured as a percentage of the size of the normal ocellus that develops on the opposite wing of the animal. (After Nijhout, 1980b)

neath. The relative positions of forewing and hind wing at this stage are quite constant, and it is a simple matter to produce a map so that landmarks on the cuticle of the forewing can be used to find specific areas on the underlying hind wing with fair precision. Having located the presumptive center of the large hind wing eyespot, I discovered that cauteries of that location failed to have any effect on the development of the eyespot. In fact, cauteries anywhere within the presumptive field of the eyespot had no effect, which initially suggested that the hind wing eyespot must be determined well before pupation, much earlier than the eyespot on the forewing. However, when cauteries were placed outside the fields of the eyespot, small eyespots developed around the sites of cautery (Fig. 5.10).

Injuries on the hind wing of *P. coenia* apparently act as pattern organizers. That the small eyespots that form around these injuries are homologous to normal eyespots is indicated by the smooth fusion of corresponding dark and light pigment rings that occurs when normal and induced eyespots are close together (Fig. 5.10). This response to cautery on the hind wing is exactly the reverse of what happens on the forewing, where cautery inhibits eyespot development.

There are two additional ways in which the developmental response to injury of eyespots on the hind wing of *P. coenia* differs from that of the forewing eyespots. First, when young pupae are subjected to a temperature shock, the eyespots on the forewing are diminished in size, whereas those on the dorsal hind wing are often enlarged and grossly distorted (Figs. 5.21 and 5.22; Nijhout, 1984). Second, when a portion of the wing imaginal disk is ablated during the larval stage, the forewing eyespot is, once more, reduced in size, but the eyespot on the dorsal hind wing is enlarged and often develops instead as an arc-shaped figure that is open toward the cut edge of the wing (Fig. 5.11; Nijhout and Grunert, 1988).

These three distinctive modes of response of the hind wing pattern find a common explanation if we assume (1) that the normal

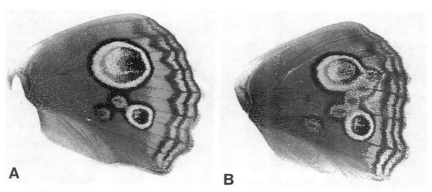

A   **B**

Figure 5.10. The hind wings of *Precis coenia* respond to cautery by developing small supernumerary ocelli around the site of injury. *A* and *B* show the patterns that developed after cells were killed by cautery at two and four places, respectively. (After Nijhout, 1985a)

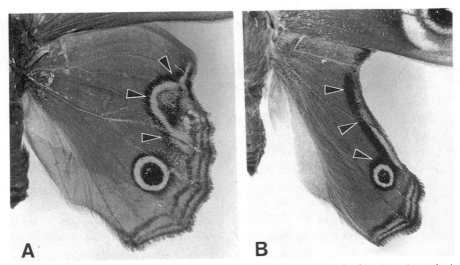

A   **B**

Figure 5.11. Surgical removal of a portion of the hind wing imaginal disk of *Precis coenia* results in the "opening up" of the nearest ocellus so that the outer ring of the ocellus (*arrows*) becomes an omega-shaped figure (*A*) or, in more severe cases, forms an arc parallel to the defective wing margin (*B*). (From Nijhout and Grunert, 1988)

eyespot on the hind wing arises, not around a source of an inductive signal as does the fore-wing eyespot, but around a relative sink for such a signal, and (2) that injured cells at the site of cautery become a relative sink with properties similar to those at the center of a hind wing eyespot (Nijhout, 1985a). Why this should be the case is shown in Figure 5.12. We'll assume for the sake of the discussion that pattern develops in response to different concentrations of one or more morphogenetic substances (morphogens) and that these substances can diffuse across the wing. The brown background color of the wing then represents the pattern that develops in the presence of an intermediate level of a morphogen. A normal eyespot forms where this morphogen is locally depleted by some catabolic activity. If cell death provokes a similar catabolic destruction of the morphogen, perhaps as an incidental side effect of wound healing, then eyespotlike patterns would be expected to form around any injury.

If the injury is small and pointlike, the induced pattern will be round and a near-perfect mimic of a normal eyespot. But if the injury is extensive, as in the case of a partial ablation, so that it cannot be surrounded by wing tissue, then an arc- or band-shaped pattern will form parallel to the injury (Fig. 5.11). The response to temperature shock (Fig. 5.22) indicates that additional sites of catabolism of the morphogen have arisen in the vicinity of the normal eyespot, though in a broad and essentially random pattern that differs from individual to individual.

Nijhout (1985a,b) has suggested that the results of all cautery experiments, including the classical work of Kühn and von Engelhardt (1933) discussed above, can be interpreted on the basis of the morphogen-sink hypothesis. Thus, instead of presenting a simple obstacle to the propagation of a morphogenetic signal, cauteries prevent the local propagation of that signal by acting as a sink for the morphogen. Computer simulations have shown that this hypothesis readily explains even the most difficult of the Class I pattern alterations in *Ephestia kuehniella* (Fig. 5.13; Nijhout, 1985a). Pattern induction around a site of injury is not restricted to *Precis coenia*. Cautery near the forewing margin of *Hyalophora cecropia* (Saturniidae) induces perfect circular patterns (Nijhout, unpub. data), and cauteries on the hind wings of *Maniola jurtina* induce circular eyespotlike patterns, just as in *P. coenia* (V. French, pers. com.).

What does all of this tell us about the developmental physiology and homology of the border ocelli? Among other things, it demonstrates that even in a single species, border ocelli can arise by two different mechanisms. Yet, on comparative morphological grounds, the border ocelli on forewing and hind wing appear to be homologous, and in some species of the genus *Precis* the forewing and hind wing ocelli are virtually identical in size, shape, and coloration (e.g., in *P. atlites*). On comparative morphological grounds, there is no basis for distinguishing between the border ocelli on forewing and hind wing, and we can be as certain about their homology across the butterflies as we can about any system of

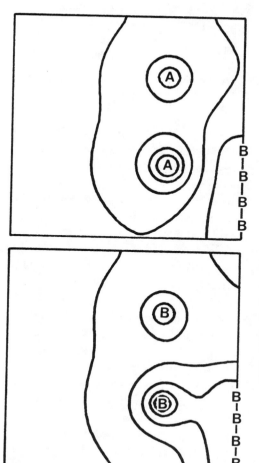

Figure 5.12. Computer simulation of the effect of a local wound along the wing margin on the morphology of a nearby ocellus. If we assume that ocelli arise around sources (*A*) of a diffusible substance (*top panel*), then if the defect produces a local sink (*B*), the ocellus near the injury will be diminished in size, particularly on the side facing the injury. If, on the other hand, we assume that ocelli arise around sinks (*bottom panel*) and the injury is a sink as before, then the outer rings of the ocellus open up and expand along the site of injury. If *B*s are assumed to be sources instead, then *A*s would have to be assumed to be sinks in order to obtain the same results as shown in the bottom panel.

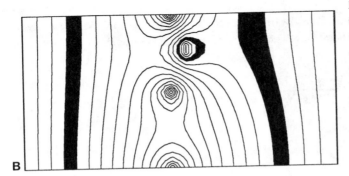

Figure 5.13. Computer simulation of cautery experiments in *Ephestia kuehniella.* The bands of the central symmetry system are assumed to be produced at certain threshold values of a diffusion gradient emerging from three sources along the middle of the wing. Cautery was simulated as a small local sink for the diffusion substance just off the midline. Depending on the location of this sink and on the threshold at which the bands develop, it is possible to get distortions of the band (*A*) as well as isolated ring-shaped patterns (*B*), like many of those in Figure 5.1. (After Nijhout, 1985a)

homologies in the animal kingdom. Yet the processes that give rise to these two types of eyespots appear to be different and, in some sense, the complement of one another.

We are thus faced with an interesting dilemma. If homology requires identity of mechanism, then forewing and hind wing eyespots would not be considered homologous. But if the criterion for homology is continuity of developmental information (Van Valen, 1982), then forewing and hind wing eyespots may be homologous to the degree that they use the same morphogenetic signal and threshold mechanism. It seems possible, for instance, that in the hind wing the eyespot arises around a sink in a morphogenetic background that is everywhere above a threshold, and on the forewing it arises as a source of the same material, acting in a background that is everywhere below a threshold. The unusual response of the dorsal hind wing

to cautery could then simply be due a very low threshold for pattern induction on this surface, so that a variety of nonspecific stimuli (such as injury) can mimic the normal inductive signal. This would be analogous to the diverse nonspecific inducers (such as pinpricks and lithium ions) that can provoke normal development in various classical embryological systems (Horstadius, 1973; Gilbert, 1985). Van Valen (1982) discusses an interesting parallel problem in the identification of homologies in mammalian molars whose cusps have diverse developmental and evolutionary origins yet whose homology as molars is generally accepted.

That we occasionally find very small circular pigment patterns organized around cauteries on the ventral forewing (Nijhout, 1985a) suggests that the different wing surfaces (forewing, hind wing, dorsal, ventral) may differ only quantitatively in their response to cautery. Thus we are left with the apparently inevitable conclusion that in *P. coenia,* border ocelli form around both sources and sinks of a morphogenetic signal. In a purely formal sense, a sink for a substance is just as effective as a source for long-range signaling, and it is impossible to tell the difference between the two on a priori grounds. We also need to consider the possibility that physiological activities other than catabolism could give rise to the appearance of a sink. It is probable, though I have not yet investigated the possibility rigorously, that a source of an inhibitor to a morphogen would have the outward appearance of a sink for that morphogen. Systems that are under simultaneous positive and negative control are commonplace in physiology, and if this proves to be the case here too, then sources on the forewing and hind wing might differ simply in the relative rates of production of two such antagonists.

It can be concluded from the foregoing that the system of border ocelli develops around a row of sources (or sinks) of pattern determination, of which there is potentially one in each wing cell, located on the wing cell midline. When such a source is absent or inactive in a given wing cell, no ocellus will develop there. As even a casual survey of pattern diversity in the butterflies shows, more often than not the border ocelli deviate significantly from being the perfect circular figures we have dealt with so far (Fig. 2.19). Because circles are the simplest patterns that can be specified in a system that uses point sources for pattern determination (Nijhout, 1978, 1985b), it follows that in order to specify patterns that deviate from perfect circularity, it is necessary to provide the system with additional information. Such information is provided by additional sources of determination at the wing veins and the wing margin. A model mechanism that appears to

be able to account for virtually the entire diversity of shapes found in the border ocelli (and in the parafocal elements, below) will be the subject of Chapter 7.

The Parafocal Elements and Submarginal Bands

The developmental response of the parafocal elements and the submarginal bands will be discussed together because the two pattern elements behave similarly in the experiments that have been done to date. Simple cautery experiments have only slight effects on the shape of the parafocal element on the ventral hind wing of *Precis coenia* (Nijhout, unpub. data), but the scope and significance of these shape alterations have not yet been analyzed. Almost all of the information about the developmental origin of the parafocal elements and submarginal bands comes from experiments in which small portions of the wing imaginal disk were ablated during the larval stage. Early experiments of this sort on *Samia cynthia* (Henke, 1933a), *Papilio machaon* (Magnussen, 1933), and *Lymantria dispar* (Henke, 1943), and more recent ones by Nijhout and Grunert (1988) on *Precis coenia,* give a relatively consistent picture of the responses of these two pattern elements to partial ablation of the wing. If the operation is done late in larval development, within one or a few days of pupation, the adult wing is simply found to have a portion missing, and no alteration of the pattern is noticeable other than a distortion imposed by tissue contraction around the wound site. If the surgery is done earlier—for instance, more than 3 to 5 days before pupation in *P. coenia* or at any time before cocoon spinning in *S. cynthia* (Henke, 1933a; Nijhout and Grunert, 1988)—a system of submarginal bands and parafocal elements develops on the foreshortened wing parallel to the new margin (Fig. 5.14). The parafocal elements and submar-

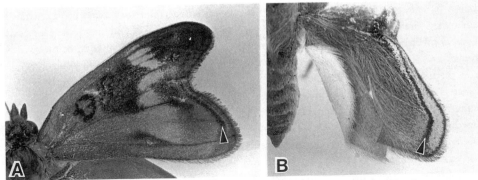

Figure 5.14. Development of parafocal elements of the forewing (*A*) and hind wing (*B*) after surgery on the wing disks of *Precis coenia*. The parafocal elements develop even on a very reduced wing, provided that the surgery is done more than 3 to 5 days before pupation. The parafocal elements always run parallel to the cut margin and, on very abbreviated wings, tend to develop as a single uninterrupted line. The parafocal element can be identified, even in these aberrant patterns, by its distinctive color. (After Nijhout and Grunert, 1988)

ginal bands appear to be the only portions of the pattern that regulate to the new position of the wing margin (Nijhout and Grunert, 1988). All other pattern elements are simply interrupted or eliminated by partial ablation of the wing.

One notable difference is seen between normal parafocal elements and those that develop along the margin of a partial wing. In the latter case the parafocal elements often form a nearly continuous band parallel to the wing margin and do not have the scalloped and wing-cell-compartmentalized appearance of normal parafocal elements (compare the parafocal elements in Fig. 5.14 with normal parafocal elements in Fig. 5.10). The lack of compartmentalization appears to have two separable causes. First, in many instances the venation pattern of a damaged wing becomes severely altered, so that fewer veins reach the wing margin. The pattern of parafocal elements in such wings clearly shows that they form a continuous band in the absence of wing veins and that the latter are responsible for the compartmentalization of these pattern elements. In *Precis coenia*

there is the additional observation that in very small wings (that is, after a large ablation) the parafocal elements always develop as an uninterrupted line, even where the veins reach the new margin, suggesting that the basal portions of the wing veins do not have the same pattern-restricting properties as their more distal parts (Nijhout and Grunert, 1988). Variation in the pattern-inducing activity along the length of the wing veins is also seen in many dependent (venous stripe) patterns (Fig. 2.9C).

It appears then that the parafocal elements and submarginal bands arise by an inductive property of the wing margin some time during late larval life. The wing veins have an inhibitory effect on this induction, and this accounts for the fragmented and wing-cell-restricted distribution of these pattern elements. It is also clear, however, that additional wing-cell-specific determinants must exist for these two pattern elements, because in most species the pattern elements are severely dislocated and have very different morphologies in each wing cell.

Aberrant Venation

Abnormal venation patterns, induced experimentally or occurring fortuitously, reveal the important function of the wing veins in controlling the morphology of several pattern elements. We just saw that the submarginal bands and parafocal elements in *Precis coenia* form continuous bands parallel to the wing margin in the absence of wing veins. Naturally occurring abnormal venation patterns have similar effects on the parafocal elements. Figure 5.15 illustrates two cases, in *Speyeria aphrodite* and *Battus philenor,* in which one of

the veins of the hind wing is shortened. These two species have parafocal elements with a much more complicated structure than those of *P. coenia,* but it is clear that here too the wing veins partition otherwise continuous patterns. It is as if the veins inhibit pattern development in their vicinity. We will see in Chapter 7 that this effect can be explained by assuming that veins are sinks for a pattern morphogen. Figure 5.16 illustrates two cases, in *Druryeia antimachus* and *Papilio polymnestor,* in which a supernumerary crossvein has developed. Here too it is clear that the new vein has inhibited pattern from

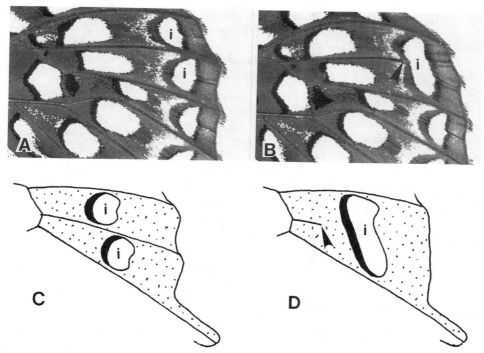

Figure 5.15. Pattern aberrations that accompany wing vein aberrations. Two instances of the effect of shortened wing veins (*arrows*) in (B) *Speyeria aphrodite* (Nymphalidae: Heliconiinae) and (D) *Battus philenor* (Papilionidae). Normal patterns are shown in (A) and (C), respectively. In each case the parafocal elements (i), which form rounded figures centered in each wing cell, have become fused. This indicates that the wing veins are involved in partitioning this pattern element. Both specimens from the author's collection.

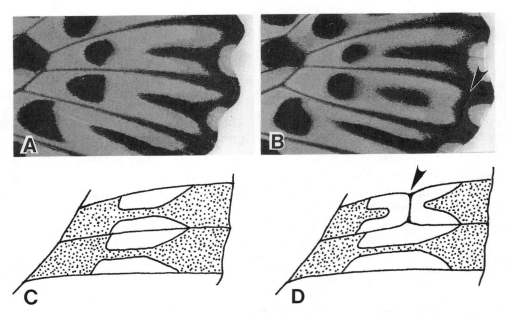

Figure 5.16. Pattern aberrations that accompany the presence of supernumerary wing veins (*arrows*) in (*B*) *Druryeia antimachus* (Papilionidae) and (*D*) *Papilio polymnestor* (Papilionidae). Normal patterns are shown in (*A*) and (*C*), respectively. In each case the effect of the extra vein is to inhibit the development of the intervenous stripe in its vicinity. Specimen A from the U.S. National Museum of Natural History; specimen B from the British Museum (Natural History), illustrated in Lewis (1987).

forming in its vicinity. The inhibitory effect, however, appears to be restricted to the intervenous pattern. In *Druryeia* the supernumerary vein has no effect on the morphology of the submarginal portion of the pattern. Because that portion of the pattern is induced by sources at the wing margin, it seems that the veins do not have the same role in pattern development as the wing margin does.

Sources and Boundaries

The various types of perturbation experiments described above give evidence for the existence of organizing centers and inductive sources in a few specific areas of the wing: the wing veins, the wing margin, and the wing cell midlines. Sources on the wing cell midline are point sources and are responsible for

determination of the border ocelli. In keeping with the elucidation of serial homology in the central symmetry system and basal symmetry systems and the fact that in some species these symmetry systems are represented by small circular patterns constrained to single wing cells (Fig. 2.5), it is most parsimonious to assume that the sources that determine the symmetry systems also reside on the wing cell midline and that, primitively, they too are point sources. It is also clear that wing veins act as boundaries to the compartments for pattern formation. It is possible that in many cases the wing veins are just passive boundaries that prevent inductive signals from passing between wing cells. But the results of some of the surgical experiments, the observation of venous stripes, and the effects on the color pattern of fortuitous venation pattern aberrations found in nature all indicate that the veins may also participate in pattern formation. In some cases the veins appear to have inductive properties (for instance, in the formation of venous stripes; Fig. 2.9), and in others they appear to inhibit the pattern locally (Fig. 5.16).

The wing margin is also an inductive source for pattern formation, but it clearly has different properties from those of the wing veins. In many cases the properties of the wing margin appear to be the reverse of those of the veins. The bordering lacuna that forms the wing margin is continuous with the lacunae of the venation system, and one might therefore expect them to have similar properties. The morphology of the bordering lacuna and the behavior of the cells around it, however, clearly distinguish it from the wing veins (Dohrmann and Nijhout, 1990). Bard and French (1984) have shown by computer modeling that some pattern aberrations can be explained if it is assumed that the wing margin acts as a sink for a gradient substance produced by the ocellar foci.

Figure 5.17 illustrates these various hypotheses for the distribution of pattern-organizing sources. They are placed on a wing whose venation system corresponds to that found in the larval imaginal disks at the time of pattern determination, but whose outline corresponds to that of a normal adult wing to facilitate comparison with the ground plan diagrams. A significant portion of pattern determination is thus likely to occur through the activity of, and interactions among, the organizing centers shown in Figure 5.17. A model of how certain interactions among such a simple array of organizing centers can account for the tremendous diversity of shapes within the various classes of pattern elements will be developed in Chapter 7.

The Response to Temperature

Changing the temperature at which development is allowed to take place offers the simplest and hence the most widely used method of perturbation analysis. Temperature methods have been particularly widely used in insects because, as small terrestrial poikilo-

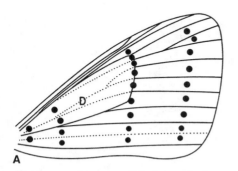 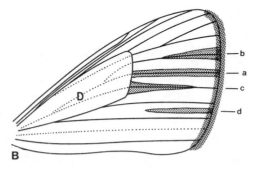

Figure 5.17. Summary of the locations of potential pattern-inducing centers in the wing, revealed by the various experiments and observations presented in this chapter. *A*, the possible positions for point sources. Primitively these occur in rows, one for each of the symmetry systems. *B*, the possible forms of extended venation-dependent sources. Three types of venous organizing centers occur: The entire vein can be active (*a*), or the activity can be restricted to the distal portion (*b*) or the proximal portion (*c*) of the vein. In any given species, all wing veins are likely to be of only one of these types. In addition, organizing centers may occur on each wing-cell midline (*d*) and along the distal wing margin.

therms, insects are naturally subject to varying temperatures during their development. Many such studies are designed to document the phenological response to seasonally varying temperatures (Danilevski, 1965; Beck, 1980; Tauber et al., 1986) and the degree of canalization of the phenotype (or its converse, phenotypic plasticity) in the face of a varying environment (Shapiro, 1976, 1980a, 1981). In butterflies, temperature studies have been especially attractive because a variety of protocols of temperature change can have spectacular and often specific effects on the development of the color pattern. The developmental effects of varying temperatures can be classified into three physiologically distinct categories (Table 5.1), which unfortunately have not always been recognized by experimenters. In some cases, reports of how the experiments were done are sufficiently ambiguous that we cannot determine with certainty which of the three responses was observed. The distinction is nontrivial because each of the responses has a different developmental origin and evolutionary significance.

Table 5.1

Differential diagnosis of the effects of temperature changes on the development and morphology of butterfly color patterns

Response	Sensitive period	Rate of development during temperature shift	Effective temperature range and duration	Effect of photoperiod
Norm of reaction	During pattern determination	Normal; proportional to temperature	10° to 35° C; best effect if duration is as long as period of pattern determination	None
Phenocopy	During pattern determination	Development ceases during temperature shock	−2° to +5° C for one or more days, or 37° to 42° C for minutes up to an hour	None
Seasonal polyphenism	Prior to pattern determination	Normal; proportional to temperature	10° to 35° C, mimicking seasonal temperature characteristics	Acts synergistically with appropriate temperature and can often substitute for temperature

The simplest response to altered temperature is what is known as the norm of reaction (Goldschmidt, 1955; Lewontin, 1974). The norm of reaction simply describes the normal phenotype that develops under a specific set of environmental conditions. Biochemical reactions and physical processes naturally go at different rates when temperatures change, and if the rates of all reactions involved in a particular developmental process do not change proportionally (e.g., with the same Q_{10}) and there is no other type of compensation, then the details of that process are bound to change, and so is the morphological outcome. The phenotypic change associated with the norm of reaction is expected to occur smoothly and continuously with changing temperature. When alterations of a particular environmental variable have no effect on a given character, that character is said to be canalized, that is, buffered against that variation (Waddington, 1942, 1956b; Rendel, 1967). Many of the early classical temperature experiments of Merrifield (1890, 1891, 1892, 1893, 1894), Standfuss (1896), and Fischer (1895, 1907) documented the norm of reaction, and not phenocopy, as is often stated.

The second response to altered temperatures is a phenocopy, a developmental aberration that occurs after exposure to an environmental stress, such as a near-lethal high or low temperature, a variety of noxious chemicals, or radiation. In many species, such developmental aberrations closely resemble the phenotypes produced by specific mutations (such as crossveinless and bithorax phenocopies in *Drosophila*); hence the term "phenocopy" (Goldschmidt, 1935a,b). Near-freezing temperatures for a few days or weeks, or sublethal heat (37° to 42° C, depending on the species) for a few minutes to about an hour, can result in dramatic altera-tions of the pattern. This response is of interest because the altered pattern occasionally bears some resemblance to the pattern of a closely related species, as discussed below.

The third response to altered temperature during development is the expression of a discretely altered morphology known as a seasonal polyphenism. Seasonal polyphenisms are adaptive alternative morphs that, in nature, develop facultatively in response to the differences in temperature and photoperiod regimes that characterize different seasons (usually summer and fall in temperate climates, and wet and dry seasons in tropical ones). In seasonal polyphenisms the temperature change has no direct effect on pattern development but serves as a token stimulus to effect a discrete developmental switch to an alternative morphology. This effect of temperature can be distinguished from the previous two effects because it can usually be enhanced and occasionally be mimicked by an appropriate change in photoperiod.

A differential diagnosis of these three types of responses to temperature is given in Table 5.1. A fuller discussion of the phenocopy response is presented below, and discussion of seasonal polyphenisms will be postponed until Chapter 6. Few studies have been done on the norm of reaction of the color pattern in Lepidoptera, none of them sufficiently systematic to lead to useful conclusions about the developmental physiology of pattern formation.

A century ago Merrifield (1890, 1893, 1894) showed that temperature shocks, applied late in larval development, can cause significant and reproducible alterations of the color pattern in several species of moths and butterflies. In the intervening years the application of cold and heat shock during the late larval and early pupal period has become one of the most commonly used procedures in the experimental analysis of wing pattern development and evolution (Fischer, 1895; Standfuss, 1896; Kühn, 1926; Köhler and Feldotto, 1935; Goldschmidt, 1938, 1940; Shapiro, 1974, 1976, 1982, 1983b; Nijhout, 1984). This experimental procedure is attractive because the color patterns that develop after temperature shock sometimes show surprising similarities to the normal patterns of different geographic races of the species or to those of closely related species. Furthermore, there appears to be a kind of convergence of patterns, so that the temperature-shocked patterns of many species resemble each other more than do the normal patterns of those species.

Both of these observations have provoked the strong suspicion that the study of temperature-induced pattern aberrations might shed light on the evolution, genetics, and development of color patterns. The results of Standfuss (1896), for instance, showed that heat shock of *Iphiclides* (then called *Papilio*) *podalirius* from Switzerland could produce specimens with patterns that resembled those of the subspecies (or race) *zanclaeus* from Sicily. Likewise, heat shock of *Papilio machaon* from Central Europe produced some specimens resembling the subspecies *sphyrus* from Syria and others that resembled the subspecies *centralis* from Turkestan. Most peculiar and suggestive is the observation that cold shocks of central European *Aglais* (then called *Vanessa*) *urticae* produced specimens resembling a subspecies from northern Scandinavia (*polaris*), whereas heat shocks of this species produced individuals resembling the race *ichnusa* from Sardinia (Goldschmidt, 1938). This nice parallelism in which cold shocks provoke phenotypes of northern races and heat shocks those of southern ones was once used in support of Lamarckian speculations about evolution (see discussion in Goldschmidt, 1938) but now appears to be merely coincidental,

although the possibility that some temperature-shocked phenotypes may have become stabilized through genetic assimilation cannot be discounted.

More recently Shapiro has taken up a reinvestigation of the ecological and evolutionary significance of the phenocopy response in a number of species of Nearctic butterflies. He has shown that cold-shocked pupae of the mourning cloak (*Nymphalis antiopa*) from California develop into individuals with phenotypes identical to those of Alaskan populations with a much reduced yellow border on the wing and a substantial enlargement of the blue marks that in this species are the homologs of the parafocal elements (Shapiro, 1976). A parallel modification can be induced in *Papilio zelicaon*, a North American species of the *machaon* group. Cold-shocked pupae of *P. zelicaon* from California develop into adults with wing patterns identical to those of *P. machaon* from Alaska (Shapiro, 1976). In general, Shapiro (1976) points out, many cold-shock phenocopies tend to resemble not only the phenotypes of more northern races of the species but also those of local cold-season morphs, if the species is multivoltine and polyphenic, as well as those of univoltine high-altitude populations of the same species. Such observations suggest that an interesting evolutionary relation may exist between phenocopies and the cold-climate forms of a species (see "Polyphenisms" in Chapter 6).

Some pattern phenocopies that can be induced in the laboratory also occur spontaneously in nature and are frequent enough that they have acquired semiofficial names among lepidopterists. Examples are the *elymi* form of *Vanessa cardui*, the *ahwashtee* form of *V. virginiensis* (e.g., Comstock, 1927; Shapiro, 1974, 1983b, 1984b), and the *nigrosuffusa* form of *Precis coenia* (e.g., illustration in Scott, 1986). The last case is possibly of evolutionary interest because *nigrosuffusa* forms occur as a natural population in northern Mexico and southern Texas and Arizona. The case of *nigrosuffusa,* however, presents an unsatisfactorily resolved problem. I have seen specimens from Arizona identified as *nigrosuffusa* that are clearly fairly normal *P. genoveva,* and others from Texas that are clearly aberrant *P. coenia* and identical to the phenocopies that can be produced in the laboratory. Comstock (1927) illustrated an aberration of *P. coenia* (ab. *schraderi*) and a *nigrosuffusa* that are barely distinguishable. Likewise, Hafernik (1982) illustrated examples of *nigrosuffusa* that are clearly *P. coenia,* but Scott (1986) classified *nigrosuffusa* as a race of *P. evarete* and illustrated it as a specimen identical to a *P. coenia* phenocopy. A typical example of a laboratory-induced phenocopy of *P. coenia* is shown in Figure 5.18, in which it is compared with the normal patterns of several American species of *Precis* and with a *nigrosuffusa* from Arizona. The *nigrosuffusa* resembles *P. genoveva* more than it resembles any of the other forms. The *P. coenia* phenocopy also clearly resembles *nigrosuffusa* more than it resembles normal *P. coenia.* A correct interpretation of the evolutionary significance of *P. coenia* phenocopies will have to await the development of a much better understanding of the cladistic relations among the American species in the genus.

Naturally occurring aberrations of many kinds of butterflies are reported regularly in specialized journals on the Lepidoptera. Perhaps the best compilation of aberrations is that of Russwurm (1978) for the British butterflies. It is highly probable that all of Russwurm's aberrations are the result of temperature shock rather than of a recurring mutation. Many of the aberrations illustrated by Russwurm are similar to those found among North American butterflies and thus appear to illustrate common responses to tempera-

ture shock for various taxa. None of the aberrations is, to my knowledge, established anywhere as a distinct population.

As we will see below, temperature shocks induce a broad range of pattern aberrations, and these can usually be ranked in a single morphological series of progressive divergence from the wild type. It is noteworthy that in each of the cases in which temperature-shock aberrations resemble the normal color pattern of distinct geographic races or subspecies, the aberrant specimens with the closest similarities to the other subspecies are all fairly close to the wild type in the morphological series, and not among the more extreme aberrations that can be produced. It is therefore reasonable to suppose that the geographic variation observed in these species is the result of relatively few and simple genetic differences that might easily have been fixed by genetic assimilation (Waddington, 1953, 1956a), although they could as easily have been produced by the fixation of a few new mutations. The implications of and limitations on each of these possible mechanisms of pattern evolution will be discussed below after we have introduced the concept of phenocopy and explored the nature of the response to temperature shock in a bit more detail.

Goldschmidt (1938), who coined the term "phenocopy," recognized that the color pattern aberrations that develop after temperature shock in butterflies are phenomenologically identical to the phenocopies of *Drosophila.* Studies on phenocopy in *Drosophila* (Gloor, 1947; Hadorn, 1955; Milkman, 1966) have revealed several general properties of the phenocopy response: (1) Each type of phenocopy has a sensitive or critical period during which it can be induced. Critical periods may be brief or quite long and can occur anytime during embryonic, larval, or pupal development, depend-

Figure 5.18. Pattern associations among some of the New World species of *Precis*. *A*, normal *P. coenia*; *B*, cold-shock phenotype of *P. coenia*; *C*, *P. evarete*; *D*, *P. genoveva*; *E*, form *nigrosuffusa* from Arizona.

ing on the phenocopy. (2) During the critical period, many different types of trauma, such as heat shock, cold shock, etherization, or x-irradiation, induce identical or nearly identical phenocopies. (3) Trauma applied at any one point in development does not induce a phenocopy in all individuals so treated, nor do all individuals that respond develop the same aberrant phenotype. Thus we could say that phenocopies in general exhibit incomplete penetrance and variable expressivity. The variability in expressivity results at least

in part from the often broad overlap of the critical periods for different phenocopies.

The genetic or developmental mechanisms of phenocopy induction have not yet been elucidated, but characteristics of the response suggest at least some of the things that might be involved. The close resemblance of many phenocopies to known mutations at single genetic loci suggests that the trauma somehow interferes with the normal expression of certain genes. Temperature shocks could, for instance, interfere with the transcription of genes whose products are necessary for normal development during the critical period. Alternatively, the temperature shock could effect an irreversible denaturation of one or more proteins whose activity is essential during the critical period. The work of Mitchell and Lipps (1978) on *Drosophila* suggests that heat-shock proteins may be involved in the phenocopy response. Heat-shock proteins are almost universally synthesized after an organism is exposed to a temperature shock or other trauma. Some heat-shock proteins inhibit general protein synthesis and appear to be part of a mechanism to shut down cellular activity temporarily, perhaps as a defense mechanism to survive the trauma. Mitchell and Lipps (1978) have shown that when protein synthesis resumes after the heat-shock proteins disappear, not all genes become active simultaneously, and the researchers hypothesize that phenocopies arise because one or a few genes do not recover function in time to participate in ongoing development. If gene reactivation after trauma is random, as the data of Mitchell and Lipps suggest, that could account for the considerable variability of penetrance and expressivity that are traditionally observed in *Drosophila* phenocopies.

The temperature-induced pattern aberrations of butterflies differ from the typical *Drosophila* phenocopy response in one significant way, namely that the variability in expressivity is extraordinarily regular. Whereas in *Drosophila* several different types of phenocopies can be induced during a given critical period, in butterflies the pattern aberrations differ only in degree and can always be arranged in smooth morphological series ranging from little or no difference from the wild type to highly aberrant (Nijhout, 1984). Figure 5.19 shows such a morphological series for the ventral hind wing of *Vanessa cardui*.

Detailed studies of the comparative mor-phology of color pattern phenocopies have been done in a few species of butterflies, namely *Argynnis paphia* (Kühn, 1926) and *Precis coenia, Vanessa cardui,* and *V. virginiensis* (Nijhout, 1984). Several detailed studies have also been done in the moths *Ephestia kuehniella* (Feldotto, 1933; Wuhlkopf, 1936), *Abraxas grossulariata* (Kühn and von Engelhardt, 1936), and *Ptychopoda seriata* (Kühn and von Engelhardt, 1944). Together these studies reveal a fairly consistent set of characteristics of the phenocopy response to temperature shock:

1. Not all pattern elements are affected by temperature shock at any one time. Pattern elements that are determined during the larval period are not altered by temperature shock during the pupal stage.

2. The bands of the central symmetry system shift toward the axis of symmetry. In the most extreme cases the proximal and distal bands become fused and may even disappear altogether.

3. Marginal and submarginal pattern elements tend to be lost.

4. Border ocelli and parafocal elements become fused along the wing cell midline. In extreme cases both of these pattern elements become fused into a single rounded or elongated figure (Figs. 5.19 and 5.20). This effect is particularly prominent in temperature shock aberrations in the Lycaenidae (see illustrations in Russwurm, 1978).

5. The overall color of the wing is darkened, and the contrast in coloration between pattern and ground is diminished.

6. The boundaries of most pattern elements become much less sharply defined than in the wild type. In extreme cases it becomes difficult to detect the boundary and shape of a given pattern element. Together with the loss of contrast between pattern and ground, this response dulls the overall ap-

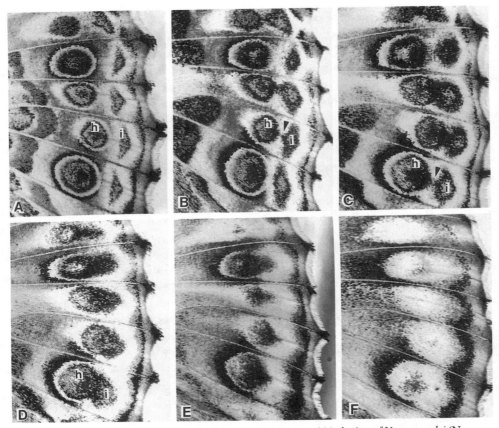

Figure 5.19. Cold-shock-induced pattern aberrations in the ventral hind wing of *Vanessa cardui* (Nymphalidae: Nymphalinae), arranged in a morphological series from normal pattern (*A*) to progressively more severe aberrations of the ocelli (*h*) and parafocal elements (*i*). Along this morphocline these two elements fuse (*B–D*), gradually merge into a single figure (*D* and *E*), and finally vanish in the most severely affected individuals (*F*). Note that wing cells respond differentially. (From Nijhout, 1985b)

pearance of the butterfly and produces a pattern that appears to be smeared (Figs. 5.19–5.22).

7. The dimensions of many pattern elements are changed. This response is difficult to summarize, because some pattern elements become enlarged and others become smaller, and the response varies among pattern elements and among species. In addition, the increasing fuzziness of pattern element boundaries in affected animals occasionally makes it difficult to define the

size of a pattern element in a way that is readily comparable with measures taken on normal individuals. In general, however, the bands of the central symmetry system become broadened and the border ocelli become diminished in size, but many exceptions exist that are of sufficient interest to merit brief discussion below.

In *Argynnis paphia* and *Ptychopoda seriata*, and to a lesser degree in *Vanessa cardui* and *Abraxas grossulariata,* the width of the bands of the central symmetry system responds in a complex way to temperature shock. In the least-affected individuals, that is, in individuals in which most other pattern elements are present in nearly their normal form and position, the width of the bands of the central symmetry system is increased, sometimes considerably. In the most affected individuals, by contrast, the bands of the central symmetry system are reduced to small dotlike patterns (Fig. 5.20) or are absent entirely. This response is clearest in *Argynnis paphia*, from which one gets the impression that the diminution of the band is due to some encroaching activity of the wing veins that occurs only in the more highly affected individuals (Fig. 5.20).

The most unusual response is seen in the

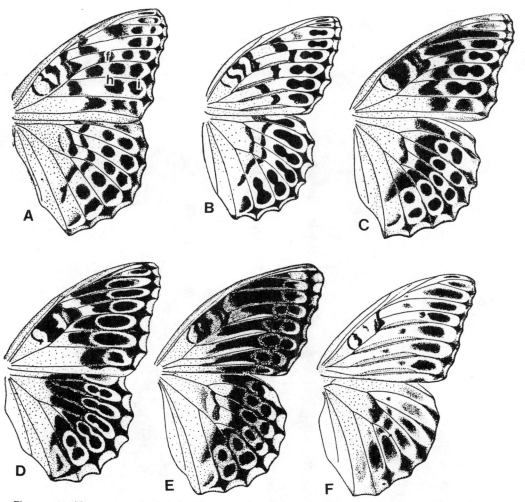

Figure 5.20. Temperature-shock-induced pattern aberrations in *Argynnis paphia* (Nymphalidae: Heliconiinae). *A*, normal pattern; *B–F*, various degrees of pattern alteration. Although the aberrations in each of the elements of the pattern can be arranged in a simple morphological series, the overall pattern cannot. Evidently each pattern element is independent in its response to temperature shock.

border ocelli of *Precis coenia*. Here the large ocelli on the forewing are invariably diminished in size (Fig. 5.21), and in the most affected individuals they are reduced to small indistinct dots (Nijhout, 1984). The large eyespots on the dorsal hind wing respond differently. In the majority of individuals, the size of the eyespot is increased, but the increase is not radially symmetrical. The eyespot has a tendency to open up toward the wing margin (yielding patterns somewhat like those seen in response to partial ablation of the wing) and to become extended along the wing veins (Fig. 5.22). In the most extremely affected individuals the eyespot becomes a large irregular blotch that can cover as much as half the surface of the wing. The response of the hind wing eyespot to temperature shock appears to involve a significant random component, because no 2 of the more than 100 animals with hind wing pattern aberrations that I and my co-workers have produced look quite alike. Whatever this random component is, it appears to vary at the level of the individual animal and not at the level of the individual wing. In all cases, the shape and size of the altered pattern elements that develop in response to temperature shock are symmetrical on the two wings (Nijhout, 1984).

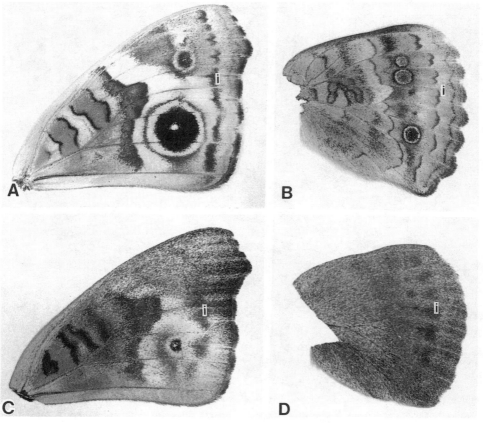

Figure 5.21. Pattern aberrations of the ventral wing surfaces of *Precis coenia* in response to cold shock. *A* and *B*, normal wings; *C* and *D*, cold-shock phenocopies. Parafocal elements (i) are indicated.

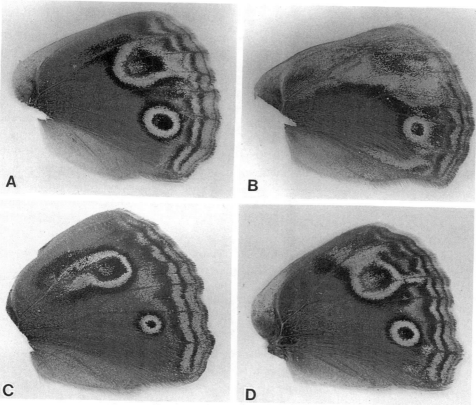

Figure 5.22. Phenocopy response to cold shock in the hind wing eyespots of *Precis coenia*. A through D illustrate four different responses to a cold shock given early in the pupal stage and show that the response of the large eyespot is erratic. This spot can become stretched toward the wing margin (A and B) or toward the base of the wing (C). In some specimens, preferential spread along the wing veins is evident (D). Usually the eyespot is enlarged, expanding preferentially along the wing veins (but see Fig. 5.18B).

In *Precis coenia* and *Vanessa cardui,* and apparently also in *Argynnis paphia,* the degree of aberration among different pattern elements is reasonably well correlated, so that individuals with highly aberrant border ocelli also tend to have highly aberrant central symmetry bands. But the correlation is never perfect; each individual of an experimental cohort expresses a unique combination of degrees of effect among its pattern elements. We may provisionally conclude that temperature shock has an independent effect on each pattern element.

The most puzzling feature of all of these pattern phenocopies is the stochastic character of the response. Perfectly synchronized animals, exposed together in the same container to an identical temperature shock, always develop a broad range of aberrant phenotypes. Most of the experimental animals either are unaffected by the shock or express only a mild pattern aberration, and progressively smaller percentages develop the more extreme aberrations. Nijhout (1984) showed that in *P. coenia* and *V. cardui* the frequency distribution of the various degrees of pattern aberration (e.g., those shown in Fig. 5.19) after a standard cold shock exhibited a Poisson distribution, a type of distribution obtained when a particular effect or state comes about through the accumulation of random and rare events (Sokal and Rohlf, 1981). What does this imply about the physiology of pattern formation after temperature shock? The Mitchell and Lipps (1978) model provides a random component in that the transcription of individual genes appears to be reactivated in a random order upon recovery from heat shock. This observation could explain why a diversity of phenocopies is often observed after a standardized cold shock in *Drosophila,* but does it also explain the long, smooth, and uninterrupted series of progressively more aberrant patterns we see in the butterflies? The continuous character of these transformation series suggests that the underlying causes of the response differ quantitatively rather than qualitatively, as they often do in *Drosophila.* Furthermore, the aberrations of most pattern elements can be arranged in a single unbranched morphological series, suggesting that variation in only a single causal parameter is involved.

Thus, in the absence of further information, it seems reasonable to conclude that the phenocopy response of a given pattern element is due to changes in a single determinant of pattern that is altered to different degrees in response to trauma. Quantitative variation in the concentration of a single gene product could come about if the rate of transcription of that gene were variable after trauma or if transcription of that gene were initiated at different times during the recovery period in different individuals. This latter possibility would be in accord with the Mitchell and Lipps model. If it is the correct explanation for the origin of phenocopy series such as the one shown in Figure 5.19, the developmental abnormalities we see must be the result of a progressive change either in the concentration of single gene products (or something made or altered by those gene products) or in the timing of their appearance.

Both possibilities have interesting implications for understanding the manner in which color patterns can evolve. If the first possibility is correct, phenocopies reveal the kinds of pattern variations that are possible after small quantitative variations in the activity of a gene product, such as might result from mutations. In this case, temperature shocks would provide a convenient way to mimic the effects of mutations, which is exactly what the term "phenocopy" implies. If the second possibility is correct, phenocopies reveal the consequences of a genetic heterochrony (i.e., differences in the relative timing of gene activation), which results from alterations in the mechanism that regulates gene activity. Such a heterochrony could come about through changes in a regulatory gene or through changes in the cell physiological processes that provoke differential gene expression. Temperature shock could thus mimic an alteration in a mechanism that regulates gene expression during pattern development, and the phenocopies that develop would reveal the consequences of small quantitative differences in the cellular mechanisms of gene regulation. Both possibilities seem equally feasible, although the first one, mimicry of mutations, is the only one that mimics what is a priori an inheritable change of evolutionary interest. The second possibility could but need not involve mimicry of a change in a gene or gene product. If it does not, as would be the case if the phenocopy mimicked changes in a physical or chemical parameter, then the phenocopy could still be of evolutionary interest because it could open the way for genetic assimilation of the aberrant pattern.

The genetic assimilation of phenocopies has been studied in *Drosophila* by Waddington (1953, 1956a), who showed that by selecting individuals that developed a particular type of phenocopy after temperature shock, and repeating the shocking and selection procedure for many generations, a stock of flies could eventually be developed that exhibited the aberrant phenotype without exposure to temperature shock. The reason for the success of such a genetic assimilation of a phenocopy is that the temperature shock essentially provides a method of identifying and selecting in favor of individuals in which the (wild-type) character in question was most sensitive to environmental variation, that is, least canalized (Waddington, 1942, 1953; Rendel, 1967). Thus genetic assimilation has the characteristics of any selection process on a complex character that can move the selected phenotype well beyond the range of phenotypes encountered in the ancestral population (Futuyma, 1986), differing only in that temperature shock is used to identify the most "promising" individuals to be used for breeding. Rachootin and Thomson (1981) discussed the evolutionary significance of genetic assimilation and defined the conditions under which a phenocopy can become fixed as the normal phenotype in a population. In essence, what is required is that (1) the population must be exposed to environmental conditions that induce a phenocopy repeatedly and (2) the aberrant phenotype must have a selective advantage, or increased fitness, in that environment.

Thus there are clearly several plausible genetic and developmental reasons why temperature-shock-induced phenocopies might resemble the color patterns of different geographic races of a particular species or the patterns of closely related species. It is not unreasonable to assume that phenocopies reveal some of the evolutionary potentials of color patterns. It is also clear, however, that unless we learn a great deal more about the mechanisms underlying the phenocopy response and the potential for their genetic assimilation, phenocopies provide no more insight than might be obtained from the study of induced mutations or of the norm of reaction (Goldschmidt, 1938; Lewontin, 1974) of color patterns. Until then, the greatest usefulness of phenocopies probably lies in revealing which kinds of pattern transformations are relatively easy to accomplish (easy in the sense that they involve changes in only a single determinant), but that may ultimately tell us more about the developmental physiology of patterns than about their evolution. It is thus unlikely that phenocopy patterns represent ancestral or atavistic pattern traits, as has sometimes been suggested. The mutual resemblance of phenocopy patterns among closely related species appears to be due to the fact that the phenocopy in each case represents a similar simplification of the color pattern (loss of elements, fusion of elements, fading of boundaries, or darkening; never an addition of a pattern element). What is ultimately most peculiar about the phenocopy response is that it involves so few modes of change, and that these modes are so similar among a broad diversity of taxa. Enough cases have been observed to suggest that it is unlikely that the restricted diversity in this response is simply a sampling artifact.

Morphometrics

Perturbation experiments of the types reviewed above shed light on the developmental physiology of pattern formation. When such perturbations result in specific and reproducible alterations of pattern that can be ordered in time or space, then this information can be used to deduce the characteristics of the processes responsible for pattern formation. If time and space parameters of the perturbation can be manipulated, it may also be possible to deduce something about the temporal and spatial dependencies among various components of the pattern. We have seen how such experiments have provided information about the relative timing of the development of different elements of the color pattern and about the dependence of certain elements of the color pattern on physical features of the wing disk, in particular the wing veins, the wing margin, and the inductive centers we called the foci. Perturbation experiments also give us a glimpse at a relation that is of immense significance for understanding the evolution of patterns, namely which elements (or aspects) of the pattern develop independently from one another. Portions of a pattern that develop independently should also be free to evolve independently.

Border ocelli and the central symmetry system clearly develop independently from one another, as do border ocelli and parafocal elements. The border ocellus in one wing cell also develops independently from the border ocelli in adjoining wing cells, but we cannot make a parallel statement about central symmetry systems and parafocal elements. The perturbation experiments done to date simply do not have the resolution, nor are the species that have been studied particularly suitable, to address the question of wing cell autonomy of these elements of the color pattern. However, other means for establishing the independence (and interdependence) of pattern elements exist, namely through the statistical analysis of individual variation in the elements of the color pattern.

Although morphometrics and the analysis of morphometric data are out of vogue among developmental biologists today, these methods are experiencing considerable interest among evolutionary biologists because they form the basis of the quantitative genetics approach to the evolution of complex characters. Characters that share one or more developmental and genetic determinants will exhibit a certain degree of correlation in their form (pleiotropy) and in the variation of their form. Knowledge of the correlation and covariation among different characters is therefore crucial for understanding and predicting the response to natural selection. The analysis of phenotypic and genetic covariance matrixes provides insight into the manner in which genetic interactions can constrain the possible modes of evolutionary change (Cheverud, 1984, 1988). Because certain correlations among characters can be inferred to have a developmental basis, a thoughtful analysis of such correlations can also be used as a tool to understand the development of those characters.

Nijhout (1985c) and Nijhout and Wray (1988) provided a preliminary analysis of variation among some serially homologous pattern elements in three genera of butterflies—*Cercyonis, Smyrna,* and *Heliconius.* The sizes of homologous pattern elements in a series of adjoining wing cells were measured to determine the degree of covariation of homologs within and among individual butterflies: If in a given individual a pattern element in one wing cell is exceptionally small, do its homologs in the other wing cells of that individual also tend to be small, and vice versa? To eliminate covariation caused by differences in the overall size of individuals (larger individuals will have larger pattern elements than smaller individuals, but that is not an interesting observation), the relative sizes of pattern elements in homologous wing cells were rank-ordered separately among all the individual animals tested, and rank correlation analyses were performed on these ordered data. The results of these analyses showed that there was relatively little covariation in the sizes of these pattern elements (Fig. 5.23) and that in most cases the sizes of homologous elements in adjoining wing cells were not correlated at all. Thus, if an ocellus in one wing cell of a given animal was unusually large for the species, the ocelli in other wing cells were not necessarily unusually large and could, in fact, be quite small relative to their homologs in other individuals.

The rank correlation analyses showed that individual variation in the color pattern occurs at the level of the pattern element, not at the level of the individual animal. Independent variation of pattern elements means that at least some of the processes that determine the dimensions of these pattern elements must have varied independently in each wing cell. The significance of this finding for the evolution of color patterns is that if variation of one pattern element in one wing cell is independent (or largely so) of variation of other elements in other wing cells, then each pattern element in each wing cell is free to evolve independently of the others. The enormous range of variations on the nymphalid ground plan discussed in Chapter 3 illustrates that such independent evolution has indeed occurred.

But variation of pattern elements among wing cells is not completely independent. Some degree of covariation is evident in Figure 5.23, and it would be interesting to determine whether there is any regularity or pattern in the correlations that do exist among pattern elements. If the variation in some element or region of the color pattern is strongly correlated with variation in another, this would suggest that the two share a portion of their genetic and developmental determinants. Such elements would not be as free to evolve independently as those whose variations are more weakly correlated.

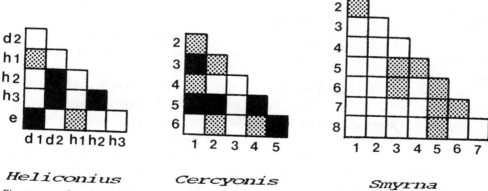

Figure 5.23. Correlation matrixes of pattern variation in three species of butterflies. The correlations measured are among different several pattern elements (letter codes as in Table 2.1) in different wing cells on the forewing of *Heliconius cydno* (Nymphalidae: Heliconiinae). In *Cercyonis pegala* (Nymphalidae: Satyrinae) the correlations indicated are of the diameters of six border ocelli on the hind wing. In *Smyrna blomfildia* (Nymphalidae: Limenitinae) the correlations indicated are of dimension of the central symmetry system in seven adjoining wing cells on the hind wing. *White,* no significant correlation among the measures; *stippled,* the hypothesis of no correlation is rejected at the 95% level; *black,* the hypothesis of no correlation is rejected at the 99% level. (After Nijhout, 1985c, and Nijhout and Wray, 1988)

Rank correlation analysis does not provide the necessary resolution and statistical sensitivity to detect finer degrees of correlation among pattern elements. Kingsolver and Wiernasz (1990) and Wiernasz and Kingsolver (1990) have done a more detailed statistical study of the variation and developmental organization of the melanic portions of the wing pattern of pierine butterflies, using analysis of variance and factor analysis. Their results showed that melanization of the venous patterns (darkly pigmented veins in the basal portion of the wings) on dorsal and ventral hind wings are most strongly correlated. They calculated broad-sense (additive plus nonadditive) genetic correlation with a median value of .95 for the venous stripes on the ventral hind wing. The venous pattern on the forewing is also strongly correlated with that on the hind wing, but this correlation is less than the within-hind-wing correlation. Bands of the central symmetry system on the forewing are also strongly correlated with each other but are not correlated with their homologs on the hind wing, and they are negatively correlated with the venous stripes. The venous pattern at the tips of the wing veins (where veins interact with the wing margin) is uncorrelated with the basal venous stripes and likewise uncorrelated with melanization of the central symmetry system. The analyses of Kingsolver and Wiernasz indicate that in pierines the wing is divided proximo-distally into three regions—basal, medial, and marginal—in which pattern variation is independent. No independence was apparent in the degree of melanization of homologous pattern elements in different wing cells; all serial homologs on a wing surface varied in unison.

The studies of Kingsolver and Wiernasz were directed at analyzing the development and evolution of those portions of the pierid wing pattern with known adaptive significance. The basal melanin pattern functions in thermoregulation, whereas the marginal pattern is used for sexual signaling (Wiernasz, 1989; Kingsolver and Wiernasz, 1990). These portions of the color pattern thus have very different functions, and they clearly develop and evolve independently of each other. An even more detailed analysis of pattern variation has been performed on the wing pattern of *Precis coenia* by Paulsen and Nijhout (1990). This study focused primarily on variation and correlations in the size and position of the border ocelli and parafocal elements and used partial correlation analysis and principal component analysis to estimate the correlations among the pattern elements from different areas of the wing. A partial summary of their results is given as a correlation matrix in Figure 5.24. It can be seen by inspection of Figure 5.24 that the correlations among the majority of elements are not significantly different from zero (white squares). Most of the significant correlations are located near the main diagonal of the correlation matrix and are the correlations among homologous pattern elements.

The following conclusions can be drawn from the results of Paulsen and Nijhout and the summary in Figure 5.24. Eyespot diameters tend to be significantly correlated among each other, though the correlations are generally less than .5. Eyespot positions and parafocal element positions also tend to be significantly (though, again, not strongly) correlated among themselves. But there is no correlation among eyespot size, eyespot position, and parafocal element position. The position of the distal band of the central symmetry system was likewise uncorrelated with any of the other characters. Eyespot sizes on one wing surface are more strongly correlated with each other than they are with eyespots on other wing surfaces; the next strongest correlation is between dorsal and ventral

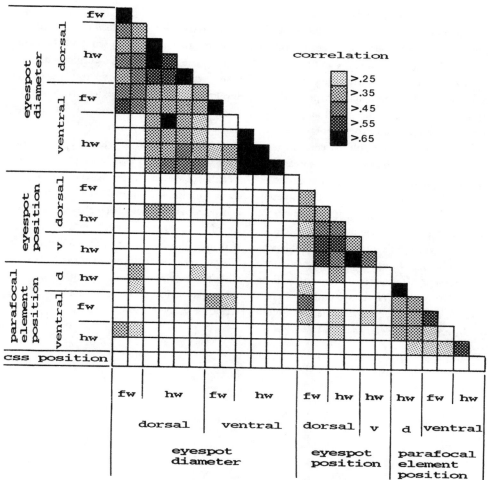

Figure 5.24. Correlation matrix for the variation of various pattern elements in *Precis coenia*. Most correlations are not significantly different from zero (*white squares*). For the data set in this figure, critical values for rejection of the null hypothesis of no correlation are $p > .95 = .24$, and $p > .99 = .36$. Various degrees of correlation are indicated by shades of gray. Coefficients of determination (correlation^2, which is a measure of the proportion of the variation in one variable that can be explained by variation in the other) are very low for almost all cases. *css*, central symmetry system; *fw*, forewing; *hw*, hind wing; *d*, dorsal; *v*, ventral. (After Paulsen and Nijhout, 1990)

sides of the same wing; the weakest correlation is between forewings and hind wings. The confidence intervals around the estimates of correlation are too broad to determine whether this same hierarchy of correlations also applies to the positions of eyespots and to parafocal elements. The correlations among eyespot diameters in homologous wing cells on different wing surfaces appear to be no higher than among those in non-homologous wing cells.

These results show that the statistical analysis of morphometric data can be used to deduce many of the same features of pattern determination that are normally approached only by means of perturbation studies. The extension of correlation analysis to path analysis (Wright, 1968; Bookstein et al., 1985) should allow us to formulate and study hypotheses about causal relations among various determinants of pattern size, shape, and position that can supplement and amplify hypotheses derived from experimental manipulations. So far we are able to confirm (and predict) that the development processes of the central symmetry system, border ocelli, and parafocal elements are almost completely independent of each other. The portions of each of these pattern elements that occur in different wing cells also develop independently, but not completely so. Trivially, the development of all pattern elements will be influenced by an array of systemic and gross environmental variables, but these can often be eliminated from the analysis by an appropriate statistical treatment of the data (Bookstein et al., 1985; Paulsen and Nijhout, 1990). The hierarchies of correlation must have a cause, probably in the use of common developmental and genetic determinants. Those elements and features of the pattern that are more highly correlated either share more determinants or share determinants with stronger effects than those that are more weakly correlated. If the analysis is extended to comparisons between "normal" patterns and patterns that have been altered by traditional perturbation, as has been done for normal and cold-shock patterns by Wiernasz and Kingsolver (1990), it may eventually be possible to develop a powerful set of experimental and statistical tools for the analysis of pattern development and evolution.

The Genetics of Color Patterns

Studies on the genetics of color patterns have taken their material primarily from naturally occurring genetic polymorphisms. Among these, melanisms and mimicry systems have provided the most interesting material. Nearly all of the investigations have been done in the context of ecological genetics and mostly constitute attempts to understand the adaptations and evolution of various mimicry systems. Almost without exception, these studies have disregarded the system of homologies among elements in the color pattern. Thus, although we know a great deal about the modes of inheritance of various spots and patches of color on the wings of an assortment of species, there has never been an attempt to interpret the identity and homologies of these parts of the pattern. Without knowledge of homology, there is no basis for generalization and, more important, no basis for critically judging how color patterns and mimicry systems evolve. Part of the task of this chapter will therefore be to interpret the results of earlier studies in terms of alterations of the elements of the nymphalid ground plan. In those cases in which it is possible to interpret the action of pattern genes in terms of alterations of pattern elements, the complicated and qualitative phenotypic effects of many genes will be seen to be relatively simple quantitative variations on existing forms.

In the second part of this chapter, we will consider the nature of color pattern polyphenisms. Unlike polymorphisms, which are due to genetic differences between individuals, polyphenic animals are capable of expressing two or more different color patterns in genetically identical individuals. Polyphenisms are generally believed to be adaptations to seasonally varying conditions, though the adaptive significance of many of

Chapter 6

Genetics, Mimicry, and Polyphenisms

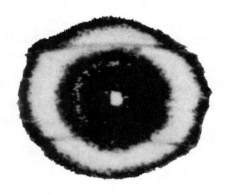

them is not fully understood. Just as in the case of polymorphisms, the interpretation of polyphenisms is much simplified when they are seen as alternative forms of the elements of the nymphalid ground plan.

Melanisms

Melanic forms and races are common in the Lepidoptera. Although the well-known industrial melanisms appear to be restricted to moths, the butterflies exhibit a variety of melanic polyphenisms and polymorphisms. For instance, melanic polyphenisms induced by day length or temperature have been described in several pierids (Watt, 1968, 1969; Hoffmann, 1973; Douglas and Grula, 1978), and melanic sexual polymorphisms are common in many papilionids (Clarke and Sheppard, 1955, 1959, 1972). As will be argued below, in many cases of mimicry in butterflies, changes in melanism of the mimic confer much of the resemblance to a model.

Among the Lepidoptera, three morphologically distinct types of melanism are possible. One consists of a general darkening of the background color of the wing, the second

involves a broadening of existing dark pattern elements, and the third consists of a darkening of the pattern elements. All three mechanisms thus achieve an overall darkening of the wing, one by replacing a light background color with a dark one, the others by covering up a light-colored background with enlarged or darkened pattern elements. All three types of melanism can occur as genetic polymorphisms or environmentally induced polyphenisms in different species of butterflies. The dark female form of *Papilio glaucus* (a presumed mimic of *Battus philenor*) is an example of melanization of the yellow background color (Fig. 6.1). In this species the normal narrow black stripes of the color pattern remain easily discernible on top of the darkened background. The melanization in several species of *Colias*, a seasonal polyphenism, is likewise due to a darkening of the background color (Watt, 1968, 1969). In these cases the melanism presents itself as an increase in the proportion of randomly peppered black scales in the yellow background.

The dark female of *Papilio polyxenes* (also believed to mimic *B. philenor*), on the other

A **B**

Figure 6.1. Polymorphism of *Papilio glaucus* females. The melanic morph is believed to be a Batesian mimic of *Battus philenor* in North America. Ventral surfaces show that the melanism of the mimic arises by a darkening of the normal yellow background color. The dark pattern stripes of the ground plan are identical in nonmimetic (A) and mimetic (B) morphs.

hand, is darker than the male of its species because the yellow background portion of the wing pattern is much reduced in size by a broadening of the black area of the pattern (Fig. 6.2A,B). The remaining small patches of yellow background are as bright in the melanic female as they are in the nonmelanic male. The sexual dimorphism of *Pontia protodice* and related pierids likewise comes about through differences in the sizes of dark pattern elements. Here the more melanic female form appears to be the primitive condition, whereas the reduction in size (and number) of pattern elements in the male is the derived condition (A. M. Shapiro, pers. com.).

The dark (e.g., *naresi*) and light (*pallescens*) forms of *Hypolimnas bolina* (Fig. 6.10A,C) are an example of a change in coloration of pattern elements. In this species the *pallescens* form is light tan and the *naresi* form is deep black, yet the two do not differ significantly in the morphology of their pattern elements. Another example is the seasonal polyphenism of *Tatochila mercedis* (Shapiro, 1980b, 1984a,b), which consists of a darkening of venous stripes. Here the pattern stripes have sparse black scales in the light morph and are bold and solid black, but not wider, in the dark morph.

These three types of melanism (darkening of the background color, broadening of dark pattern elements, and darkening of light-colored pattern elements) have each been shown to be controlled by single genetic loci in nearly all cases in which the genetics has been studied, yet they come about through different developmental mechanisms. Darkening of the background color can occur through one of two mechanisms: (1) by a switch in the pigments synthesized (for instance, the replacement of a light pterin pigment by a darker pterin, an ommochrome, or melanin) in all scales, or (2) by a change in

the frequency of dark versus light scales. Both mechanisms are known to occur. The first accounts for the differences in background color in different mimetic races of *Papilio dardanus,* and the second is responsible for the melanism of female *P. glaucus* (Fig. 6.1) and various *Colias* species. The first mechanism is relatively easy to understand, both genetically and physiologically, because it involves a simple switch in biochemical pathways, which could be accomplished by the activation or inactivation of specific enzymes. The second case is more difficult to understand mechanistically because it involves discrete switching of biochemical pathways at the single-cell level. In the melanisms of *P. glaucus* and *Colias,* individual scales are either completely black or yellow and are arranged in an apparently random distribution. The genetic or environmental switch somehow effects a shift in the proportion of each scale type. This type of discrete peppering of scales with different colors is seen in several other contexts as well—for instance, in many pattern heterozygotes and in normal color gradients. A possible mechanism to account for this effect is discussed in Chapter 7.

The second mechanism, a broadening of dark pattern elements, involves a change in the position of the boundary between pattern and ground. This could come about by changing the rate or timing of propagation of the signal that establishes this boundary or by altering the threshold of response to this signal. In Chapters 7 and 8 we will see how each of these developmental alterations could be effected by changes at single genetic loci. The third type of melanism, a change in the color of most or all pattern elements, is probably caused by a switch from phaeomelanin (brown) to eumelanin (black) synthesis. Biochemically this appears to be a relatively simple switch, because it occurs as phenotypic alternatives throughout the vertebrates and invertebrates. Unfortunately, the underlying biochemistry has so far resisted elucidation. Most likely it involves a change in the use of substrates, not a change in the enzyme (Kayser, 1985; Nijhout, 1985b). Exactly how the alteration of a single gene could accomplish such a switch is unclear.

As noted above, melanisms are often attributable to changes at a single locus. Moreover, when the genetic control of the melanism is simple, the melanic form always appears as the dominant condition. The heredity of the melanism of *Papilio glaucus* has been worked out by Clarke and Sheppard (1959, 1962), Clarke et al. (1976), and Scriber and Evans (1986). Almost without exception, yellow females produce only yellow daughters, and black females produce only black daughters, irrespective of the origin of the male (which is always yellow, no matter what the phenotype of its mother). This mode of inheritance is diagnostic of sex-linked inheritance in Lepidoptera, in which females are the heterogametic sex (XY) and males the homogametic sex (XX). Thus the black color is likely controlled by a locus on the Y chromosome that either has no homolog on the X chromosome or has a homolog on X that is somehow suppressed.

The genetic control of melanic forms in the *Papilio machaon* complex (Fig. 6.2) has been examined by Clarke and Sheppard (1955). They studied the genetics in interspecific crosses of four species in this complex: *P. machaon, P. polyxenes, P. zelicaon,* and *P. brevicauda.* These species are sufficiently closely related that interspecific hybrids between them produce viable F₁ progeny, and though the F₁ individuals are sterile among themselves, they produce viable offspring when backcrossed to one of the parents. Among these four species, *P. polyxenes* is polymorphic, having a melanic female and a

yellow male, and *P. brevicauda* is melanic and monomorphic. Both of the other two species are yellow and monomorphic. Clarke and Sheppard (1955) interpreted the differences in these color patterns as being due to differences in the "ground color," presumably meaning the background coloration. The background color of all four species is, however, bright yellow, and the differences in their color patterns are almost entirely due to differences in the widths of the pattern elements. In fact, pattern diversity in the entire *machaon* complex worldwide appears to involve species-specific changes in the relative widths of black bands on forewing and hind wing, coupled with variation in the degree of darkening of the wing veins.

The melanic forms (i.e., with wide bands) of both *P. polyxenes* and *P. brevicauda* are inherited as single dominant genes in crosses with *P. machaon.* Clarke and Sheppard (1955) showed that when the black gene of either *P. polyxenes* or *P. brevicauda* was placed in a *P. machaon* genetic background (by hybridization and backcrossing to *P. machaon*), the offspring have a sexual color dimorphism similar to that of *P. polyxenes.* These results were taken to indicate that (1) the same gene controls the melanic form in both *P. polyxenes* and *P. brevicauda,* (2) the sexually dimorphic expression of the melanism in *P. polyxenes* is due to the presence of modifier genes not active in the monomorphic *P. brevicauda,* and (3) *P. machaon,* which is normally nonmelanic and monomorphic, has a gene complex that produces sexual dimorphism (like that of *P. polyxenes*) in the presence of the dominant melanic gene.

Mimicry

Batesian mimicry, the resemblance of a palatable species to an unpalatable one, and Müllerian mimicry, the evolution of similarity between two or more unpalatable species,

Figure 6.2. Species in the *Papilio machaon* complex. *A*, female *Papilio polyxenes*; *B*, male *P. polyxenes*; *C*, *P. brevicauda*; *D*, *P. machaon*; *E*, *P. zelicaon*.

have traditionally been regarded as two complementary modes of convergent evolution, involving different evolutionary strategies and pathways. Turner (1984), in his recent synthesis of ideas on the relation between mimicry and palatability, concluded that the pathways by which these two mimicry systems are established may not be as different as is commonly believed. He pointed out that

the initial phase in both Müllerian and Batesian mimicry consists of a single mutation of large effect that brings the less protected species into reasonably close similarity with the better-protected one. In Müllerian mimicry, evolution then proceeds by gradual mutual convergence to some intermediate pattern by the evolution of modifier genes with smaller effects on the pattern. Both co-mimics are unpalatable, and both derive increased protection from predation by the evolution of a common warning signal. In Batesian mimicry, by contrast, the model is placed at a relative disadvantage because it loses the protection afforded by a 100% correlation between unpalatability and an easily recognized pattern. The model should therefore evolve away from the mimic (Brower and Brower, 1972). Evolution then proceeds by the mimic's somehow evolving faster toward the model than the model can evolve away, and close mimicry is maintained by accumulation of the small effects of modifier genes in the mimic.

Mimicry is a common adaptation among the butterflies, and several species, such as *Papilio dardanus* and *P. memnon,* are well known for the spectacular polymorphism that has evolved as different geographic races have simultaneously come to mimic an array of different model species in their respective regions. The evolutionary theory of mimicry has been tested and refined largely by breeding experiments with these polymorphic species. Most of our knowledge of the genetics and evolution of butterfly mimicry systems comes from the dedicated and sometimes heroic genetic studies of E. B. Ford, C. A. Clarke and P. M. Sheppard, and J. R. G. Turner. In the sections that follow, I will briefly review the genetics of color patterns in these mimicry systems as elucidated by these workers and will attempt to interpret the effects of the pattern genes they have discovered in terms of our present knowledge of the homology and developmental physiology of patterns.

Papilio dardanus

Papilio dardanus is widely distributed across sub-Saharan Africa. It is sexually dimorphic everywhere except in Madagascar, and the females have evolved color patterns that mimic several different species of unpalatable danaids, mostly species in the genera *Amauris, Bematistes,* and *Danaus* (Fig. 6.3). *Papilio dardanus* is differentiated into at least six geographic races that differ subtly in the color patterns of the monomorphic males. But in each of those races the females are polymorphic and may mimic four to six different unpalatable models, depending on the region in which the race occurs. The different genotypes interbreed randomly, and the various

Figure 6.3. Polymorphism of *Papilio dardanus. A,* male pattern, which is also the pattern of nonmimetic females (form *meriones*); *B,* form *hippocoon; C,* form *cenea; D,* form *planemoides.*

mimicking phenotypes segregate quite cleanly. The varietal nomenclature in this species can be confusing because subtle geographic variants of the same mimicking phenotype are often given different names. Thus, form *hippocoon* and form *hippocoonides* not only look similar but also appear to be genotypically identical for the loci that control mimicry (Clarke and Sheppard, 1960a,b).

The genetic work of Ford (1936) and of Clarke and Sheppard (1959, 1960b) has shown that this female polymorphism is controlled by a single genetic locus with 10 alleles. These, together with the phenotypes they produce, are shown in Tables 6.1 and 6.2. Although each of the 10 alleles produces a distinctive pattern, only 4 are involved in mimicry systems. Hybrids with heterozygous genotypes either resemble a homozygote perfectly (or nearly so) or, as in the case of *salaami*, produce a phenotype distinctive enough to have been given a name. The origin and evolutionary significance of the 6 nonmimicry-producing alleles and the mechanisms responsible for their maintenance in a population, to my knowledge, have not been studied.

As many as 6 of the alleles shown in Table 6.1 may occur in a given race (geographic population) of *P. dardanus*. In each population, however, the phenotypic expression of a given allele is balanced by modifier genes specific for that population. Breeding experiments between races of different geographic origins have shown that in each race the allele(s) that convey mimicry occur within a coadapted gene complex of modifiers that not only ensure close mimicry but also reduce variability in the phenotypic expression of the major gene (Clarke and Sheppard, 1963). Each allele produces a close mimicry only if it is expressed in the genetic background within which it evolved (Clarke and Sheppard, 1960a, 1963).

The main pattern gene of *P. dardanus* (referred to as *H*) has a complex set of effects on several portions of the color pattern. The general complexity of these effects, and the diversity of effects produced by the different alleles at this locus, have led to the suggestion that *H* is really a complex locus, or supergene, consisting of several closely linked genes, each with a specific effect on the pattern. The evidence for a supergene is weak, however, because it seems to rest mostly on theoretical grounds (Sheppard, 1953) and on the interpretation of a single putative crossover event that produced the *salaami* phenotype from a single (homozygous) allelomorph rather than as a heterozygote (Clarke et al., 1968).

The specific effects of the various alleles of locus *H* on the color pattern are summarized in Table 6.2 and Figure 6.3. Each of the alleles has a distinct effect on the background color and on one or more of the pattern elements. On the forewing there appear to be three regions in which the background color is under independent genetic control. The size of the black pattern elements in each of these three regions is also controlled independently. Altering the background color and pattern size in various combinations accounts for the array of phenotypes in Table 6.1. The approximate outlines of the independent pattern areas are shown in Figure 6.4. The positions of each of the pattern elements and the derivation of some of the structural features of the pattern of *P. dardanus* are given in Figure 6.5. It should be noted that the boundaries of the three areas on the forewing that are under independent genetic control do not correspond to any known landmarks on the wing. Developmental or genetic compartmentalization of the sort found in *P. dardanus* has so far not been found in any other species of butterfly. The interpretation of the homologies of pattern elements in Figure 6.5 rests

Table 6.1

The genetics of polymorphism in *Papilio dardanus*

Female form	Principal allele	Known genotype(s)*	Model(s)
Mimics			
hippocoonides	h	hh	*Amauris niavius*
cenea	H^c	H^cH^c, H^ch, H^cH^{na}	*Amauris albimaculata, A. echeria*
trophonius	H^T	H^TH^T, H^Th, H^TH^c	*Danaus chrysippus*
planemoides	H^{Pl}	$H^{Pl}H^{Pl}$, $H^{Pl}h$	*Bematistes poggei*
Nonmimics			
natalica	H^{na}	$H^{na}H^{na}$, $H^{na}h$	
leighi	H^L	H^LH^L, H^LH^{na}, H^LH^c, H^Lh	
niobe	H^{Ni}	$H^{Ni}H^{Ni}$, $H^{Ni}h$, $H^{Ni}H^c$, H^TH^{Pl}	
Yellow	H^y	H^yH^y, H^yh, H^yH^c	
Pale-*poultoni*	H^{pp}	$H^{pp}H^{pp}$, $H^{pp}h$	
Bright-*poultoni*	H^{bp}	$H^{bp}H^{bp}$, $H^{bp}h$, $H^{pp}H^T$	
salaami	—	H^LH^T	
Imperfect heterozygotic mimics			
cenea-like	—	H^cH^{bp}, H^cH^{pp}, H^cH^{Ni}	
trophonius-like	—	H^TH^{na}	

Apparent dominance relations†

$H^c > H^{na}$	$H^T \approx H^{na}$
$H^T > H^c$	$H^{bp} \approx H^c$
$H^{pp} > H^T$	$H^{pp} \approx H^c$
$H^L > H^c$, H^{na}	$H^T \approx H^{Pl}$
$H^y > H^c$	$H^{Ni} \approx H^c$

Note: The known genotypes of the named female forms are shown, together with the species mimicked by each. Most of the phenotypes can be produced by several different genotypes. Not all forms of *P. dardanus* are known mimics.

Source: After Clarke and Sheppard, 1960a.

*The phenotypic effects of these genes are given in Table 6.2.

†The symbol > indicates dominance of allele at left over allele at right; the symbol ≈ indicates additive or codominant alleles. Allele *h* appears to be recessive to all the others.

Plate 1. The color pattern of butterflies is a mosaic of colored scales. Each scale bears only a single color, which is determined by an interaction of developmental and genetic processes. Shown is a close-up of the hind wing of *Graphium marcellus* (Papilionidae).

Plate 2. Demonstration of structural colors in *Morpho rhetenor* (Nymphalidae: Morphinae). The blue wing color is produced by the constructive interference of light reflecting from evenly spaced ridges on the vanes of the scales. When a drop of acetone is put on the wing, it appears green. The reflected light must now pass through acetone instead of air, and the higher refractive index of acetone causes constructive interference to occur at longer wavelengths. Hence the reflected color is shifted to green. When the acetone evaporates, the normal blue metallic color reappears.

Plate 3. Differences in pigmentation, without altering the position or morphology of the pattern elements, can have dramatic effects on the appearance of a butterfly. The ventral wing patterns of two closely related species, *Asterope (Sallya) amulia (top)* and *A. occidentalium* (Nymphalidae: Limenitinae), illustrate this mode of pattern divergence.

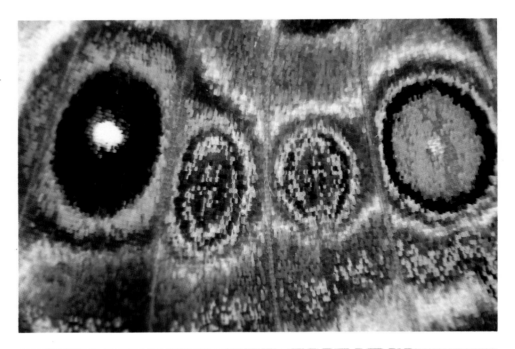

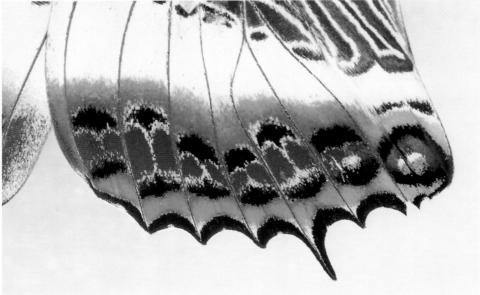

Plate 4 A, The morphologies of homologous pattern elements in adjoining wing cells can evolve independently, as illustrated by the divergence in form of the border ocelli on the ventral hind wing of *Smyrna blomfildia* (Nymphalidae: Limenitinae). B, Border ocelli and parafocal elements on the ventral hind wing of *Charaxes brutus* (Nymphalidae: Charaxinae), illustrating variation in size in each wing cell (see also Fig. 3.11).

Plate 5. Müllerian mimicry in *Heliconius* (Nymphalidae: Heliconiinae). Pairs of co-mimics from various geographic areas are arranged side by side; *left column, H. melpomene; right column, H. erato. From top to bottom,* specimens from southern Ecuador, southern Brazil, northern Ecuador, western Brazil, and Peru.

Plate 6 A, Seasonal polyphenism of *Araschnia levana* (Nymphalidae: Nymphalinae). *Top,* spring form (*levana*); *bottom,* summer form (*prorsa*); *middle,* intermediate form produced in the laboratory. Specimens courtesy of P. B. Koch, University of Ulm, West Germany. B, Seasonal polyphenism of *Precis octavia* (Nymphalidae: Nymphalinae). *Top,* dry-season form; *bottom,* wet-season form; *middle,* intermediate form collected in nature.

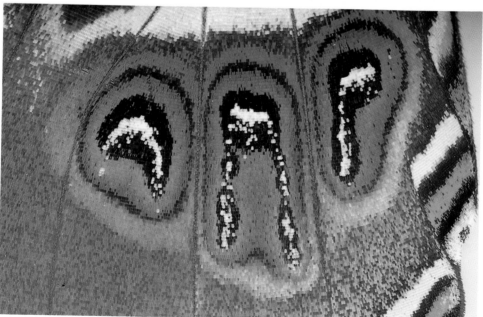

Plate 7 A, Male *Pieris napi microstriata* (*left*) from California and *Tatochila vanvolxemii* (*right*) from central Argentina, showing convergently evolved seasonal polyphenisms: *top,* cold-season form; *bottom,* warm-season form. Animals in each pair are sibs reared under different regimes of photoperiod and temperature (Photo by Samuel W. Woo, courtesy of A. M. Shapiro). B, Arc-shaped foci in *Morpho hecuba* (ventral hind wing), also showing partial fragmentation (Nymphalidae: Morphinae).

Plate 8. Independent evolution of dorsal and ventral color patterns in *Baeotus baeotus* (Nymphalidae: Charaxinae).

Table 6.2

Phenotypic effects of various alleles of *Papilio dardanus*

Pattern		Expression by major allele								
	b	H^c	H^T	H^L	H^{na}	H^TH^L	H^{pp}	H^{bp}	H^{Ni}	H^{Pl}
Background color*										
Forewing discal bar	1	1	1	3	2	1	1	2	1	1
Forewing band	1	1	1	3	2	3	2	5	4	1
Forewing posterior blotch	1	2	4	3	3	4	2	5	4	1
Forewing submarginal spots	1	1	1	3	2	3	2	5	4	1
Hind wing	1	3	4	2	2	4	2	5	4	1
"Pattern" on forewing†										
Discal bar	M	M	M	M	L	S	S	S	M	L
Band	L	S	L	L	L	M	L	L	L	L
Posterior blotch	L	S	L	S	L	L	L	L	L	L
Submarginal spots	L	M	M	M	S	S	S	S	M	M
"Pattern" on hind wing†										
Rays	L	L	M	L	L	S	S	S	M	L
Black band	M	L	S	L	L	S	S	S	M	L
Submarginal spots	M	M	S	M	M	M	M	M	S	L

Note: Each allele controls a combination of background colors and pattern sizes simultaneously, which has led to the supposition that the H locus constitutes a supergene. This tabulation allows for a convenient comparison of the differential effects of each allele on various features of the color pattern. Dominance relations of the alleles, insofar as they are known, are given in Table 6.1.

Sources: After Clarke and Sheppard, 1959, 1960b, 1963.

*Arbitrary color scale: 1 (white) to 5 (dark orange-brown).

†Arbitrary size scale, relative to mean: S = small, M = medium, L = large. All these patterns, except the hind wing rays and black band, actually represent background color. When such a colored band is large, for example, it means that the (black) pattern elements that flank it are small or narrow. The identities of the pattern elements of *P. dardanus* are given in Figure 6.5.

on studies of the comparative morphology of patterns in related papilionids and corresponds to the general relations found in the Nymphalidae.

The effects of the various alleles on the pattern elements and background coloration seem to be purely quantitative. The background color varies from white to orange-brown, but the lighter shades of buff and orange appear to be merely dilutions of the darker shades. The yellow of the males (and of the monomorphic Madagascan female) lies approximately halfway along the scale of color intensity found among the various forms, but it is clearly a different pigment, because it fluoresces under ultraviolet light, and the other background pigments do not (Clarke and Sheppard, 1963). Thus the H^y allele may code for a type of pigment different from that coded by the other nine alleles. No chemical analysis has been done on the pigments, however, so the interpretation given here must be provisional. The modifications of the black pattern elements are also quanti-

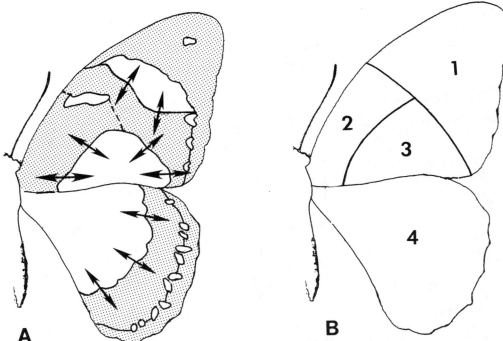

A **B**

Figure 6.4. The pattern diversity among the several mimetic races of *Papilio dardanus* arises from two types of pattern formation. The black elements of the color pattern can increase or decrease in size, as shown by the arrows in *A* (compare with Fig. 6.3), or the background color of the wing can change independently in the four areas shown in *B*. In addition, the intervenous stripes, or rays, on the hind wing (not shown) also vary among mimetic forms (see Table 6.2). Changes in the width of the pattern elements on the forewing can allow large or small areas of background to show. The pattern of these colored areas depends on the pattern of expansion of the pattern elements. Often expansion is stronger along wing veins, and that can leave a heavily spotted background pattern, as in Figure 6.3C.

tative, consisting of variations in the width of existing bands. Widening of black bands constricts the background color that shows through and can have dramatic effects on the overall appearance of the pattern and the closeness of the mimicry. The size and form of the diagonal pale forewing stripe between elements **f** and **g–h**, for instance, is controlled by expansion of these two elements. Expansion of these elements is preferentially along the wing veins so that in the races with the smallest forewing stripe (which is background), this stripe becomes constricted into a series of elliptical patches of background

color. The large patch of background color along the posterior border of the wing varies in size by expansion or contraction of elements **c** and **d** proximally, and of element **f** distally. Thus in both cases, constriction of the background patches occurs equally from all sides and cannot be attributed to preferential expansion of one pattern element. It is possible that the *H* locus simply affects a threshold parameter that determines the position of the pattern-ground boundary over a large portion of the wing.

Although the *H* locus affects pattern over the entire forewing and hind wing, the man-

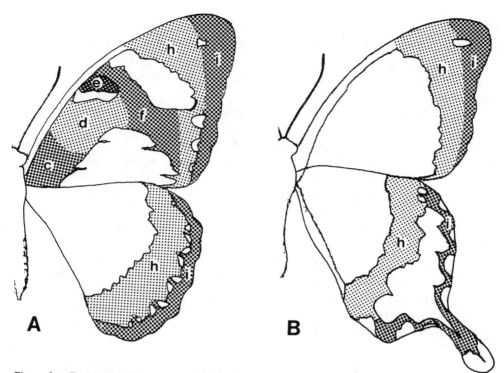

Figure 6.5. Derivation of the patterns of mimetic (A) and nonmimetic (B) *Papilio dardanus*. The nonmimetic form (*meriones*) has a much abbreviated pattern consisting only of elements h and i (pattern element codes are described in Table 2.1). In the mimetic forms, most of the elements of the nymphalid ground plan are added, and the variation in the size of these elements (together with alterations of the background color) enables mimicry to a variety of models.

ifestations of the effects of its alleles appear to be compartmentalized. Background color as well as pattern elements are individually and independently controlled in four different areas (one on the hind wing and three on the forewing; Fig. 6.4). Thus, if *H* is a supergene, it must have at least eight independent functional units.

It is worth noting that the pattern produced by what is presumably the ancestral allele, *H^y* (the nonmimetic yellow-and-black *meriones* phenotype), is a very abbreviated version of the pattern theme that is manipulated by the other nine alleles. The derivative patterns are not merely modifications of the an-

cestral pattern but appear to be additions to it (Fig. 6.5), and in the case of the background color, as noted above, a completely different pigment has been substituted. The *H^y* pattern has no expression of elements c and d in the discal cell, nor of element f on the forewing. In addition, pattern element h on the hind wing is much reduced in width and lies well proximal of the position it occupies in all the derivative patterns. The color patterns of closely related species of *Papilio* (e.g., *P. phorcas*, *P. hesperus*, and *P. echerioides*) resemble the derivative mimetic patterns of *P. dardanus* much more closely than the malelike (and presumably primitive) nonmimetic pattern. This observation suggests that the derivative patterns of *P. dardanus* are atavisms.

Successful mimicry in *P. dardanus* depends not only on changes in color pattern but also on changes in wing shape. None of the models, for instance, have tails on their hind wings, whereas nonmimicking *P. dardanus*, and its relatives, do. All mimicking females are tailless. The tailless condition is controlled by a single dominant gene in some populations, and in others tail length is multifactorially controlled, but with a predominant effect from a single locus. Clarke and Sheppard (1960a) have shown that modifiers can have a considerable effect on the expression of tail length, and they argue that cases of good dominance are probably derived. The

gene with the main effect on tail length is designated T (t for recessive) and is autosomal and unlinked to the gene for color pattern (Clarke and Sheppard, 1960b).

Papilio memnon

The Southeast Asian *Papilio memnon* is a polymorphic species whose forms are Batesian mimics of several *Aristolochia*-feeding papilionids (Fig. 6.6). As in the case of *P. dardanus*, only the female is mimetic and polymorphic, and several female morphs occur in each local population; some females may be malelike, whereas others may mimic one or several different models. Different female morphs may segregate from a single brood (depending on parentage), and breed-

Figure 6.6. Polymorphism of *Papilio memnon* females. *A*, male form, for comparison; *B*, form *achates;* *C*, form *alcanor;* *D*, form *anceus;* *E*, form *trochila;* *F*, form *laomedon;* *G*, form *butlerianus.* *e*, epaulette.

ing experiments among the various forms have revealed that the major differences between forms are controlled by alternative alleles at a single locus. In *P. memnon* the case for a complex locus, or supergene, is stronger than in the previous species, because Clarke et al. (1968) and Clarke and Sheppard (1971) have documented several probable cases of crossing-over within the locus.

Clarke et al. (1968) and Clarke and Sheppard (1971) have studied the genetics of 17 distinct named forms, 6 of which came from two or three different geographic locales. The researchers found that the major gene controlling the differences in pattern among these forms has clear dominant-recessive relations among its many alleles. As in *P. dardanus,* this accounts for the relatively simple segregation of mimetic forms in crosses between the various forms within a single population. However, when forms from different geographic regions are crossed, there is a classical outbreak of variance in the F_2, and a breakdown in dominance and precision of mimicry among the segregants. Thus, as in *P. dardanus,* each of the allelomorphs produces good mimicry with low variance only if it occurs within the gene complex in which it evolved. Presumably, the mimicry has been refined by different sets of modifier genes in each population.

The mimicry locus in *P. memnon* has more-complex functions than that in *P. dardanus.* On the basis of rare crossover events, Clarke et al. (1968) were able to distinguish three closely linked loci within the supergene. Locus *T* affects the presence or absence of tails on the hind wing (tail presence is dominant, but the expression is markedly affected by modifier genes in many populations). Locus *W* affects the color pattern on forewing and hind wing; at least nine alleles for this locus were found among the 17 forms studied. Locus *B* affects the coloration of the body, and

three alleles were recognized. Two additional unlinked loci control the mimicry pattern: locus *P*, with two alleles, which affects the presence or absence of tails in races on the island of Palawan (its relation to locus *T* is unclear), and locus *Y*, which affects a pale yellow background coloration of the hind wing (yellow dominant over white). In a subsequent paper, Clarke and Sheppard (1973) reported nine additional forms of *P. memnon* within which they documented several more rare putative crossover events in the mimicry locus. These allowed them to separate the *W* locus into three distinct functional loci: *E*, *F*, and *W* (Table 6.3).

Locus *E* affects the color of the epaulette, a small brightly pigmented triangular region at the base of the forewing (Fig. 6.6D,E) whose color mimics the bright and distinctive thorax colors of models. Two alleles were recognized, E^w and E^s, whose phenotypes are white (to pale orange) and scarlet epaulettes, respectively. Scarlet is usually dominant over white, but many intermediate hues, including a bright yellow, are produced, depending on the genetic background. Locus *F* affects the color pattern on the rest of the forewing. The forewing pattern of *P. memnon* is simple, consisting of venous and intervenous stripes (Fig. 6.6). The various forms differ primarily in the relative brightness of the background, in the width of the intervenous stripes, and in whether or not they have a bright forewing patch. The criterion given to evaluate this last feature is unclear, but it appears to consist primarily of a brightening of the background color in the wing cells anterior to vein M_3 (e.g., Fig. 6.6E,F). In the most extreme forms (such as *P. memnon* f. *esperi*), it also involves a reduction of the width and length of the intervenous stripes in this portion of the wing. Two alleles were recognized among the nine forms studied, *F* and F^l, the latter associated with expression of the forewing patch

Table 6.3
Genetics of female polymorphic forms of *Papilio memnon*

Female form	Supergene					Recessive or codominant sympatric supergenes*				
	T	W	F	E	B	T	W	F	E	B
achates (*distantianus*)	T	W^d	F	E^s	B^Y	t	w	F^l	E^s	b^y
						t	w	F^l	E^s	b
						t	w	F	E^w	B^y
						t	w	F	E^w	b
laomedon (yellow tip)	t	w	F^l	E^s	b^y	t	W^i	F	E^s	b
laomedon	t	w	F^l	E^s	b	t	W^i	F	E^s	b
trochila (yellow tip)	t	w	F^l	E^s	b^y	t	w	F	E^w	B^y
trochila	t	w	F^l	E^s	b	t	w	F	E^w	B^y
isarcha	t	W^i	F	E^s	b	t	w	F^l	E^s	b^y
						t	w	F^l	E^s	b
anceus	t	w	F	E^w	B^y	t	w	F^l	E^s	b^y
						t	w	F^l	E^s	b
anceus (black body)	t	w	F	E^w	b					
titania	T	W^i	F	E^s	b					
anura	t	W^d	F	E^w	B^Y					
gerania	t	W^g	F^l	E^s	B^y					
ityla	t	W^g	F^l	E^s	b					

Note: The phenotype is controlled by a supergene of five closely linked genes (*T W F E B*), each with two or more possible alleles. Each of the forms is characterized by a particular combination of alleles in the supergene. Most populations have two or more arrangements of the supergene that differ in the alleles that are held in linkage.

*Each of these combinations, when homozygous, yields one of the forms in the left column.

and nearly completely recessive to F in most (but not all) genetic backgrounds. The more narrowly defined locus W affects the pattern on the hind wing. Clarke et al. (1968) and Clarke and Sheppard (1971) focused their attention primarily on the genetic control of the size (and pigmentation) of the white central region of the hind wing, which they called the window (Fig. 6.6B,C). Four alleles were recognized among the nine forms stud-ied: W^g, W^i, W^d, and w. The white window occupies the proximal half of the hind wing in W^d, the distal half in W^g, and an intermediate position in W^i; the window is eliminated by the fully recessive allele w. The dominance relations among the three other alleles have not yet been clarified.

Clarke and Sheppard (1971) proposed the following provisional arrangement for the loci in the supergene, based on their interpretation of rare crossover events: *T W F E B*. They point out that it is perhaps no coincidence that the three loci affecting the wing pattern are adjacent to each other in the supergene. These loci, however, appear to affect very different aspects of the color pattern. It is tempting to postulate that *F* and *W* control homologous expressions of a white patch on forewing and hind wing, respectively, but the anatomical locations of the forewing patch and the window are sufficiently different to cast doubt on such homology. Likewise, the epaulette is not related in an obvious way to either of the other two patterns. Table 6.3 gives the genotypes for the various mimetic and nonmimetic forms studied by Clarke et al. (1968) and Clarke and Sheppard (1971).

The most interesting phenomenon in pattern inheritance that emerged from the studies on *P. memnon* received no attention from Clarke et al. (1968) and Clarke and Sheppard (1971). It involves the inheritance of the wing cell patterns of the hind wing. Unlike the forewing, which has a nearly identical pattern of venous and intervenous stripes in all forms of *P. memnon* (and also in most of the species closely related to it), the patterns on the hind wing differ greatly among the many forms. Figure 6.7 gives a semidiagrammatic summary of the wing cell patterns of the various forms. As can be seen, the patterns vary from simple intervenous stripes similar to those on the forewing, to single or multiple spot patterns, with spots of various sizes and shapes. When two forms with different hind wing patterns are crossed, the hybrids bear patterns intermediate between those of the two parents but with a number of potentially interesting dominance effects. Figure 6.7 illustrates several such instances. The patterns of the various forms (and those of the hybrids between them) can be arranged in a morphological series, as shown in Figure 6.8. That

these patterns fall into a relatively simple series as readily as they do suggests that we are observing quantitative changes in one or a few variables along the morphocline.

Papilio polytes

Papilio polytes is widely distributed across India and Southeast Asia. As in the case of the previous two species, *P. polytes* has monomorphic males and several female forms (Fig. 6.9). One female form is nonmimetic and resembles the male, whereas three are mimetic and, as in the case of *P. memnon*, resemble several *Aristolochia*-feeding swallowtails. Some 20 varietal names are established for this species, but most of these appear to be geographic variants of only four genotypes (Clarke and Sheppard, 1972). The nonmimetic form occurs sympatrically with either one or two of the mimetic forms across most of the range of this species.

Fewer differences occur among the four forms of *P. polytes* than among the forms of the previous two species. The color pattern characters that distinguish each of the forms are listed in Table 6.4. The major differences reside in the pattern and background color of the hind wing. The forewing, with a pattern of intervenous stripes, is essentially identical in all forms except form *romulus*, in which the intervenous stripes are shortened in two areas on the wing. Together with a lighter background color in those areas, these shortened intervenous stripes give the forewing of form *romulus* the appearance of having two bright diagonal stripes (making the forewing pattern virtually identical to that of its model, *Pachliopta hector*). The pattern differences outlined in Table 6.4 appear to be controlled by alleles at a single locus. The dominance relations among these alleles are shown in the lower portion of Table 6.4. Clarke and Sheppard (1972) suggested that the differences

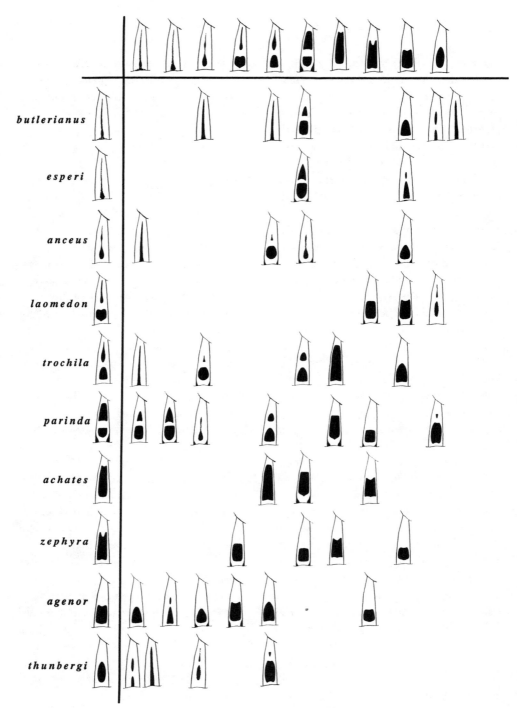

Figure 6.7. Diversity and inheritance of the wing cell patterns on the hind wings of female *Papilio memnon*. The patterns of 10 forms of *P. memnon* are shown along the two axes, and the patterns of the hybrids between some of them, produced by Clarke at al. (1968) and Clarke and Sheppard (1971), are shown in the body of this matrix. Individual variability in the parental patterns as well as in the hybrid patterns is not captured by this diagram.

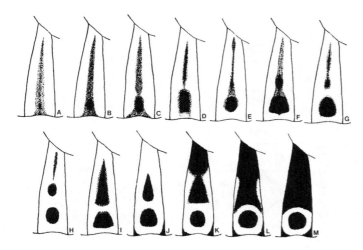

Figure 6.8. Morphocline of the patterns of *P. memnon* (*A–J*) and its sibling species, *P. rumanzovia* (*K–M*). The patterns are arranged to illustrate the full range of intermediates between intervenous stripes (*A*) and ocelli (*M*). *A*, form *retorina;* *B*, form *butlerianus;* *C*, form *esperi;* *D*, male; *E* and *F*, form *anceus;* *G* and *I*, form *trochila;* *H*, form *laomedon*. (From Nijhout, 1985b)

between form *polytes* and form *theseus* may be due only to differences in polygenic modifiers and not to a difference at the pattern locus.

Clarke and Sheppard (1972) found three specimens that had the forewing pattern of form *polytes* and the hind wing pattern of form *romulus*. They suggested that this unusual segregation was probably due to crossing-over within the pattern locus, which would imply that the pattern locus constitutes a supergene. If this interpretation is correct, then forewing and hind wing patterns are controlled independently and by different loci within the supergene, as in *P. memnon*.

In their paper on the genetics of *P. polytes*, Clarke and Sheppard (1972) also illustrated and described the offspring of interspecific

Figure 6.9. Polymorphism of *Papilio polytes*. *A*, the male and nonmimetic female form, *cyrus; B*, form *polytes; C*, form *romulus*.

Table 6.4

Phenotypes and dominance relations of the female forms of *Papilio polytes*

| Female form | Forewing | | Hind wing | | | | |
| | Intervenous stripe | Edge spot | Background color between elements | | Separation between elements | Width of dark base (probably elements |
			d and f	f and h	f and h	c and d)
cyrus	Complete	Yes	White	—	Fused	Small
polytes	Complete	No	White and red	Red	Small	Medium
theseus	Complete	No	Red	Red	Small	Medium
romulus	Shortened	No	Red	Red	Large	Large

Dominance relations

  romulus > *cyrus, polytes, theseus*
  theseus > *cyrus*
  polytes > *cyrus*
  polytes = *theseus*

crosses between *P. polytes* and seven other species of *Papilio.* In nearly all cases the offspring have color patterns that are nearly linear intermediates between those of the two parental species (though a few cases show partial dominance of one of the parental types), suggesting that pattern differences between those species are under polygenetic control and that the effects of these genes are additive.

Hypolimnas bolina

The nymphalid *Hypolimnas bolina* is widely distributed, from Madagascar through India and Southeast Asia to New Guinea, Australia, and Micronesia. Females are polymorphic in most locations throughout its range, and males are identical and monomorphic throughout (Fig. 6.10). Clarke and Sheppard (1975) recognized four basic female forms and three additional forms intermediate be-

tween pairs of the basic forms. One of the female forms (*euploeoides*) is a mimic of various *Euploea* species (Danainae). The other three female forms are nonmimetic.

In contrast to the previous species, color pattern and mimicry in *H. bolina* is controlled by two unlinked loci, *E* and *P.* For neither locus is there evidence for a supergene. Table 6.5 gives the genotypes and phenotypes for the various forms of this species.

Locus *E* affects the size of the dark pattern elements on both forewing and hind wing. Two alleles are known. The dominant allele, *E*, causes an enlargement of all pattern elements in the central region of the wing. This obliterates nearly the entire background, resulting in a very dark butterfly. This allele also increases the size of the white M-shaped lunules near the margin in each wing cell. The recessive allele, *e*, produces narrower areas of pattern, so that a band of background appears on both forewing and hind wing between pattern elements f and **g**. The recessive allele also decreases the size of the submarginal lunules.

Locus *P* affects the color of the pattern elements. Three alleles are recognized at this locus. The recessive allele, *p*, produces a pattern that is completely black. A dominant allele, *P*, causes all pattern elements to be expressed in a very light brown, giving rise to the form *pallescens*. Another dominant allele, P^n, causes a patch of brown to appear near the posterior margin of the forewing. It is not clear at present whether this brown patch represents background or a local change in

Figure 6.10. Polymorphism of *Hypolimnas bolina*. A, male form (*naresi*); B, form *nerina;* C, form *pallescens*.

Table 6.5
Genotypes and phenotypes of the female forms of *Hypolimnas bolina*

| Form | Genotype(s) | Color of pattern elements | Forewing | | | Hind wing | |
			Lunules	White bar	Brown spot	Lunules	White bar
euploeoides	*EEpp* *Eepp*	Black	Large	Absent	Absent	Large	Absent
pallescens	*eePP* *eePP^n*	Brown	Large	Present	Present[a]	Large	Small
nerina	*eeP^{n}P^n* *eeP^{n}p*	Black	Small	Present	Present	Small	Large
naresi	*eepp*	Black	Small	Present	Absent	Small	Small

Note: Two unlinked genes (*E* and *P*) control the pattern differences.

Source: After Clarke and Sheppard, 1975.

[a]This phenotype cannot be ascertained, because the overall color of *pallescens* is brown.

the color of a pattern element. In the latter case, the pattern element in question would have to be i. We deduce this because in *euploeoides-pallescens* intermediates (genotype *E- P-*), only element h on both forewing and hind wing is pale, and the position of this pale element on the forewing corresponds precisely with that of the brown spot produced by P^n (see illustrations in Clarke and Sheppard, 1975). Allele *P* appears to be dominant to P^n.

Clarke and Sheppard (1975) note that the form *euploeoides* (genotype *EE pp*) mimics different species of *Euploea* in different parts of its range. Unlike the previous cases in *Papilio,* this mimicry polymorphism is brought about not by multiple alleles at a single locus but by differences in polygenes that modify details of the main pattern, such as the size and shape of the white submarginal lunules and the amount of iridescent blue on the forewing. The linkage among these modifier genes, and whether any of them have large effects on the pattern, is unknown.

Supergenes and Regulatory Loci

The inheritance of the color patterns of three of the species discussed above is believed to involve coordinated changes at a supergene. Supergenes are clusters of two or more genes that are tightly linked and affect several different components of the pattern. Supergenes are presumed to have evolved either from previously unlinked genes brought together by chromosomal rearrangements, or through gene duplication. The rationale for the evolution of supergenes is that they provide a mechanism for ensuring that specific alleles from several different loci are stabilized in one or more sets of particularly favorable combinations; their close linkage then assures that the alleles will not segregate often. In the case of Batesian mimicry, the need for

a coordinated transformation of several aspects of the pattern and the need to prevent favorable combinations from being lost by recombination are presumed to be the driving force behind the evolution of a supergene for color pattern.

It is generally accepted that the work of Clarke and Sheppard, outlined above, provides strong evidence for the involvement of supergenes in the evolution of mimetic patterns in butterflies (Charlesworth and Charlesworth, 1975; Futuyma, 1986). In most cases, however, little or no crossing-over is detectable within the pattern locus, and thus mapping by traditional techniques cannot be done. The inference that a pattern locus is a supergene rests on three observations and arguments: (1) the existence of one or a few specimens with unusual combinations of characters (which were inferred, not proved, to be due to crossing-over within the locus), (2) hyperpolymorphism at the locus in question (at least 10 and 9 alleles in *P. dardanus* and *P. memnon,* respectively), and (3) the effect of each allele on several apparently unrelated aspects of the pattern. Among these three observations, only the first (crossing-over within the locus), if unambiguously documented, can serve as proof for the existence of a supergene. Therefore, among the four cases of Batesian mimicry discussed above (which constitute the four best-studied systems), there is no evidence for a supergene in *Hypolimnas bolina,* extraordinarily weak evidence in *Papilio dardanus* and *P. polytes,* and quite convincing though still circumstantial evidence in *P. memnon.*

The theoretical genetics of Batesian mimicry and the evolution of supergenes have been explicitly worked out by Charlesworth and Charlesworth (1975, 1976a,b). Throughout their work, as well as all the previous speculations about the evolution of supergenes by Clarke and Sheppard (1960b,

1971) and Clarke et al. (1968), the (reasonable) supposition has been that supergenes are simple linear arrangements of discrete genes, each with a characteristic product and effect, that differ from other genes only by their proximity and perhaps some special mechanism preventing recombination at that site. Recent work on *Drosophila,* however, gives us a different view of complex loci (loci with many apparently different functions). For instance, Mendelian genetic studies of the Bithorax complex have documented the existence of five closely linked loci, whose positions can be mapped with some accuracy (Lewis, 1963, 1978). Molecular genetic studies have revealed, however, that this portion of the Bithorax complex produces only a single RNA transcript, and only about 1% of this transcript is eventually translated, the remaining 99% being introns (noncoding intervening sequences) that are eliminated during RNA processing (Duncan, 1986). The original transcript can be processed into several different mRNAs, depending on the presence of insertions and deletions at specific sites in the gigantic introns. The introns, though not coding for protein, appear to regulate RNA processing somehow. Our current understanding is that these critical insertion and deletion sites on the introns, and not the actual coding regions of the gene, are what were originally recognized as the Mendelian loci giving the various Bithorax effects. Along similar lines, the Antennapedia complex, with at least nine apparent Mendelian loci, produces only four transcripts; and the Abdominal region of the Bithorax complex, with at least eight apparent loci, produces only one or two transcripts (Duncan, 1986; Gehring and Hiromi, 1986).

These findings drastically alter the classical interpretation of the structure and function of complex loci and the modes of evolution of supergenes. We must consider the possibility that in the *Papilio* mimicry systems the apparent supergenes are functionally, as well as evolutionarily, single loci that produce different types of proteins depending on how their RNA product is edited. By analogy to the Bithorax complex, such a model could readily account for the large number of allelomorphs present at these loci.

The Bithorax gene, however, is not a structural gene but a regulatory gene. That the *Papilio* supergenes may also constitute regulatory genes, whose function it is to control the activity of a number of subservient structural genes, has been noted on several occasions by Clarke and Sheppard. At present this appears to be the general view of the function of complex loci and of loci that control complex characters. The presence of a homeobox in several complex loci in *Drosophila* lends support to this view. (A homeobox is an evolutionarily conserved base sequence that appears to code for a DNA-binding domain on a protein; its presence is considered prima facie evidence for a gene-regulatory function.) It is essential to recognize, however, that whether a particular genetic effect is complex depends on whether its production involves many independent interacting variables, not on whether the effect is visually complicated. It is therefore worth examining whether the effects of the color pattern loci in the Batesian mimicry systems discussed above are sufficiently complex to warrant the supposition that the pattern loci are regulatory.

Each of the four cases of mimicry discussed above is brought about by changes in the background coloration of the wing and by changes in the width and/or position of the black bands of the pattern. These changes may be homogeneous over the entire wing or may involve only a restricted region of the wing surface, as in *Papilio dardanus.*

In *P. dardanus* the differences in background coloration of the mimics consist of quantitative differences in the amount of a brown pigment. In *P. memnon* the background coloration of forewing and hind wing varies from off-white to gray-blue and consists of quantitative variations in the proportion of dark to light-colored scales. In addition, some forms of *P. memnon* have a yellow hind wing background (controlled by a single locus not linked to the pattern gene; see the section "*Papilio memnon,*" above), and others have a red background color on the distal third of the hind wing. In *P. polytes* too, differences in forewing coloration are due to differences in the proportions of light and dark scales, and the background color of the hind wing is either white or red. *Hypolimnas bolina* demonstrates no polymorphism in background coloration among the forms.

Differences in the width and position of black pigment bands constitute the other main variable in the Batesian mimicry systems discussed above. Among the various forms of *P. dardanus* the hind wing differs only in the width of the broad dark border band that occupies the distal third of the wing and presumably constitutes element f. In addition, the forms may differ in the length of the intervenous stripes, although these usually co-vary with the width of the border band. Clarke and Sheppard (1960b) have noted segregation of band width and length of intervenous stripes in some of their interform crosses. This means either that these two characters are genetically unlinked or that their expression is greatly affected by additional unlinked (modifier) genes. A correct expression of both the width of the border band and the intervenous stripes is necessary, however, for visually effective mimicry. Differences in the forewing pattern consist entirely of coordinated changes in the widths of elements f and h, which affect the

anterior background band, or of elements f and d, which affect the posterior background band. In *P. memnon,* changes in the forewing pattern consist of alterations in the width of the intervenous stripes and of variation in the color of the epaulette. The epaulette is a pattern element at the base of the wing that probably corresponds to pattern element c. This element is black and confluent with element d in most papilionids, but in closely related species, such as *P. rumanzovia,* element c is red. *Papilio memnon* differs from *P. rumanzovia* in the expression of element c. In the former, only the portion of element c that is within the discal cell is expressed in red, and the portion outside that cell is black. Epaulettes like those of *P. memnon* are also present in *P. deiphobus, P. lowi, P. alcmenor,* and *P. thaiwanus.*

Differences in the hind wing patterns of the various forms of *P. memnon* constitute a special case of considerable developmental and evolutionary interest whose implications are more fully discussed in Chapters 7 and 8. The point here is that the major differences between the forms can be readily explained as being due to variation in a single developmental parameter. In three of the forms of *P. polytes* the forewing patterns differ only slightly in width and length of the intervenous stripes. The white diagonal bands that characterize the forewing pattern of form *romulus* are produced by localized narrowing of the intervenous stripes. The hind wing patterns of *P. polytes* forms differ mostly in the width of element f and to a lesser degree in the width of elements d and g. The differences in these three pattern elements are coordinated. All three are narrowest in forms *romulus* and *theseus* and widest in form *cyrus.* The pattern differences among the various forms of *Hypolimnas bolina* likewise involve simple changes in the width of pattern elements. Here, elements **d** and **f** are involved, and changes in their width are coordinated on both forewing and hind wing.

It should be clear from the foregoing that most of the wing pattern differences that characterize the various forms of the species under consideration are simple quantitative variations on existing patterns and colors. These consist of changes in size, in color, or, in the case of background, in the proportion of different scale types. From a developmental viewpoint, these are all simple changes. Changes in color can come about through the activation (or suppression) of a single enzyme (thus, a single gene), and changes in the intensity of a color could result from simple changes in enzyme activity due to mutations that slightly alter the structure of an enzyme. A change in the proportion of colored scales can be due to a quantitative change in a stochastic switching mechanism at the cell level (see Chapter 7). Finally, changes in the size of a pattern element can be effected either through a change in the rate at which pattern determination spreads across the wing surface, a change in the time at which it starts to spread, or a change in a threshold for pattern determination (see Chapter 7). The third possibility is most likely to be the case where size changes of several pattern elements occur in a coordinated fashion. Both rate and threshold changes could be readily attributable to quantitative changes in single gene products.

None of the color pattern changes in these Batesian mimicry systems are sufficiently complex to cause us to suspect that the pattern genes controlling them are regulatory genes controlling networks of subservient genes. All changes can be accounted for by relatively straightforward (and mostly quantitative) changes in the products and activities of standard structural genes. This is not

to say, however, that the pattern changes by which each mimic is derived from a nonmimicking form are simple. It is clear that in several instances the pattern locus affects several features that will, in most likelihood, prove to be attributable to different gene products. For example, the simultaneous control of the width of a pattern element and the color of the background in *P. dardanus* is likely due to two independent genes, as is the simultaneous control of tails and color pattern in *P. memnon*. In general, control of wing form is likely to be independent of the control of color pattern, and control of pattern shape is likely to be independent of the control of pattern (or background) color. On the other hand, homologous pattern elements on forewing and hind wing could easily be under common genetic control (see the section "*Heliconius*," below), as could the coloration of wing and body pattern (as in *P. memnon*). Whenever several pattern elements change in a coordinated fashion among three or more mimicking forms, as on the hind wing of *P. polytes* and the forewing of *H. bolina*, it should be assumed that a single genetic variable is involved. Thus, in most of the cases discussed above, it is unlikely that more than two or three independent variables are responsible for the observed diversity of pattern.

Probably the most difficult case to interpret is the subdivision of the forewing of *P. dardanus* into three areas in which both pattern and background appear to vary independently (Fig. 6.4 and Table 6.2). There is either separate genetic regulation of both variables in each area, requiring six genes within a supergene, or a dynamic compartmentalization of the wing by a wave function (for example) whose value in each area affects the quantitative expression of a single pattern gene and a single background gene. In the latter case, we would have to hypothesize the existence of the two genes plus whatever genes are involved in establishing variation in the wave function.

In conclusion, the pattern diversity in the Batesian mimicry systems that have been studied provokes a strong suspicion that the systems are controlled by supergenes and that these supergenes consist of very few linked loci of ordinary structural genes. It must be emphasized, however, that the existence of supergenes for color pattern formation in Batesian mimicry has yet to be critically demonstrated in many of the species in which they are assumed to occur.

Heliconius

The work of P. M. Sheppard, J. R. G. Turner, L. E. Gilbert, and their co-workers on the color pattern genetics of several species of *Heliconius* has provided us with the most complete picture to date of the modes of pattern evolution and genetic control over pattern formation. The genus *Heliconius* constitutes about 55 species of forest-dwelling butterflies of tropical and subtropical America. All of the species are distasteful to predators and display warning coloration. Several species in this genus have diverged into a spectacular series of Müllerian mimicry rings that provide us with one of the best examples of a recent massive adaptive radiation. These Müllerian mimicry rings may have their origin in the forest refugia where many previously widespread species survived the South American droughts of the last glacial maximum (about 20,000 years ago). It is believed that contraction of the ranges of a number of previously widespread species into many small refugia caused many different combinations of species to be brought into close and prolonged contact. As a consequence of the combination of differences in species

composition and differences in relative distastefulness of those species, and as a consequence of founder effects and drift within the refugia, many of these isolated populations converged on different mimetic patterns (Brown et al., 1974; Turner, 1982).

Two major pattern complexes emerged among this diversity of species combinations, each containing many independently evolved mimicry rings. In the so-called tiger, or silvaniform, pattern complex, *Heliconius* butterflies have mostly a yellow-and-orange wing pattern with relatively small patches of black. Species within this pattern complex have formed Müllerian mimicry rings with each other and with several species of Ithomiini, Pieridae, Papilionidae, and some moths. In the second group, the black-and-red pattern complex, the wings are predominantly black with bold red, yellow, and white bars, spots, and stripes. The majority of species of *Heliconius* belong to this second pattern complex. The remarkable thing about the many mimicry rings within this pattern complex is that two species, *H. melpomene* and *H. erato,* are participants in nearly every one of them. Throughout tropical America it is possible to find pairs and triplets of species of *Heliconius* that have nearly identical color patterns in any one geographic area but whose patterns differ dramatically from one region to another (Turner, 1982, 1984). The resemblance of color patterns of *H. melpomene* and *H. erato* in some areas is sufficiently good as to challenge even experienced workers (see Plate 5). The species are easily told apart, however, on the basis of internal, genitalic, and pupal anatomy. In addition, they have different mating systems, with *H. erato* being a pupal mater (Gilbert, 1984), and the two cannot interbreed. There are no fertility barriers in crosses between members of the same species that belong to different mimicry complexes, and this has enabled the development of a rich understanding of the Mendelian genetics of color patterns in *H. erato, H. melpomene,* and closely related species (Sheppard et al., 1985; L. E. Gilbert, unpub. data).

Perhaps the most remarkable finding to emerge from studies of the coevolved radiation of *H. melpomene* and *H. erato* is that it appears to have a fairly simple and comprehensible genetic basis. The vast majority of the pattern alterations that evolved within this radiation involve changes at relatively few genetic loci; the differences in color pattern that characterize most forms of *H. melpomene* and *H. erato* are due to differences at no more than three or four loci with major effects (Turner and Crane, 1962; Turner, 1984; Sheppard et al., 1985). In each geographic area each species is monomorphic, and the pattern is stabilized by homozygosity at the loci concerned. About 22 loci involved in pattern formation have been documented in *H. melpomene,* and 17 in *H. erato.* These are discussed in some detail below.

Figure and Ground Revisited. Before we embark on an exploration of the genetics of pattern formation in *Heliconius,* it will be useful to sensitize ourselves once more to the distinction between pattern and background. Unless we utilize a consistent convention when referring to those portions of the wing pattern we call pattern, and those we call ground, we risk confusion of purpose and of cause and effect. For instance, growth of a pattern element is necessarily correlated with a diminution of the remaining background. Hence an allele that causes an increase in a physiological activity associated with the growth of a pattern element could appear, if the definitions of pattern and ground were switched, to cause a diminution of pattern instead. Unfortunately, in virtually the entire literature on color pattern genetics in *Heliconius,* the convention of pattern and ground is

exactly the reverse of the one we must use to accurately understand variation and origin of pattern. The convention used in the past is to interpret the *Heliconius* wing as an essentially black surface with a red and yellow pattern on it. This is a reasonable supposition because in many races and species the wing is mostly black, with the red and yellow marks coering only a small portion of the wing surface. Figure 6.11 illustrates the nomenclature that has been used for the various colored parts of the *Heliconius* pattern.

However, the elements of the nymphalid ground plan are generally expressed as dark patterns on a light background, which forces a reinterpretation of the pattern structure of *Heliconius* (Nijhout and Wray, 1988). We saw in Chapter 3 that the diversity of color patterns in *Heliconius,* as well as in many other taxa, can be most easily and consistently explained on the assumption that dark portions of the wing constitute pattern whereas light-colored portions are background. We also saw, quite unexpectedly, that large black areas of the wing are often complex, being made from the fusion of several adjoining pattern elements. When we use the conventions derived from the nymphalid ground plan (Figs. 2.17 and 3.18C), it is necessary to treat the black portions of the wing as pattern and the colored portions as background. So far, this convention is a simple reversal of the assumptions of Sheppard et al. (1985) and others. However, for rasons that will become clearer as we proceed, it appears that in many species the red portions of the color pattern also constitute pattern (Nijhout and Wray, 1988). After extensive experience with hybridization of many species of *Heliconius,* L. E. Gilbert (pers. com.) has concluded that most if not all red portions of the pattern in *Heliconius* constitute pattern elements, not background (see also Chapter 3 on homologies in the color patterns of *Heliconius*). In the

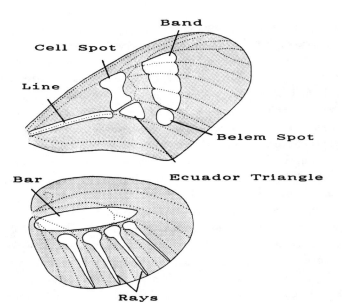

Figure 6.11. Nomenclature for the color pattern of *Heliconius*. These named areas of the wing pattern represent the background color of the wing. Their presence, shape, and position are affected by the differential expansion of the black portions of the pattern, which are homologs of the elements of the nymphalid ground plan. (From Nijhout et al., 1990)

discussion that follows, we will therefore assume that both red and black constitute pattern, and yellow and white are background.

We must now deal with the problem of how to handle the many standardized names for various patches of color that former workers considered to be pattern but that we now know to be background (compare Figs. 3.19 and 6.11). Named genetic variants and races are described almost exclusively in terms of their effects on the shape of these pieces of background. The background patches, of course, owe their existence and form entirely to the size and shape of the black and red pattern elements that establish their boundaries. Variation in these background patterns is in reality caused by complementary variation in the black portions of the pattern. Thus, to understand the actual effects of a specific gene on variation in the color pattern, it is necessary to translate its more visually obvious effects on the background portions of the pattern into a statement of its (often much less obvious) effects on specific elements of the ground plan. This I have attempted to do in

the sections that follow. The phenotypic effects of the genes listed in Tables 6.6 and 6.8 will be described in terms of their traditionally recognized effects on the background, and an interpretation of the effects of these genes on the pattern elements of the *Heliconius* ground plan will be given. Then, as we proceed, I will restrict discussion to the effects on pattern elements alone, because it is only in terms of the alterations of pattern elements that the mechanism of action of these genes can be accurately understood. Reference to the *Heliconius* ground plan (Fig. 3.18C) will be useful, if not essential, for nearly all of the following discussion of *Heliconius*.

Wing Pattern Genes. Tables 6.6 and 6.8 list most of the currently known and named loci that influence the color patterns of *Heliconius melpomene* and *H. erato,* respectively. The loci are grouped by the general type of their effect. Table 6.6, for *H. melpomene,* also includes the effects of loci found in *H. cydno* and *H. pachinus.* These last two species are sometimes considered subspecies of *H. cydno,* and all three species hybridize easily with full interfertility of their offspring (L. E. Gilbert, pers. com.). The tables give only telegraphic statements about the effects of each gene; Sheppard et al. (1985) have given extensive descriptions of these genes and their modes of inheritance. Somewhat fuller explanations and rationales for the particular interpretations with respect to pattern formation are also given in the sections that follow. It should be noted that the recognition of a few of these genes is based on the observation of relatively few segregants. The voluminous work of Sheppard et al. (1985) should be consulted for a full account of the genetics in this system.

The Pattern Loci of *Heliconius melpomene.* The descriptions of the gross phenotypic effects of each of the *Heliconius melpomene* loci discussed below are taken from the work of Sheppard et al. (1985), unless otherwise noted. Descriptions of the loci derived from *H. cydno* are from the work of L. E. Gilbert as reported in Nijhout et al. (1990). The *H. cydno* genes have identical effects in *H. cydno* and *H. melpomene* backgrounds. Some of the interpretations of the effects of these genes on the elements of the ground plan are taken from the work of Nijhout et al. (1990), but most are given here for the first time.

Of the 22 known loci, 5 affect the type of pigment in which a particular pattern element is expressed (Table 6.6). Three pigments are of interest here: Blacks are all melanin; reds, browns, and oranges are xanthommatin or its reduced form, dihydroxanthommatin; and yellow is 3-hydroxykynurenine. White is either a structural color or possibly a colorless metabolite in the tryptophan pathway.

Locus *Or* (orange) switches the color of red areas of the wing between red and orange-brown. These two alternatives appear to be different oxidation states of xanthommatin, with red being the reduced form and orange the oxidized form (L. E. Gilbert, pers. com.). Sheppard et al. (1985) pointed out that the dominant phenotype (red) can be accurately scored only on fresh specimens, because it fades to orange after a period of time. How this gene alters or stabilizes the oxidation state of a small organic molecule is not known; presumably its product controls the redox environment within the scale.

Locus *B* (band) controls the color of pattern element h and switches it between red and black. Locus *B* has the same effect in a broad diversity of genetic backgrounds. When introduced into *H. hecale* (one of the tiger-patterned *Heliconius* species), for instance, it likewise causes only pattern element h to become red (R. Boender, pers. com.; L. E. Gilbert, pers. com.); the expression of the red pattern element from *H. melpomene* on the wings of *H. cydno* is illustrated by Gilbert (1984).

Locus *Wh* (white) switches color of the forewing band between white and red. Its effect is complex and appears to require the presence of the dominant allele *B*; it may be a special manifestation of the *B* locus in the genetic background of the east Ecuador stock. Thus the two loci, *B* and *Wh*, may affect alternative switches in the use of pigment biosynthetic pathways.

Loci *D* (dennis) and *R* (rays) behave much like locus *B* in that they affect a red-black switch in specific pattern elements. In each of these three loci the dominant allele produces a red color, and the recessive allele codes for black. The dennis phenotype refers to the presence of a red base on the forewing coupled with a red bar in the anterior portion of the hind wing (Fig. 6.12). These red areas correspond to the positions of pattern ele-

Table 6.6
The wing pattern genes of *Heliconius melpomene* and their phenotypic effects

Letter code	Name of gene	Main apparent phenotypic effect[a]
Or	Orange	Switches red areas to orange-brown
B	Band	Switches color of red forewing band to black
Wb	White	Switches color of red forewing band to white
D	Dennis	Switches red base of forewing and hind wing to black
R	Rays	Switches red hind wing rays to black
F	Fused	Breaks up yellow forewing band into spots
Rr	Red restricted	Breaks up red forewing band
N	Yellow band	Affects color of yellow forewing band
C	Cell spot	Removes yellow dumbbell spot at tip of discal cell
T	Triangle	Removes white Ecuador triangle
S	Short	Shortens yellow forewing band
Fs	Forewing shutter	Moves black forewing band proximo-distally
Cs	Cydno shutter	Removes black band from center of hind wing
Yb	Yellow bar	Removes yellow hind wing bar and yellow forewing line
Ps	Pachinus shutter	Removes black band near base of forewing
Cv	Convex	Alters shape of distal margin of yellow forewing band
Tc	Trinidad cell spot suppressor	Removes dumbbell spot at tip of discal cell
Ac	Anterior cell spot suppressor	Removes only the anterior lobe of dumbbell spot
Ub	None	Modifies the effects of *Yb*
Rs	None	Modifies the effects of *N* and *B*
Tr-1	None	Modifies the effects of *T*
Tr-2	None	Modifies the effects of *T*

Sources: After Sheppard et al., 1985, and Nijhout et al., 1990.

[a]The effects listed are those on the patterns shown in Figure 6.12, which is the way in which the genetics of *Heliconius* has been traditionally described. Only a telegraphic description is given. For a fuller analysis and interpretation of how these genes appear to affect the elements of the nymphalid ground plan, see text. The effect of only one allele is given; the alternate allele has the opposite effect.

ment c on both forewing and hind wing (Nijhout and Wray, 1988), and we thus assume that gene *D* specifically switches the color of pattern element c, just as gene *B* switches the color of element h. Locus *R* affects the expression of red rays on the hind wing. The radiate phenotype consists of a pattern of red stripes along the midline of each wing cell. These stripes probably correspond to the intervenous stripes of the nym-

Locus *Yb*

Locus *D*

Locus *Ps*

Figure 6.12. Interpretation of the phenotypes produced by three genes that affect both forewing and hind wing patterns of *Heliconius melpomene*. The recessive phenotype of the *Yb* and *Ps* loci is indicated by the dark gray area; the dominant phenotype, by the expansion into the light gray area. Locus *D* changes the color of pattern element c at the base of the wings; this area is indicated by light gray (red dominant to black). The pattern in the unshaded portions of these wings is not affected by the genes in question. (From Nijhout et al., 1990)

phalid ground plan (Fig. 6.13). Intervenous stripes of various colors are a common feature of the hind wing pattern in many species of *Heliconius* (Nijhout and Wray, 1988). The simplest interpretation for the phenotypic effect of the *R* gene is that it switches the color of intervenous stripes on the hind wing between black and red. When the remainder of the wing is mostly black, the appearance is that this gene adds (or subtracts) a set of rays on the hind wing.

Loci *B*, *R*, and *D* stand out among the other pattern loci because of their apparently reversed dominance relationship. As will be evident from the discussion below, all other genes that affect color pattern have as their dominant effect an increase in the area covered by black. This effect is consistent with

the usual finding that a dominant allele is hypermorphic (i.e., produces more product or produces a phenotypic effect of greater magnitude) to a recessive one. It is therefore interesting that the three loci that effect a simple red-for-black switch (*B*, *R*, and *D*) have the red phenotype as dominant over black. This observation suggests not only that all three may operate on similar biochemical processes but also that their effect is different from that of the other genes discussed below. These three loci do not affect the shapes or sizes of pattern elements, as most of the others do, but affect only the color of pattern elements. The finding that each of the three loci switches the color of a single pattern element (actually, each switches all the serial homologs) provides an

Locus *R*   Locus *Cs*

Figure 6.13. Interpretation of the phenotypes of two genes that affect only the hind wing pattern of *Heliconius melpomene*. Locus *R* affects the color of the intervenous stripes (red dominant to black). The recessive phenotype produced by locus *Cs* is indicated in dark gray; the expansion of elements **f** (light gray) is the dominant phenotype. (From Nijhout et al., 1990)

important principle for understanding pattern evolution of butterflies in general. Evidently, single genes can control the color in which a pattern element is expressed and allow it to stand out or blend into the rest of the wing pattern. A great deal of pattern evolution in butterflies consists of simple color switches without altering the shape or position of pattern elements.

Among the loci that affect the shape of pattern elements, two (*F* and *Rr*) appear to affect pattern-inducing activity of the wing veins. Locus *F* (fused) affects the width of the colored forewing background band between elements **f** and **h**. The dominant allele causes the background to become broken up into a series of elliptical spots by encroachment of the black boundaries along the wing veins (Fig. 6.14). Locus *Rr* (red restricted) has a superficially similar effect on the shape of pattern element **h**, but this is evident only in the presence of allele *B*, which transforms element **h** from black to red. The dominant allele causes a constriction of element **h** at the wing veins. This effect must, however, be quite different from that of gene *F*, because **h** is a pattern element and not background. Gene *Rr* must somehow affect the shape of the boundary of element **h** by reducing (not enhancing) its expansion at the wing veins,

thus giving this element a scalloped appearance.

The remaining 16 loci shown in Table 6.6 affect the position, the size, or the presence or absence of particular pattern elements. In most cases different alleles of a given gene cause identical changes in all elements of a homologous series. In two cases a gene affects the morphology of homologous pattern elements on both forewing and hind wing.

Locus *N* (yellow band) affects the color and shape of the forewing band in a complex interaction with gene *B* (Fig. 6.15). The nature of this interaction, and a hypothesis to explain the complex phenotypes that are produced, are discussed below (see "The Interaction of Genes *N* and *B*"). There are two codominant alleles, N^N and N^B, the latter being slightly recessive to the former. On first sight, these alleles determine whether the forewing band will be yellow (N^N, yellow; N^B, not yellow). The model proposed below explains the effects of the *N* gene as controlling the position of pattern elements **f** and **h**.

Locus *C* (cell spot) affects the presence or absence of three colored spots of background—a dumbbell-shaped spot at the apex of the discal cell (the cell spot), the Belem spot, and the Ecuador triangle (Fig. 6.11; the name derives from the fact that this fea-

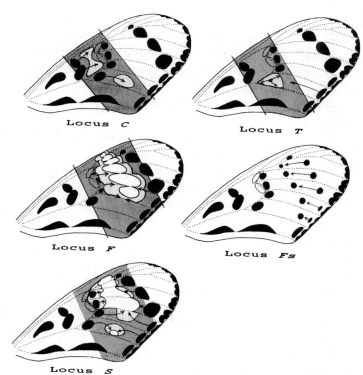

Figure 6.14. Interpretation of the phenotypes of five genes that affect only the pattern on the forewing of *Heliconius melpomene*. The recessive phenotypes are indicated by the dark gray areas; the dominant phenotypes, by expansion into the light gray areas. The pattern in the unshaded areas of the wing is unaffected by these genes. The alternative alleles at locus *Fs* cause a shift in the position of pattern element h. (From Nijhout et al., 1990)

ture is characteristic of the race from east Ecuador)—in what appears on first sight to be a complex pleiotropic effect: The dominant allele suppresses the cell spot and the Belem spot, and the recessive allele suppresses the Ecuador triangle. Nijhout et al. (1990) have shown that this combination of effects is readily explained by a lateral movement of pattern element d in several wing cells. If the recessive allele codes for a proximal position of element d, then the position of these elements would allow the background to show through at two places: in the apex of the discal cell (the cell spot) and in the middle of the forewing (the Belem spot). The dominant allele causes elements d to be shifted distally, covering up these two background areas but uncovering a piece of background in the crotch of wing cell Cu_1–Cu_2, the Ecuador triangle (Fig. 6.14).

Locus *T* (triangle) controls the presence (recessive) or absence (dominant) of the white triangular background area in the crotch of wing cell Cu_1–Cu_2, the Ecuador triangle. This triangle occurs in the place normally occupied by pattern element d in this wing cell (Fig. 6.14), and either the inactivation of this pattern element or a shift in its position could account for the phenotype. This effect differs from that of gene *C* in that it is restricted to this single wing cell. None of the other serial homologs of d is affected.

Locus *S* (short) causes a shortening and narrowing of the yellow background band on the forewing. This effect is due to a widening of the adjoining pattern elements d, f, and h in the presence of the dominant allele (Fig. 6.14). This effect could come about in one of two ways: (1) by an intrinsically controlled increase in the size of each of the three pattern elements (for instance, an increase in the rate at which the determination front for each element proceeds) or (2) by a change in the threshold required for induction of black pigment in all pattern elements. Given that nearly all other genes in this system affect individual pattern elements (or ranks of serial homologs) without affecting any of the other elements, it is reasonable to suppose that gene *S* alters a response threshold that affects all pattern elements equally. This seems the more parsimonious explanation of the observed phenotypes. The dominant allele has been found only in east Ecuador and controls the so-called split, or division of the forewing band from the cell spot (Emsley, 1965; Sheppard et al., 1985). The split phenotype appears to be due to an enlargement of element **f**, causing a larger separation than usual between the cell spot (background between **d** and **f**) and the band (background between **f** and **g**). Sheppard et al. (1985) noted that the short and split phenotypes are always co-inherited and are probably due to a single gene. The fact that the split phenotypic effect is also consistent with a threshold hypothesis supports this assumption.

Locus *Fs* (forewing shutter) affects the position of a broad black band on the forewing (Fig. 6.14). The two alleles, Fs^p (proximal position of the band as in *H. pachinus*) and Fs^c (distal position of the band, as in *H. cydno*), are codominant, and in the heterozygote the band position is intermediate between the two extremes. This band represents pattern element **h**, and no other elements of the color pattern are affected by this locus. The *Fs* gene is therefore responsible for shifting the position of a rank of serial homologs of a single pattern element along a proximo-distal axis. Each individual element moves only within its own wing cell. In combination with gene *B*, we now have a way to control both the

color and the position of a row of homologous pattern elements.

Locus *Cs* (cydno shutter) adds (dominant) or removes (recessive) a black pattern element that covers (or uncovers) a large central white area of background on the hind wing of *H. cydno* (Fig. 6.13). The area occupied by this pattern element corresponds perfectly to the position of element **f** in *H. melpomene*. It is thus reasonable to suppose that gene *Cs* affects the activation or inactivation of the entire row of pattern elements **f** on the hind wing.

Locus *Yb* (yellow bar) affects the yellow bar on the anterior portion of the hind wing and the yellow line in the proximal portion of the forewing (Figs. 6.11 and 6.12). Absence of the bar and line is the dominant effect. Both bar and line are background areas in the field normally occupied by pattern elements **c** and **d**. The recessive effect appears to be due to a reduction in the size of both pattern elements. In the case of the forewing line, the extent of **c** and **d** is reduced in the antero-posterior direction to reveal a thin line of yellow background along the posterior margin of the discal cell and the anterior margin of wing cell Cu_2–2A. The *N* locus affects the basal portion of the forewing line, possibly by having a minor modifier effect on pattern element **c**.

Locus *Ps* (pachinus shutter) affects the size of the entire series of **d** elements on the forewing of *H. cydno*. The recessive allele reduces the size of element **d**, and the dominant allele causes a substantial increase in the size of this pattern element out to the distal edge of the discal cell (Fig. 6.12). In the recessive, presumably hypomorphic condition, element **d** is missing entirely from the crotch of wing cell Cu_1–Cu_2, whereas this pattern element is large in the presence of a dominant allele. This observation suggests that in the recessive condition the unexpressed element **d** may be present at a subthreshold level. The *Ps* locus also affects a pattern element along the anterior margin of the hind wing, the dominant allele producing an increase in the size of this element so that it covers about half of the large central background area of the hind wing (Fig. 6.12). The anterior portion of the hind wing is occupied by a confluence of elements **c** and **d** (Fig. 3.18), and the boundary between the two is unclear. Thus it is always difficult to ascribe any pattern changes in this region of the hind wing to a specific pattern element. But given that the *Ps* locus affects only pattern element **d** on the forewing, it is reasonable and parsimonious to assume that it affects only element **d** on the hind wing as well. Thus, here we may have a case in which a single locus controls the size of a whole row of serial homologs of a pattern element on both forewing and hind wing. This contrasts with locus *T* (above), which probably controls the expression of only a single element (in one wing cell) of this same series on the forewing.

Locus *Cv* (convex) affects the shape of the distal margin of the yellow forewing band. Sheppard et al. (1985) described the alternate phenotypes as being either concave (dominant) or convex (recessive). Their illustrations suggest that this gene affects the shape of the proximal boundary of pattern element **h**. In the concave form the boundary of **h** within the anterior three wing cells is nearly a straight line, and in the convex form this boundary is scalloped, with a pronounced expansion at the wing veins. Thus we could provisionally assume that this gene controls the shape of the boundary of element **h** by altering some property of the wing veins in that region.

Locus *Tc* (Trinidad cell spot repressor) has been described only from the Trinidad race.

Its dominant allele suppresses the dumbbell-shaped spot in the discal cell (the cell spot) in animals of the genotype N^NN^B and N^NN^N, either in its entirety or in only the posterior half of this spot, depending on the cross. Locus *Ac* (anterior cell spot repressor) has also been described only from Trinidad. As its name suggests, the dominant allele of this locus suppresses the anterior lobe of the dumbbell-shaped cell spot. Both *Tc* and *Ac* act independently from locus *C*. The differential effect of these two loci on the posterior and anterior portions of the cell spot, respectively, is interesting because pattern element **d** has two homologs within the discal cell that arise when the base of the media vein bisects the discal cell early in development but atrophies during pattern formation. It is easy to see that locus *Tc* could be concerned mainly with the posterior element **d**, whereas locus *Ac* clearly affects only the anterior elements **d** within the discal cell.

The remaining four loci could all be classified as modifiers of effects already discussed. Locus *Ub* is a modifier of *Yb*. The phenotypic effect of the recessive allele is to enhance the yellow hind wing bar. It appears that the homozygous recessive *ubub* causes a partial loss of dominance of *Yb*, so that the *Ybyb* heterozygote has the yellow hind wing bar expressed as a shadow bar, consisting of a mixture of yellow and black scales. In the presence of the dominant allele *Ub*, *Yb* is nearly fully dominant over *yb*.

Locus *Rs* modifies the interaction of the *N* and *B* loci, so that a narrow red line (presumably element **h**) is present in N^BN^B *bb* genotypes, which would otherwise be completely black in the region of pattern element **h**.

Loci *Tr–1* and *Tr–2* are modifiers of the *T* locus. The dominant alleles of both suppress the Ecuador triangle. Thus homozygosity for the recessive alleles at both loci is required for development of the Ecuador triangle.

The Interaction of Genes *N* and *B*. When red-banded ($N^B N^B$ *BB*) and yellow-banded ($N^N N^N$ *bb*) *Heliconius melpomene* are crossed, an unusual and unexpected assemblage of characters is produced (Turner, 1971a,b; Sheppard et al., 1985). In the F_1 generation, all animals have a forewing band that is not intermediate in color but is in fact half red and half yellow. The split is roughly down the middle, with the distal half red and the proximal half yellow. In the F_2, six distinct phenotypes segregate as shown in Figure 6.15. On first sight it looks like this might be a relatively straightforward case of epistasis between the *N* and *B* loci, but the interpretation is confounded by a need to account for the fact that in $N^N- B-$ genotypes the red portion of the band is distal to the yellow,

whereas the position of the red band in $N^B N^B$ *BB* homozygotes is largely proximal to the position of the yellow band of $N^N N^N$ *bb* homozygotes.

The following model accounts for this morphological feature as well as for the pattern of phenotypic segregation in the F_2. We must start with the understanding that yellow represents background, and red is pattern. Comparative morphology (Nijhout and Wray, 1988; Nijhout et al., 1990) and breeding experiments (L. E. Gilbert, pers. com.) indicate that the broad red band of $N^B N^B$ *BB* corresponds to pattern element **h**. The red band extends into the distal portion of the discal wing cell, which is well proximal to the expected range of variability of pattern element **h** and is thus likely to involve more than just element **h**. For the purpose of initial analysis, however, we will assume that only pattern element **h** is red and then exam-

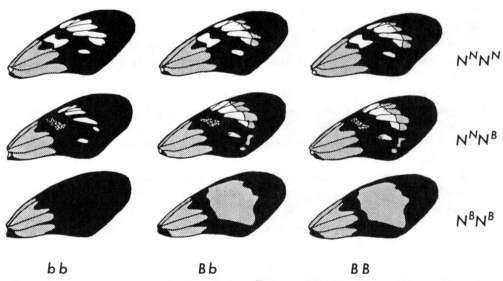

Figure 6.15. Inheritance of the color of the forewing background band of *Heliconius melpomene*. Two genes, *N* (with codominant alleles N^N and N^B) and *B* (with dominant allele *B* and recessive allele *b*), control the major features of this phenotype and appear to interact epistatically to produce the six phenotypes shown. *White areas,* yellow; *gray areas,* red. In cell spot, gray indicates that the yellow mark is variable in size and color intensity. (From Turner, 1984; reprinted with permission of New York Zoological Society)

Table 6.7

Phenotypic effects of the N and B genes of *Heliconius melpomene*

Genotype	Width of h	Color of h	Position of h
$N^N N^N$ BB	Narrow	Red	Medial
$N^N N^N$ bb	Narrow	Black	Distal
$N^B N^B$ BB	Wide	Red	Proximal
$N^B N^B$ bb	Wide	Black	Proximal[a]
$N^N N^B$ BB	Medium to narrow	Red	Medial
$N^N N^B$ bb	Medium to narrow	Black	Medial

Note: This interpretation is based on the epistatic interactions among the alleles at these two loci as illustrated in Figure 6.15. The results of Turner (1971a) can be interpreted in terms of changes in the width, color, and position of pattern element h. When element h is narrow and distal, it allows a portion of the yellow background to appear; this produces the impression that gene N controls the yellow background band.

[a]Proximal position is presumed but cannot be ascertained, because the whole wing is black.

ine the consequences of that assumption. After describing the model that emerges from this simple assumption, we will examine the origin and significance of this extreme proximal extent of the red band.

Table 6.7 lists the phenotypes associated with each of the genotypes, in terms of their effect on pattern element h. The simple interpretations that emerge from this table are that locus B affects the color of element h (red dominant over black) and that locus N affects the width of element h (the two alleles, N^N and N^B, codominant). This observation accounts for the apparent effect of locus N on the width of the yellow portion of the forewing band (Sheppard et al., 1985), because the width of this band is complementary to the width of element h. In addition, Table 6.7 shows that allele N^N encodes a distal and N^B a proximal position, with a possible slight epistatic effect of B in the $N^N N^N$ homozygote. Thus three genetic effects, on color, width, and position of element h, account for the principal aspects of phenotypic variation in the F_2.

The shading of the cell spot in $N^N N^B$ het-

erozygotes in Figure 6.15 indicates that this mark is variable; it is usually smaller and less brightly yellow than in $N^N N^N$ homozygotes and is occasionally absent (Turner, 1971a). The cell spot is absent in $N^B N^B$ homozygotes. As indicated above, the cell spot is an area of background between pattern elements d and f. Its "elimination" therefore probably comes about through enlargement or displacement of element d (as in the effect of locus C) or f. In $N^B N^B$ $B-$ genotypes (broad red band), the portion of the wing normally occupied by the cell spot is red and constitutes the extreme proximal extent of the red band. Thus in $N^B N^B$ $B-$ the position of the cell spot has been invaded by a pattern element that is red and confluent with pattern element h. These observations clearly support the conclusion that pattern element f is involved in the elimination of the cell spot and in the broadening of the red forewing band. This means that the size of element f is controlled by locus N, being small in $N^N N^N$ homozygotes, larger and variable in $N^N N^B$ heterozygotes, and large in $N^B N^B$ homozygotes, which parallels the effect of the N locus on element h.

The color of element f would be controlled by locus B but with an epistatic effect of the N^N allele, so that red is expressed only in $N^B N^B$ homozygotes. This is unlike the effect of the B locus on element h, and because it is without precedent among any of the other genetic effects on wing pattern in *Heliconius*, we cannot look elsewhere for support of this interpretation. The possibility that the broad red forewing band is a composite of elements f and h is, however, part of an internally consistent interpretation of the joint phenotypic effects of loci N and B.

The main technical difficulty with the preceding interpretation is that it appears to violate the assumption of the nymphalid ground plan that pattern element f cannot be within the discal cell. It is worth noting, however, that pattern element f also violates this assumption in the hind wing of *Heliconius* (Nijhout and Wray, 1988), and it may be that this is a derived character for the genus. This violation of a morphological rule is easy to understand in developmental terms, because the discal cell becomes established late in development, usually during the later stages of color pattern determination. Consequently, no causal or necessary relationship exists between the position of the discal crossveins and color pattern, and the exclusion of pattern element f from the discal cell is probably a coincidence, albeit an extraordinarily regular one. The only pattern element in this region of the wing that is dependent on the wing venation is element e (which forms on the discal crossvein; Fig. 2.17), and this pattern element is noticeably absent from the *Heliconius* ground plan (Fig. 3.18C). In fact, *Heliconius* is possibly the only genus in which pattern element e is lacking, and its absence may be taken as circumstantial evidence that in this genus the discal crossveins form after pattern determination is over. If the discal crossveins of *Heliconius* in-

deed form so late in development, then the inclusion of all or part of element f in the discal cell is only mildly noteworthy.

The Pattern Loci of *Heliconius erato*. Sheppard et al. (1985) have described 17 genetic loci that affect the color pattern of *H. erato* (Table 6.8). Six or 7 of these loci (*Wh, Or, R, D, Ly, Cr,* and possibly *Yl*) are probable homologs of those found in *H. melpomene,* and the others represent genetic functions absent or as yet undescribed in *H. melpomene*. None of the loci of *H. erato* appear to have phenotypic effects that are completely identical to those of *H. melpomene*. Even those genes that are quite certainly homologous between the two species have slightly different phenotypic and pleiotropic effects. Presumably these differences in phenotypic expression are due to small divergences in function and to differences in the genetic backgrounds in which these genes occur. In general, the genetic loci of *H. erato* appear to have less sharply defined effects on the phenotype and to have many more pleiotropic effects than those of *H. melpomene*. In many cases it is impossible to deduce the extent of parallelism in the genetics of pattern formation among the two species, because parallel crosses (i.e., between the same pairs of geographic races of both species) have not been done. Sheppard et al. (1985), in fact, reported only a single parallel cross (*H. m. thelxiope* × *H. m. nanna* and *H. e. amazona* × *H. e. phyllis*) among the many interracial crosses they did on both species.

The discussion that follows, like the description of the loci in Table 6.8, presents first the genes that are possible homologs of those already seen in *H. melpomene,* followed by those that have been described for *H. erato* only. The emphasis of this discussion will be on drawing parallels and distinctions among the phenotypic effects of genes in the two

Table 6.8

The wing pattern genes of *Heliconius erato* and their phenotypic effects

Letter code	Name of gene	Main apparent phenotypic effect[a]
Or	Orange	Switches red areas to orange-brown
Wh	White	Switches color of red forewing band to white
D	Dennis	Switches red base of forewing and hind wing to black
R	Rays	Switches red hind wing rays to black
Rt	Raylets	Removes small red hind wing rays
Ly	Broken band	Breaks up yellow forewing band into spots
Yl	Yellow line	Removes yellow hind wing bar and yellow forewing line
Cr	Cream rectangles	Removes white marks from ventral hind wing and diminishes expression of bar and line
Ybs	Yellow bar suppressor	Diminishes width of yellow hind wing bar and yellow forewing line
Y	Yellow band	Switches color of yellow forewing band to red
Ro	Rounded	Affects shape of distal portion of red forewing band
Bf	Bar forward	Bends tip of hind wing bar backward
Fb	Forward bar	Bends tip of hind wing bar forward
Cs	Costal spot	Removes small red spot at base of ventral forewing
St	Split band	Breaks forewing band into two spots on ventral surface
Ur	Upper split	Breaks forewing band into two spots on dorsal surface
Sd	Short band	Affects shape of forewing band

Source: After Sheppard et al., 1985.

[a]The effects listed are those on the patterns shown in Figure 6.12, which is the way in which the genetics of *Heliconius* has been traditionally described. Only a telegraphic description is given. For a fuller analysis and interpretation of how these genes appear to affect the elements of the nymphalid ground plan, see text. The effect of only one allele is given; the alternate allele has the opposite effect.

species and, where possible, on the manner in which pattern elements are affected by each locus.

Locus *Or* (orange) appears to be the same as the *Or* locus of *H. melpomene*. It probably controls the oxidation state of xanthommatin, yielding a stable red or orange-brown pigment with its alternate alleles.

Locus *Wh* (white) has the same phenotypic effects and occurs in the same geographic areas as the *Wh* locus of *H. melpomene* and

thus probably represents a homologous locus. The expressivity of the recessive allele appears to be much more variable than that in *H. melpomene*, although this has not been investigated in parallel crosses.

Locus *D* (dennis) in *H. erato* has an identical phenotypic effect as the *D* locus in *H. melpomene*. The red bar on the hind wing, however, is smaller than in *H. melpomene* and follows the outline of the discal cell nearly perfectly. In *H. melpomene* the hind wing bar crosses the discal cell nearly perpendicularly. The difference between the two species is that in *H. melpomene* this locus affects all of elements c, whereas in *H. erato* it affects only the portions of these elements within the discal cell.

Locus *R* (rays) appears to be the same as the *R* locus of *H. melpomene*. The dominant allele adds a radiate pattern of red intervenous stripes on the hind wing. These rays differ from those of *H. melpomene* in that they tend to be a little longer, and they lack the widening at the proximal end that often gives a nailhead appearance to the rays of *H. melpomene*.

Locus *Rt* (raylets) affects the development of small short rays on the hind wing. These raylets probably represent the bases of the hind wing rays (Sheppard et al., 1985). The effect of this gene does not parallel that of gene *R*, however, because here the recessive allele encodes the raylet character. Thus the *Rt* locus probably has a different effect on pattern formation from that of the *R* locus. For instance, the *Rt* locus could modulate a pattern-inducing activity of the veins of the hind wing, partially or completely obliterating the expression of red intervenous stripes. This possibility could be tested by observing the interaction between the *R* and *Rt* loci, because it predicts that raylets would be expressed only in the presence of a dose of the dominant allele for *R*.

Locus *Ly* (broken band) may be the same as the *F* locus in *H. melpomene*. The dominant allele at this locus causes the yellow forewing band to be strongly indented and interrupted at the wing veins. The homozygous recessive has a broader band with a smoother outline.

Locus *Yl* (yellow line) has the same primary phenotypic effect as the *Yb* locus of *H. melpomene*. The dominant allele removes a yellow background line along the posterior border of the discal cell in the forewing. This gene also affects expression of the yellow hind wing bar in the same way as the *Yb* locus. In addition, this gene has pleiotropic effects on the shape of the red forewing band (when present) that have no parallel in *H. melpomene*. The effects on the red forewing band cannot be attributed to changes in the same pattern elements that are involved in controlling the disappearance of the yellow line and bar. Sheppard et al. (1985) suggested that nothing in their data argued against the possibility that two closely linked loci are involved. They suggested a hypothetical locus (called *T* but not related to the *T* locus of *H. melpomene*) for the control of the band morphology but no recombinants have yet been found that might prove the existence of two loci. Several alleles of the *Yl* locus, derived from different geographic areas, have quantitatively different effects on the line or bar, but these differences are slight.

Locus *Cr* (cream rectangles) may be a modifier of *Yl*. The dominant allele weakens the expression of both the yellow line and the yellow bar, and the recessive allele either strengthens the expression or is required for the normal expression of the line and bar. This gene is probably homologous to gene *Yb* of *H. melpomene* (Mallet, 1989). A pleiotropic effect of this gene (the effect from which it derives its name) is on a series of whitish marks along the apical margin of the ventral hind wing. These white marks (cream rectangles) do not obviously correspond to a standard pattern element. They may be remnants of the white marginal background pattern that is characteristic of the Ecuadorian form *cyrbia* and probably represent the absence or diminution of pattern element i. This effect seems to be entirely unrelated to the phenotypic effects of this locus on the line and bar. Again, the possibility of two closely linked genes cannot be excluded.

Locus *Ybs* (yellow bar suppressor) may also

function as a modifier of the effects of the *Yl* locus and interacts epistatically with the *Cr* gene to affect the width of the yellow forewing line and hind wing bar. The dominant allele suppresses the line and bar, whereas the recessive allele enhances both, but only in the presence of the correct genotypes for *Cr* and *Yl*. Thus its phenotypic effect is much like that of the *Ub* locus in *H. melpomene*.

Locus *Y* (yellow band) affects the color of the forewing of *H. erato*. The dominant allele produces a red band, and the homozygous recessive has a yellow band. This effect is reminiscent of the combined effects of the *N* and *B* loci in *H. melpomene*. The genetics here is quite different, though. When two races of *H. melpomene* are crossed, one with a yellow forewing band and the other with a red one, the F_1 possess bands that are half yellow and half red, and the F_2 segregate as shown in Figure 6.15. The parallel cross in *H. erato* gives a different result (tested in the cross *amazona* × *phyllis*). The F_1 of such a cross all have a red forewing band, and the F_2 segregates as 3 red : 1 yellow, indicating that a single factor, with red dominant over yellow, controls the color of the forewing band in this species. Thus the *Y* locus of *H. erato* behaves like the *B*, *R*, and *D* loci of *H. melpomene* in having red coloration of a pattern element as a dominant effect. But it differs from these in that the recessive allele allows the yellow background color to show, whereas the recessive alleles of the other loci produce a black pattern element. To remain consistent with the preceding interpretation of the genetics of *H. melpomene*, it would be necessary to assume that the *Y* locus of *H. erato* affects the presence of pattern element **h** and that this pattern element is red (perhaps specified by an independent and yet undiscovered gene) in the genetic background in which this gene has been studied. Thus in the presence of **h** (dominant) the forewing band is red, and in the absence of **h** (recessive) the yellow background shows. This interpretation also preserves the positive relation between dominance and hypermorphic pattern development that we found throughout *H. melpomene*. Judging from the phenotypic effect, it is unlikely that the *Y* locus is homologous to either the *N* or the *B* gene of *H. melpomene*.

Locus *Ro* (rounded) affects the shape of the distal portion of the red forewing band. The phenotypic effect, as described by Sheppard et al. (1985), is reminiscent of that of the *Cv* locus of *H. melpomene*, except that the latter exerts its effect only on the yellow background forewing band.

Loci *Bf* (bar forward) and *Fb* (forward bar) both affect the shape of the tip of the yellow hind wing bar. The dominant allele *Fb* and the recessive allele *bf* both cause the tip of the hind wing bar to be curved forward. The complementary alleles at each locus cause a backward turn of the tip of the bar. The dominant allele of the *Fb* locus has been found in the Ecuador race only. The shape of the tip of the bar appears to be controlled largely by the expansion of black at the anterior margin of the tip. Pattern elements **c** and **d** occur there, so these loci may affect the size of either or both of these elements in the most anterior wing cells.

Locus *Cs* (costal spot) affects the presence of a small bright red spot on the ventral forewing near the thorax. This spot is characteristic of most races of *H. erato*. The dominant allele is necessary for the expression of this red spot. The costal spot does not correspond to the position of a known pattern element. Element **c** would be the only one present in this area, but there is no independent evidence from comparative morphology that this pattern element can be as small and as well defined as the costal spot.

Loci *St* (split band), *Ur* (upper split), and

Sd (short band) all affect the morphology of the forewing band in a manner related to the phenotypic effect of locus *S* in *H. melpomene.* The split band phenotype refers to the race from east Ecuador (*notabilis,* a co-mimic of *H. m. plesseni*), which has two somewhat rounded spots of background color (half white and half reddish pink) on the forewing, representing background areas between pattern elements **d** and **h**, separated by a black "gutter," element **f**. Two loci, *St* and *Ur*, affect this trait, both having apparently identical effects but on different wing surfaces: *St* acts only on the ventral pattern, and *Ur* acts only on the dorsal pattern. The other main difference between the phenotypic effects of these two loci is that their dominance relations are reversed; also see Sheppard et al., 1985). It is possible that an epistatic interaction exists between these two loci, so that the split phenotype is expressed on both dorsal and ventral surfaces only in the presence of the dominant allele, *St*, and the recessive allele, *ur*, and that the latter has no effect on the dorsal pattern in *stst* homozygotes.

In the short band phenotype, the distal white forewing band does not extend beyond vein M_3 (dominant over the long phenotype, which extends well into wing cell M_3–Cu_1). In the east Ecuador race of *H. melpomene* these two phenotypes cosegregate and appear to be due to different alleles at a single locus, yielding only two possible phenotypes: short split and long entire. In *H. erato,* however, three phenotypes involve these traits: long entire, long split, and short entire. This suggests that recombination between the two traits must be possible. Because no recombinants were ever recovered in laboratory crosses (Sheppard et al., 1985), the split band and short band phenotypes must be due to separate but tightly linked loci. This might, of course, also be true of *H. melpomene,* so that the *S* locus of melpomene may be homolo-

gous to the *Ur St* loci of *H. erato.* The main problem with this interpretation is to reconcile the dominance relationships among the genes. The east Ecuador race of *H. erato* must be *sd Ur St,* whereas that of *H. melpomene* must be *S.* Yet the *H. melpomene* gene affects the dorsal pattern, which in *H. erato* is affected by *Ur* and *sd,* whose dominances do not match. It is possible that we are dealing with homologous loci and that differences in modifier genes have altered the dominance relations of their alleles in the two species, but this remains to be investigated. According to Mallet (1989), the *Sd* gene of *H. erato* is homologous to the *S* gene of *H. melpomene.* Mallet (1989) has also suggested that the effects ascribed by Sheppard et al. (1985) to the *Ur* gene could be explained by an interaction of the *Sd* and *St* genes.

Linkage Groups. In *Heliconius melpomene* the linkage groups have been established for the following loci: *D R B, N Yb, S T, C,* and *Or.* In *H. erato* the following linkage groups have been established: *Y D R, Yl St Sd,* and *Ly.* Insufficient data are available to ascribe the remaining loci either to these or to additional linkage groups with certainty (Sheppard et al., 1985). Both species of *Heliconius* have 21 potential linkage groups (Suomalainen et al., 1971). The tight linkage between *R* and *D* genes results in the cosegregation of hind wing rays and dennis phenotypes with only rare recombinations. Mallet (1989) has suggested that this linkage constitutes a supergene. The evolution of a supergene, however, is predicated on the capture of favorable allelic combinations in a polymorphic population, and that is not the case here. All pattern genes in *Heliconius* occur as homozygotes, and each population is monomorphic for one allele only. Finding linkage under such conditions is not evidence for the evolution of a supergene, because such linkage has no adaptive value in a monomorphic system. Linkage between two genes that affect prominent aspects of the color pattern would be expected to occur with a certain frequency by chance alone.

Genetic Differences among the Mimetic Races of *Heliconius*.

Turner (1986) has pointed out that the differences in the color pattern between the races of *Heliconius melpomene* and *H. erato* are of the same type and magnitude as pattern differences between species. The genetic studies of Sheppard et al. (1985) and their predecessors indicate that most of the mimetic races of *H. melpomene* and *H. erato* differ in allelic composition at four or fewer genetic loci with large effect (and at an undetermined but probably fairly small number of modifier loci, because in most cases the characters segregate cleanly).

Tables 6.9 and 6.10 present a summary of the genotypes of several races of the two species of *Heliconius* taken from the work of Turner (1984) and Sheppard et al. (1985). The point of this summary is to dramatize the enormous phenotypic differences that can be accomplished by relatively few genetic alterations in this system. If Turner's inference is correct, then an equally small number of genetic changes could underlie pattern differences between related species. Turner (1984) has used the data in Tables 6.9 and 6.10 to deduce the phylogenetic pattern of divergence of *H. melpomene* and *H. erato*. By assuming that of a pair of alleles the dominant one is likely to be derived and the recessive one ancestral or primitive (a hypothesis he calls the dominance sieve, because new dominant mutations are presumed to become established more efficiently through natural selection than do new recessive mutations), Turner was able to use phylogenetic reconstruction methods to deduce the sequence of

Table 6.9

Genotypes of various races of *Heliconius melpomene*

Race	Source	Gene											
		D	*R*	*B*	*N*	*Yb*	*C*	*Or*	*F*	*Rr*	*S*	*T*	*Wh*
1. *plesseni*	East Ecuador	*d*	*r*	*B*	N^N	*Yb*	*c*	*Or*	*F*	*Rr*	*S*	*t*	*wh*
2. *thelxiope*	Lower Amazon	*D*	*R*	*b*	N^B	*Yb*	*c*	*or*	*F*	*rr*	*s*	*T*	*Wh*
3. *aglaope*	Upper Amazon	*D*	*R*	*b*	N^B	*Yb*	*C*	*or*	*F*	*rr*	*s*	*T*	*Wh*
4. *meriana*	Guiana	*D*	*r*	*b*	N^B	*Yb*	*c*	*Or*	*F*	*rr*	*s*	*T*	*Wh*
5. *rosina*	Central America[a]	*d*	*r*	*B*	N^B	*yb*	*c*	*Or*	*f*	*rr*	*s*	*T*	*Wh*
6. *melpomene*	Venezuela	*d*	*r*	*B*	N^N	*Yb*	*c*	*Or*	*f*	*rr*	*s*	*T*	*Wh*
7. *nanna*	East Brazil	*d*	*r*	*B*	N^N	*yb*	*C*	*Or*	*F*	*rr*	*s*	*T*	*Wh*

Note: The numbers preceding the names of each race correspond to those of their *H. erato* mimics in Table 6.10 and to the diagrams in Figure 6.16.

Sources: Turner, 1984; Sheppard et al., 1985.

[a]The presumptive genotype for this race is not given by Turner (1984). It is deduced here from its color pattern. Alleles whose expression cannot be ascertained from the *rosina* phenotype alone have been assumed to be similar to those of the Venezuela population.

Table 6.10

Genotypes of various races of *Heliconius erato*

Race	Source	Gene											
		D	R	Y	Yl	St	Sd	Ly	Cr	Or	Ur	Wh	Ro
1. *notabilis*	East Ecuador	d	r	Y	Yl	St	sd	ly	Cr	Or	ur	wh	Ro
2. *amazona*	Lower Amazon	D	R	y	Yl	st	sd	Ly	Cr	or	Ur	Wh	ro
3. *lativitta*	Upper Amazon	D	R	y	Yl	st	Sd	Ly	Cr	or	Ur	Wh	ro
4. *amalfreda*	Guiana	D	r	y	Yl	st	sd	Ly	Cr	Or	Ur	Wh	ro
5. *petiverana*	Central America	d	r	Y	Yl	st	sd	ly	Cr	Or	Ur	Wh	ro
6. *hydara*	Venezuela	d	r	Y	Yl	st	sd	ly	Cr	Or	Ur	Wh	ro
7. *phyllis*	East Brazil	d	r	Y	yl	st	sd	ly	cr	Or	Ur	Wh	ro

Note: The numbers preceding the names of each race correspond to those of their *H. melpomene* mimics in Table 6.9 and to the diagrams in Figure 6.16.

Sources: Turner, 1984; Sheppard et al., 1985.

divergence of the various races of these two species. Several equally parsimonious phylogenies were obtained for each species, but among these phylogenies, two proved to be identical for the two species (Fig. 6.16). The phylogenetic trees shown in Figure 6.16 do not necessarily represent the true relations among the races, because the assumptions of the dominance sieve are questionable (Mallet, 1989) and because the trees are not congruent with the biogeographic pattern of distributions of the races. The phylogenies also do not correspond to the biogeographic pattern of Quaternary refugia, which are the presumed cause of the pattern of divergence. What is important about these phylogenies is that they focus our attention on the fact that we have an adaptive radiation in progress today. This system offers unequaled opportunities to study and reconstruct an active evolutionary radiation in which we are challenged to reconcile genetic, morphological, biogeographic, and ecological data.

Genetics in Other Species of *Heliconius*. Genetic studies on other species of *Heliconius*

consist mainly of opportunistic hybridizations. *Heliconius numata,* a member of the silvaniform (tiger-patterned) group, has two forms that differ, among other ways, in having a yellow versus a brown bar on the hind wing, in the same location as the yellow *H. melpomene* bar. Crosses between these two races show that brown is recessive to yellow. A cross of two yellow individuals of unknown genotype yielded 26 yellow and 22 brown offspring (Sheppard, 1963), which departs significantly from the 3 : 1 ratio expected if the two parents were heterozygous for a single effective locus. Sheppard (1963) noted that this may in fact represent a 9 : 7 ratio, diagnostic of a two-locus system with epistasis. This situation is thus quite different from that found in *H. melpomene* and *H. erato,* in which the yellow bar is inherited as a recessive character.

Brown and Benson (1974) reported an additional series of genetic effects that emerged from interracial crosses with *H. numata.* Unfortunately, they gave no breeding data and reported the inheritance of characters only in a very general qualitative way. Crosses between *H. n. superioris* and *H. n. silvana* reveal that the presence of element **d** in wing cell Cu_1–Cu_2 is controlled by a single gene. Presence of the spot is dominant over its absence (this is analogous to the effect of locus *T* in *H. melpomene*). These two races also differ in the position of element **h** (distal in *silvana,* proximal in *superioris*), and this position also appears to be controlled by a single gene (perhaps an analog of the *Fs* gene of *H. melpomene*). The presence of pattern element **c** in the discal cell (the black dagger phenotype) is dominant over its absence. Whether the genes that control these characters are collected in a single supergene, as Brown and Benson (1974) suggested, is uncertain because there appears to be quite a lot of recombination between them. More-extensive quantitative data with an attempt to draw more-explicit comparisons to the *H. melpomene* case could be quite enlightening.

In *Laparus* (formerly *Heliconius*) *doris* a polymorphism exists in the background color of the hind wing, which can be red, metallic blue, or metallic green. All three forms occur sympatrically and can be reared from single egg rafts (representing oviposition from a single mated female). Table 6.11 gives the data from five broods reared by Sheppard (1963). These data do not admit to a simple genetic interpretation. It is noteworthy, however, that segregation is quite clean (only a single intermediate form is reported), and that the ratios approximate those expected from backcrosses.

Implications for Understanding the Genetics and Evolution of Pattern. Perhaps the most interesting observation that emerges from analyzing the phenotypic effects of pattern genes in terms of alterations of the elements of the ground plan is that almost all changes are quantitative and narrowly lo-

calized in space. Different alleles at most loci either enlarge or diminish (sometimes to the point of disappearance) specific elements of the ground plan. The allele that enlarges a pattern element is almost without exception dominant over the one that diminishes its size. Some genes affect the expression of only a single pattern element in a single wing cell, though most affect the entire row of serial homologs on a wing. Most genes affect the pattern on only one wing surface: either forewing or hind wing, and either dorsal or ventral. Only three of the loci discussed above affect the size or color of homologous elements on both forewing and hind wing (Fig. 6.12). Those genes that do not have obvious effects on the size of a pattern element either change the color of a specific pattern element or change the color of the background.

We can use this information to speculate about the minimal genetic architecture for pattern formation. The size or width of an entire rank of serially homologous pattern elements, such as the proximal band of the central symmetry system, its distal band, the border ocelli, and the parafocal element, can be under independent genetic control and can be regulated by a single gene for each pattern element. The size of each of the serial homologs within a wing cell can also be regulated independently of the others by individual genes. The position of each element along the proximo-distal axis of the wing can be independently controlled by a single gene at the level of the entire rank of homologs and at the level of the individual wing cell. The color of each pattern element can likewise be independently controlled by single genes. In general, the characteristics of each rank of serial homologs are regulated by a few genes, with potentially only one major gene for each major feature such as size, position, and color. Then, in each individual wing cell, there is an additional level of control of

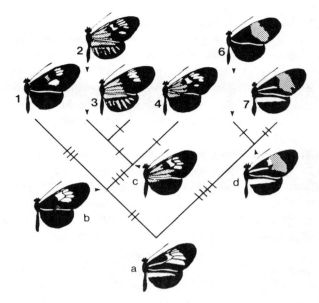

H. melpomene

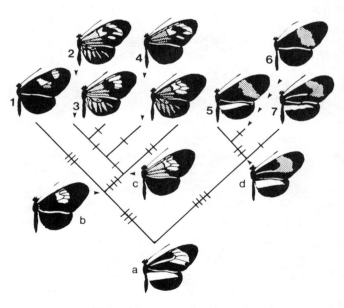

H. erato

Figure 6.16. Cladograms of hypothesized phylogenetic relations among the mimetic races of *Heliconius melpomene* and *H. erato*, deduced by applying the dominance sieve hypothesis to the genotypic data in Tables 6.9 and 6.10. The dominance sieve assumes that the recessive alleles represent the primitive condition for each gene in the tables. Each hatch mark represents a mutation in one of the genes listed in the tables. The number next to each of the terminal taxa corresponds to the numbering in Tables 6.9 and 6.10. The specimens indicated by lowercase letters are the phenotypes of hypothetical ancestors, deduced from the genotypes at the nodes. Many equally parsimonious trees can be obtained by applying phylogenetic reconstruction methods to the data in Tables 6.9 and 6.10; the ones shown here were chosen by Turner as best fitting biogeographic data. (From Turner, 1984; reprinted with permission of Royal Entomological Society of London)

Table 6.11

Distribution of the three hind wing phenotypes of *Laparus doris* from Trinidad

	Form		
Brood	Blue	Red	Green
1	68	0	0
2	25	16	0
3	40	0	0
4	12	0	10
5	86	83	1

Source: Sheppard, 1963.

the same characteristics of the same pattern elements, each again provided by a single major gene.

This genetic toolbox thus mirrors the modular structure of wing patterns derived from studies on comparative morphology. The genetic structure is also modular, with relatively little cross talk between the genetic regulations of each pattern element. This is obviously a highly speculative scenario. It is unlikely that the effects of the genes of *Heliconius* hold true across all pattern elements and wing cells. Furthermore, this simple genetic scenario does not account for the many small-effect genes (or modifier genes) that have evolved to fine-tune the various features of the pattern. Relatively few of these modifier genes have so far been found in *Heliconius,* but that is undoubtedly because our attention has been riveted on the spectacular large-effect genes that account for the main features of mimicry. The important role of modifier genes is much more easily appreciated in the *Papilio dardanus* and *P. memnon* mimicry systems, in which we have a balanced polymorphism for mimicry and in which mimicry, in fact, breaks down when the major pattern genes are placed in an inappropriate

genetic background. Until now the presence of genes with small phenotypic effects that are responsible for refining and stabilizing the mimicry has been all but ignored. The progeny of interracial crosses and backcrosses in *Heliconius* are almost always quite variable in details of their phenotypic expression (Turner, 1971a; Mallet, 1989), suggesting the interaction and segregation of many modifier and small-effect genes. Thus the genetics of quantitative variation in relation to the evolution of Müllerian mimicry might be well worth exploring.

The genetics of *Heliconius* and *Papilio* unfortunately tell us little if anything about the genetic regulation of the details of pattern shape, because the pattern elements in these species are mostly represented as spots and blotches with ill-defined shapes that blend together where they meet. None of the species have the kinds of detailed patterns whose morphoclines we analyzed in Chapter 4. Thus we have to assume the existence of an additional, yet-to-be-discovered body of genes that function in the regulation of pattern shapes in the more traditional expressions of the nymphalid ground plan. In Chapters 7 and 8 we will make some predictions about the properties of such genes.

Is Mimicry the Result of Convergent Evolution?

In almost all cases, mimicry is achieved by a melanism superimposed on a change in the background coloration. The melanism is of a straightforward kind, namely an increase in the dimensions of existing pattern elements. Occasionally, as in *Papilio dardanus,* mimicry is achieved by the addition of a previously "lost" pattern element; on first sight this might seem like an evolutionary reversal, or atavism, but it is probably a simple threshold phenomenon. With the exception of pigment switches, most of the alterations of the color pattern that effect mimicry are quantitative changes on existing patterns.

The two co-mimicking species of *Heliconius* clearly use the same pattern elements, and in some cases the same genes, to achieve a close mutual resemblance. Unfortunately, we cannot say anything with certainty about the total extent of this parallel usage of pattern elements and genes. Genetic studies with *H. erato* have almost all used source material from geographic areas and from mimetic complexes different from those that have been used in studies with *H. melpomene*. The two species are sufficiently distant phylogenetically that they have accumulated a large number of genetic differences, so one might assume that they will use different modifier genes to refine the precision of their mimicry. But the overall genetic and morphological architecture of their mimicry systems seems to be virtually identical. This means that mimicry here is not a case of convergent evolution in the classical sense, because that would require the achievement of similarity of structure using different materials. The patterns look the same because the mimics use fundamentally the same pattern elements and largely the same genes. Mimicry is convergent evolution in a phylogenetic sense, however, because the ancestors did not look similar to one another but their descendants do. Although phenotypic similarity among the mimics is achieved by the same structural materials (pattern elements), they have clearly achieved this similarity through different developmental and evolutionary pathways.

In the Batesian mimicry systems of *Papilio memnon* and *P. dardanus,* evolution has also been convergent in the phylogenetic sense described above. There is no question that most of the mimics use pattern elements that are homologous to those of their models. In a way, this is not entirely surprising, because there are relatively few elements to choose from in building the overall pattern of any species (Chapter 3). In each case the major convergence has involved a color switch and either an enlargement or a positional shift in one or two of the pattern elements. Based on what we have learned about the genetics of *Heliconius,* such phenotypic alteration need involve relatively few genetic switches. But the enormous genetic distances between models and mimics in these cases, and the absence of an equivalent genetics of pattern variation in the models, make it unlikely that we will be able to discover whether homologous genes are used in this phenotypic convergence.

Quantitative Genetics

The two-dimensional character of color patterns, the manifest variation in the elements of the pattern (Chapter 5), and the ease with which the size and position of these pattern elements can be quantified make it possible to apply the methods of quantitative genetics (Falconer, 1981) to the analysis of color pattern heredity and evolution. Few studies have explored the genetics of quantitative variability and its consequences for the evolution of color patterns. This dearth of studies on a system of such considerable previous interest to evolutionary biology, with a methodology that is experiencing considerable current interest among evolutionary biologists, is surprising. After all, the theories on the evolution of coadapted gene sets, on supergenes, and on the evolution of dominance have received some of their best support from observations on butterflies, and most theories on the evolution of mimicry and polymorphism have been developed around butterfly material: (Ford, 1945, 1971; Sheppard, 1958; Sheppard et al., 1985). The lack of research is possibly due to a general failure to appreciate that pattern elements are homologous across the butterflies and many moths. It should be possible to apply the methods of quantitative genetics to study the evolution of mimicry, crypsis, aposematism, and polyphenism in butterfly color patterns.

The eyespots of *Maniola jurtina* (Satyrinae) have been subjected to quantitative genetic analysis by Brakefield (1984) and Brakefield and Van Noordwijk (1985), who found that the number of eyespots on the hind wing varied from none to six, with a heritability of 0.66 to 0.89, depending on the comparison that was made (these values were not significantly different from each other). Eyespot sizes on the forewing had heritabilities between .59 and .80. The mean position of the eyespots on the hind wing—that is, the mean point along the line of border ocelli of all eyespots present in a given individual—had heritabilities between .35 and .57 (midparent). McWhirter and Creed (1971) have documented a progressive backward shift of this mean eyespot position in populations of *M. jurtina* in the southwest of England since 1957, which suggests that evolution of wing pattern characteristics may be measurable in nature. Pansera and Arauju (1983) studied the quantitative variation in the number of raylets (short versions of the hind wing rays) on the hind wing of *Heliconius erato phyllis*. They showed that the number of these raylets varied with a heritability of .74. No segregation pattern was evident in their crosses, and they estimated that more than three independently segregating factors are likely to account for the observed variation. A most interesting result of a quantitative genetic analysis comes from the work of Wiernasz and Kingsolver (1990) on the control of melanization in *Pieris (Pontia) occidentalis*. They have shown that genetic variation exists in the melanization response to cold shock. This response is partially adaptive, because mela-

nization affects thermoregulation. Wiernasz and Kingsolver (1990) argued that such genetic variation in the canalization of a phenotype (Shapiro, 1973, 1978, 1981) that yields variation in melanization, which in turn has adaptive consequences, is an essential prerequisite for the evolution of seasonal polyphenism.

Homeotic Transformations

The term "homeosis" refers to the transformation of the characteristics and structure of one member of a meristic series into those of another member (Bateson, 1894). Among the insects, homeoses are best known from homeotic mutants (such as those in the Bithorax and Antennapedia complexes) that transform the characteristics of one segment or developmental compartment into those of another. Nonheritable manifestations of homeosis also occur, such as the transdeterminations and somatic mutations that transform the characteristics of small clonal regions of one segment into those of another segment. For instance, portions of an antenna can be homeotically transformed into leg. The characters of such homeotic clones are always position-specific. That is, they form the pattern and structure expected at the homologous location in a different segment: Homeotic clones at the tip of an antenna look like the tip of a leg, and at the base of an antenna they look like the base of a leg. This is generally taken as evidence in favor of Wolpert's (1969, 1971) theory of positional information.

Somatic mutations that result in homeosis or in gynandromorphs (male-female mosaics) are a fairly common occurrence in the wing patterns of butterflies (Fig. 6.17). Sibatani (1980, 1983a,b, 1987a) has published an extensive documentation of several hundred cases of wing homeosis in Lepidoptera, and these probably represent but a small fraction of the cases preserved in public and private collections worldwide. Wing homeoses in butterflies are particularly easy to detect and often are striking in appearance, because the four wing surfaces (dorsal and ventral forewing and hind wing) almost always bear distinctively different patterns. Sibatani has shown that homeotic transformations occur between forewing and hind wing (most common) and between dorsal and ventral surfaces (much less common), in both directions. A clonal restriction barrier may exist in the vicinity of vein M_2 that prevents homeotic clones from the anterior compartment from penetrating into the posterior, and vice versa (Sibatani, 1980). Such a clonal restriction barrier in a nearby position on the wings of *Drosophila* has been the subject of extensive investigation (Garcia-Bellido et al., 1979).

In species with sexually dimorphic wing patterns, it is also possible to find gynandromorphic switches of portions of the wing (Fig. 6.17). Sibatani (1987b) illustrated an unusual specimen of *Calinaga buddha* (Calinaginae) in which the shape, venation pattern, and dorsal color pattern of the hind wings are transformed into those of the forewings (the ventral pattern was normal). Thus the specimen appears to have two pairs of forewings.

The homeoses documented by Sibatani range from transformation of a whole wing surface to transformation of small patches of only a few scales. Perhaps the most striking feature of wing homeoses in butterflies is their common expression as mosaics of a large number of unconnected patches of transformed tissue randomly spread across the wing. These mosaics give the impression that

Figure 6.17. Wing homeosis in *Heliconius melpomene*. This specimen came from a colony of mixed genotypes; its parentage is unknown. The mosaicism of its wing pattern is probably due to a somatic mutation. Black flecking in the white anterior margin of the hind wing is gynandromorphic (white = male, black = female), but the mosaicism for the expression of the *D* and *R* loci (dennis and hind wing rays) cannot be explained this way, because these genes are not sex-linked or sex-limited.

the homeotic transformation occurred many times independently on different parts of the wing surface. The transformed patches on one wing can differ greatly in size. Assuming that each patch is a clone, such a result indicates that the transformations occurred at many different times in development. No mechanisms have yet been proposed for the origin of mosaic homeoses in wing patterns. They are probably due to somatic mutations, and in this regard it may be noteworthy that these mosaic homeoses bear a close phenotypic resemblance to the somatic variegation patterns produced through somatic mutation by transposable elements in plants and animals.

Polyphenisms

Many species of butterflies have seasonally distinct phenotypes or morphs. In tropical climates such alternate morphs usually occur in wet and dry seasons, whereas in temperate climates distinct morphs are often found in spring and summer. Seasonal dimorphism of the color pattern is often so extreme that the two forms were originally described as distinct species. Perhaps the best known among these polymorphisms from temperate and tropical climates, respectively, occur in *Araschnia levana* and *Precis octavia* (Fig. 6.19; Plates 6A and 6B). In each of these species, both morphs can be reared from a single brood of eggs, and the morphology of any given individual is strictly controlled by the photoperiod or temperature regime under which it is reared. Thus, because both morphs are produced by the same genotype, the term "polyphenism" is preferred to describe this phenomenon, although the term "seasonal polymorphism" is quite acceptable.

Shapiro (1984a) provided an evocative summary of the evolutionary significance of polyphenisms:

An organism with two generations a year, which faces two seasonally selective regimes a year will be permanently out of phase: the genetic makeup of the "summer" generation reflects selection in a "winter" regime and *vice versa*—such an organism is like a military general always planning for the last war. The shorter the generation time relative to the environmental periodicity, the closer the organism may track its environment, and the less the load at any given time. However, polyphenism can reduce the load and the time lag of the response to zero, if we assume that all individuals are equally competent to make the correct developmental decision and that the cue(s) they respond to in the environment is (are) trustworthy.

Expression and Adaptations

Polyphenisms are widespread among the butterflies, but they differ enormously in their manifestation. Some degree of seasonal variation can be detected in perhaps as many as half of the species of Nymphalidae and in many Pieridae, though it is less prevalent (or at least less conspicuous) among the Papilionidae. In most cases, the seasonal morphs differ only in the coloration of the background or in the size and coloration of one or more pattern elements. The position or shape of pattern elements is generally not affected. In several of the North American crescents—for example, *Phyciodes tharos* and *P. morpheus*—the summer form is paler and has much less distinct markings than the spring and fall forms (Fig. 6.18). However, none of the pattern elements are displaced from their normal position, nor is their shape different in the two morphs. The main difference in the appearance of the patterns of the two forms comes about through a diminution in the width of almost all pattern elements in the summer form. Some elements are so reduced in width that they shrink to a small spot or arc that represents what was formerly the broadest part of the pattern element; other elements may be so much reduced that they disappear entirely (Fig. 6.18). Greater differ-

Figure 6.18. Seasonal poly-phenism of *Phyciodes tharos* (Nymphalidae: Nymphalinae). *A*, summer form; *B*, spring form.

entiation in the shades of background color also contributes to the much crisper defini-tion of the pattern of the spring and fall forms (*marcia*) of *Phyciodes tharos*. The inner field of the central symmetry system and the field around the border ocelli can develop a shade of brown or yellow that is considerably darker than the rest of the background color in many specimens of the spring and fall forms (though individuals vary much in this re-gard), whereas in the summer form the back-ground coloration is a nearly homogeneous pale yellow.

Similar types and degrees of pattern alter-nation characterize the seasonal polymor-phism of many anglewings (*Polygonia* spp.). In these, the summer generation has a darker

background pattern with more contrast; this enhances the expression of the ripple pattern on the ventral surface. The polyphenisms of various Pieridae likewise consist mostly of changes in the size or intensity of black pat-tern elements. In the pierid genus *Tatochila*, whose pattern consists mostly of dependent stripes paralleling the wing veins, the alter-nate morphs differ not in the width of these stripes but in the length. The pale morphs have short fragmented stripes mostly near the wing margin, whereas the dark morphs have bold stripes that extend along the entire length of the veins enclosing each wing cell (see illustrations in Shapiro, 1980b).

Even the most extreme pattern polyphe-nisms such as those of *Araschnia levana* (Fig.

Figure 6.19. Seasonal polyphenism of *Araschnia levana* (Nymphalidae: Nymphalinae). *A*, summer form (*prorsa*); *B*, spring form (*levana*).

A

B

6.19) and *Precis octavia* (Plate 6B), involve mostly alterations in the size of pattern elements and in the color of pattern elements and background. The morphology of the pattern elements—that is, their shape and relative positions on the wing—differs little if at all between the alternative morphs. In *A. levana* the spring morph (*levana*) has smaller and lighter brown pattern elements than the summer (*prorsa*) form, and the background color is a homogeneous orange-brown except near the wing tips. In the summer form, by contrast, the orange background color is restricted to the distal and proximal quarters of the wing and is much reduced by the enlargement of border ocelli and bands of the central symmetry system, and the background within the central field of the central symmetry system is white or yellowish. The

overall effect is to give the spring form an orange-and-brown appearance and the summer form a black-and-white one (Plate 6A). A virtually identical polyphenism occurs in the Oriental species *Araschnia burejana*.

The seasonal polymorphism of *P. octavia* is only slightly more complex in that two pattern elements, the distal band of the central symmetry system and the parafocal elements, can be missing entirely in the dry-season form. There is quite some variability in the expression of these two elements in the dry-season form, but, as before, the variability is only in the intensity of their pigmentation, not in their position or shape; in individuals with only a weakly developed distal band, the band appears as a thin shadow of sparse black scales on an orange background. The overall appearance of the wet-season form is orange and black, whereas the dry-season form is mostly blue with some black and orange. Orange is clearly the background color, but in the dry-season form this color is restricted to the field between the border ocelli and the parafocal elements (except in the most anterior wing cells, where this area is white), and the background color of the remainder of the wing is a dull metallic blue. In both forms the pattern elements are all a dark brown or black.

The two species *A. levana* and *P. octavia* thus offer vivid illustrations of how simple changes in coloration and dimensions of pattern elements and background can have dramatic effects on the overall appearance of the pattern. By analogy to the genetic polymorphisms discussed above, it seems reasonable to assume that these types of polyphenisms probably involve few and relatively simple genetic or developmental switches.

In most species the different seasonal morphs also have slightly different wing sizes and wing shapes. One of the more extreme examples of this is found in *Precis almana,* a subtropical species from the Indo-Australian region. This species has distinct wet-season and dry-season forms that differ in both wing shape and ventral wing pattern (Fig. 6.20). The dry-season form has a much subdued color pattern on the ventral surface, which mimics a dead leaf. This mimicry is en-

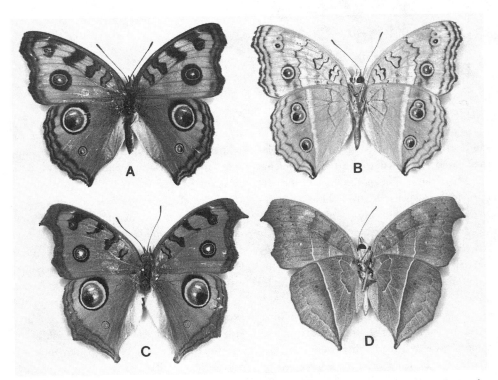

Figure 6.20. Seasonal polyphenism in *Precis almana*. *A* and *B*, dorsal and ventral sides, respectively, of wet-season form; *C* and *D*, dorsal and ventral sides of dry-season form. The dorsal patterns of the two seasonal forms are virtually identical, but the wing shapes and ventral patterns differ significantly. The ventral pattern of the dry-season form is a typical dead-leaf mimic.

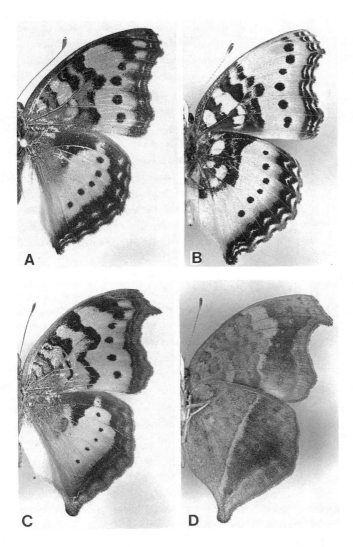

Figure 6.21. Seasonal polyphenism of *Precis antillarum* (Nymphalidae: Nymphalinae). *A* and *B*, dorsal and ventral sides, respectively, of wet-season form; *C* and *D*, dorsal and ventral sides of dry-season form.

hanced by a strongly angular wing shape resembling the outline of a leaf. The wet-season form has a ventral pattern and a wing shape that are typical of the monomorphic species in the genus (a parallel polyphenism occurs in several African species of this genus, e.g., *P. antilope* and *P. pelarga*). Interestingly, the two seasonal forms of *P. almana* have virtually identical dorsal wing patterns. The obvious dead-leaf mimicry of the dry-season morph argues strongly for an adaptive

function of this pattern polyphenism, and this is one of the few instances in butterfly polyphenisms in which the adaptive significance of one of the morphs is readily interpretable.

The dead-leaf mimicry of the dry-season form of *P. almana* is but one case of a widespread syndrome of dry-season adaptations among tropical butterflies (Figs. 6.20 and 6.21), a small sample of which is listed in Table 6.12. Brakefield and Larsen (1984)

Table 6.12
Tropical butterflies with seasonal polyphenisms

Pieridae
  Ixias pyrene
  Pinacopteryx eriphia
  Colotis danae, C. antevippe, C. eucharis, C. ione
  Gideona lucasi
  Eurema laeta, E. brigitta
Nymphalinae
  Precis almana, P. octavia, P. cuama, P. antilope,
    P. pelarga, P. ceryne, P. actia,
      P. touhilmasa, P. eurodoce, P. terea
  Salamis parhassus
Satyrinae
  Mycalesis perseus, M. mineus, M. visala
  Bicyclus vansoni, B. danckelmani, B. safitza
  Henotesia simonsii
  Ypthima inica
  Orsotriaena medus
  Melanitis leda, M. phedima, M. zitenius
Charaxinae
  Memphis glycerium
  Zaretis ellops

Note: These butterflies have a cryptic dry-season form, usually a dead-leaf mimic, and a less cryptic wet-season form. Wet-season forms have patterns that are more typical of the taxon as a whole. The adaptive significance of most of these polyphenisms is unknown.

Source: In part after Brakefield and Larsen, 1984.

studied the general characteristics of wet-dry seasonal polyphenisms in tropical butterflies and developed a model of their evolutionary significance. The researchers noted characteristic phenotypic differences between the wet- and dry-season forms of tropical Nymphalidae. In all cases studied, the dry-season form is more cryptic than the wet-season form. The wet-season forms of many Nymphalidae have a well-developed pattern of border ocelli, believed to serve as predator-deflection devices. In general, wet-season forms have a relatively bright ventral wing pattern whose coloration, contrast, and cryptic value resemble those of related monomorphic species. Dry-season forms, by contrast, commonly have their ocelli reduced to dots. In addition the ventral wing pattern of the dry-season forms always has much less contrast between pattern elements and ground than is found in wet-season forms. This general dullness of the dry-season form greatly enhances its crypsis. The adaptive significance of the dry-season pattern is clearest in those cases in which the coloration and texture of the pattern resemble those of bark or dry leaves, but Brakefield and Larsen (1984) pointed out that although the dry-season form is always duller and more cryptic than the wet-season form, in many cases in the adaptive significance of the dry-season morph is not at all obvious. Such is the case for the polyphenic Pieridae, in the tropics at least (in temperate zone pierids the adaptive value of the polyphenism is clearer, as discussed below), and for species like *Precis octavia,* whose dry-season morph is more cryptic than the wet-season form but, because it retains bright white border ocelli, does not appear to be particularly inconspicuous.

Brakefield and Larsen's (1984) explanation for the adaptive significance of seasonal polyphenism in these tropical butterflies is based on a model developed by Brakefield (1984) to account for sexual dimorphism in border ocelli patterns of *Maniola jurtina.* In this model, Brakefield proposed that the evolution of border ocelli in a butterfly species is dominated by its interaction with visually orienting predators. The size and conspicuousness of the ocelli reflect a balance between their effectiveness in allowing a butterfly to escape the attack of a predator (by serving as deflection devices for a predator's strike or as a means of evoking a startle response) and allowing it to remain undetected by a searching predator. Where this balance lies depends on the butterfly's activity pattern and choice of habitat, because these affect the probability of encountering a predator, and on the degree of match between color pattern and background.

Brakefield (1984) has suggested that in *M. jurtina,* behavioral differences between the sexes may explain the evolution of sexual dimorphism in the pattern of ocelli. This suggestion is based on the observation that females are more sedentary than males and are thus more likely to be encountered by a predator while they are sitting still. Females have a cryptic pattern on their ventral hind wing, with relatively small and inconspicuous border ocelli. But they also have a large ocellus on the ventral forewing that is normally hidden by the hind wing but can be flashed to provoke a startle response. Males are more active and thus more conspicuous to a predator. Males also have ventral hind wing patterns with much more conspicuous ocelli that are believed to act as deflection devices should the male be attacked.

The model that Brakefield and Larsen (1984) proposed for the evolution of wet-dry seasonal polyphenism is a direct extension of this cryptic balance model. They proposed that in each of the seasons a different optimal balance exists between the effectiveness of the antipredator role of ocelli and the effectiveness of crypsis of the pattern. In the wet season, when larval food is plentiful, reproductive success is optimized by an active adult life-style. In the dry season, with fewer resources available for reproduction, survival may be maximized by quiescence, or a more sedentary behavior, and in some species by aestivation. The predators that are most likely to be encountered also differ in the wet and dry seasons. In the wet season the principal predators are browsers, such as geckos, skinks, and certain birds and mammals, whereas in the dry season active visual predators, mostly birds, are more common. Both of these biotic factors should favor a high degree of crypsis in the dry season and the evolution of predator-deflection devices in the wet season.

This balance model is supported by a variety of observations, but Brakefield and Larsen have emphasized that more-detailed ecological studies, as well as comparative studies with related nonpolyphenic species, will be essential to provide rigorous tests of the model. The manner in which such contrasting evolutionary pressures can be accommodated within a single genome is discussed in the section "Control of Polyphenic Expression," below.

The only other seasonal polyphenism for which the evolutionary significance is well established is the pale-dark dimorphism of certain temperate zone Pieridae. This has been best studied in *Colias eurytheme* (Watt, 1968, 1969; Hoffmann, 1973) and *Nathalis iole* (Douglas and Grula, 1978). Both species have cool-season forms and warm-season forms that develop under short and long day lengths, respectively. The cool-season forms have a much darker ventral hind wing than the warm-season forms. This melanization is due not to a change in the intensity or color of the scales but to an increase in the proportion of black scales among the normal yellow background color. This gives the ventral hind wing a grayish green appearance. In *N. iole,* for instance, the cool-season form has three times as many black melanic scales in a standard area of the hind wing as the warm-season form (Douglas and Grula, 1978). *Colias eurytheme* and *N. iole* are both lateral baskers—they are able to increase their body temperature by exposing the ventral surface of their hind wings to the sun. Heating and cooling curves for both species show that when exposed to a source of radiant heat, the cool-season melanic forms warm up more quickly and attain a higher steady-state thoracic temperature than nonmelanic forms

(Watt, 1968; Douglas and Grula, 1978). Because butterflies have no endogenous mechanism for increasing their thoracic temperature prior to flight, their activity pattern
often depends on their ability to increase
their thoracic temperature by basking. By increasing the efficiency with which they ab-
sorb radiant heat, the cool-season forms of
these pierids have evolved a mechanism
whereby they can extend their flight season as
well as their geographic range. *Colias eurytheme* has an additional parallel polyphenism
in its dorsal wing pattern, which is orange in
the warm-season form but nearly yellow in
the cool-season form (Hoffmann, 1973,
1974). Nothing appears to be known about
the adaptive significance of this portion of the
polyphenism.

Many other temperate zone Pieridae exhibit seasonal polyphenisms that parallel
those of *C. eurytheme* and *N. iole*. All North
American *Colias* spp., *Pieris rapae, P. napi,*
and *Pontia* spp. have cool-season forms that
are distinctly darker than their warm-season
counterparts (Watt, 1968; Hoffmann, 1973;
Shapiro, 1978). Because wing pigmentation
has a demonstrable role in body temperature
regulation (Wasserthal, 1975; Douglas and
Grula, 1978; Kingsolver and Moffat, 1982;
Kingsolver, 1983, 1985, 1988), it is not unreasonable to assume that in many of these
cases the darker coloration of cool-season
forms is an adaptation.

It is worth noting at this point that the
adaptive significance of the warm-season
form in temperate zone butterflies is not at all
obvious in any of the cases mentioned so far.
Because seasonal polyphenisms are so widespread, it would seem that both seasonal
forms must have an adaptive value. Shapiro
(1976) notes that the necessity of such a
double adaptation was recognized nearly a
century ago by Weismann (1896). An adap-

tational model parallel to that proposed by
Brakefield and Larsen (1984) might be constructed for some other Nymphalidae, but
because their model is basically a predator-
interaction model that depends on variability
in eyespots, it is clearly not directly applicable to the many Nymphalidae and Pieridae
that do not have well-developed ocelli. Shapiro (1976) has done release-recapture experiments with cold- and warm-season morphs of
Pieris protodice during the summer months in
California and showed that cold-season forms
had a marginally lower survivorship than the
warm-season ones. This finding supports the
assumption that the warm-season forms must
be adaptive in summer, but the cause (as well
as the magnitude) of their enhanced fitness
remains unresolved. The simplest assumption would be that the cold-season forms
overheat too easily in the summer, but this
remains to be demonstrated and would have
to be reconciled with the evolution of a great
many sympatric taxa of dark-colored monomorphic butterflies. Comparative studies of
the phenology and behavior of polyphenic
and monomorphic taxa will be needed to
understand and resolve this problem.

Shapiro (1976) has discovered a striking
and potentially important general feature of
seasonal polyphenisms in Nearctic butterflies. He notes that the cold-season phenotypes of multivoltine butterflies usually
resemble the phenotypes of univoltine populations that live at higher altitudes or at
higher latitudes. In several species it is possible to induce cold-shock phenocopies in the
warm-season form that closely resemble the
normal cold-season and boreal phenotypes. It
seems reasonable to suppose that the close resemblance among these three phenotypic
classes is not accidental. One possibility is
that the cool-season forms and races arose by
genetic assimilation of a phenocopy (Wad-

dington, 1953, 1956a). The general requirements of Rachootin and Thomson (1981) are
met: It can be demonstrated that phenocopies can be induced, and the adaptive significance of the cold-season form can be
understood or inferred in at least some of the
cases. This does not prove, of course, that genetic assimilation played a role in the evolution of seasonal polyphenism. The close phenotypic resemblances noted above could have
come about through independent recurrence
and selection of a single mutation, or
through repeated independent selection for a
particular combination of alleles at a few
polymorphic loci. The possible scenarios are
indefinitely large and are likely to remain so
until we learn more about the ecology and
the genetic and developmental bases of some
of these polyphenic complexes.

Shapiro (1979, 1980b, 1984a) has con-
ducted extensive investigations on the evolution of the seasonal polyphenism of South
American pierids of the *Tatochila sterodice* species group. This genus is closely related to
the Holarctic Pierinae (none of which extend
into South America) of the *Pieris callidice* species group. *Tatochila* species occur in temperate to subarctic environments along the
Andes from Colombia to Tierra del Fuego.
Various degrees of polyphenism occur among
the sibling species of the *T. sterodice* group,
each of which occupies a well-defined and relatively narrow range of habitats and climatic
zones within its range (Shapiro, 1980b,
1983a, 1984a). The northern species, such as
T. macrodice, are monomorphic. In Argentina, *T. vanvolxemii* is strongly seasonally diphenic in the male (Plate 7A), and the female
is genetically dimorphic (the two alternative
phenotypes of the female resemble the alternative seasonal morphs of the male; see Shapiro, 1980b). In Chile, *T. mercedis* is polyphenic in both sexes, with continuous

seasonal variation between two extremes.
The polyphenisms in both *T. vanvolxemii* and
T. mercedis appear to be controlled mainly by
temperature, though photoperiod can exert a
modulating and perhaps synergistic effect.
Individuals reared at high temperatures have
ventral hind wings that are almost entirely
white, whereas those reared at low temperatures develop a bold pattern of black lines
parallel to the wing veins, so that the whole
venation system is beautifully outlined. In
T. mercedis a complete range of intermediates
with lighter venation patterns is found in nature and is believed to be produced by intermediate temperatures.

The *Tatochila* polyphenism is similar to
that of the Holarctic *Pieris napi,* but the pattern of loss of the venous stripes in the summer form is different. Reduction occurs
through a retreat toward the base of the wing
in *P. napi,* through a retreat toward the outer
wing margin in *T. vanvolxemii,* and as a general diffusion in *T. mercedis.* These differences
suggest that in the three species the venous
stripes of the cool-season form evolved independently (Shapiro, 1980b). Shapiro has attempted to deduce the evolution of polyphenism in *Tatochila.* He takes issue with the
suggestion by Klots (1933) that the Andean
pierine fauna was derived from the Nearctic
one via the *Pieris callidice* group (Shapiro,
1978, 1980b) and instead suggests that the
initial populations of *T. sterodice*'s ancestors
spread southward along the Andes as a monomorphic species. Populations that invaded
the drier and sunnier lowlands in Argentina
and Chile independently began to evolve
physiological responses to the more pronounced seasonality in those regions as
they differentiated into *T. vanvolxemii* and
T. mercedis. The evolution of the mixture of
sex-limited polyphenism and genetic polymorphism in *T. vanvolxemii* is a considerable

puzzlement and has no apparent parallel in any other butterflies.

Control of Polyphenic Expression

Seasonal polyphenisms of butterfly color pattern are but one manifestation of a diverse and widespread array of seasonal adaptations of insects. Most insects have evolved mechanisms to perceive and respond to environmental cues that herald the onset of a seasonal change. Among these adaptive responses are dormancy (diapause), migration, and polyphenisms such as changes in coloration, wing length, and body shape. The cues to which insects respond are almost always token stimuli. Photoperiod (day length or night length) is the most common among these, but photoperiodic cues are often modulated by other factors such as temperature, humidity, or food quality (Tauber et al., 1986). The adaptations of insects to seasonally fluctuating conditions in their environment have long been of interest to ecologists and evolutionary biologists, and in consequence a voluminous literature exists on the geographic and environmental correlates of seasonal adaptations and on the physiological mechanisms that mediate an appropriate response to the change of seasons (Danilevski, 1965; Shapiro, 1976; Beck, 1980; Tauber et al., 1986). The control of pattern polyphenisms will be discussed separately for tropical and temperate zone butterflies, because the seasonal cues in these climatic zones differ considerably.

Tropical Polyphenisms

In tropical regions there are often very distinct wet and dry seasons. Tropical insects capable of adaptive developmental responses to these seasons differ widely in the type of cue they use to predict the onset of a season. Photoperiod is not always used as a token stimulus in the tropics, because differences in day length between wet and dry seasons are generally not great. Although some tropical insects use photoperiod, many appear to use a change in food quality, humidity, or temperature. In nature, combinations of these stimuli may be used simultaneously to effect the same type of synergism seen in temperate zone insects (Tauber et al., 1986).

The control of the wet-dry seasonal polyphenism of the African *Precis octavia* has been studied by McLeod (1968), who showed that the dramatic pattern polyphenism of this species (Plate 6B) can be controlled exclusively by the temperature at which the animals are reared. When animals are reared at 27° to 32° C, all become the red, wet-season morph (*natalensis*). At lower temperatures, 10° to 16° C, nearly all animals become the blue, dry-season morph (*sesamus*). At intermediate temperatures (18° to 24° C), a variety of transitional forms between the two extremes is produced. McLeod (1968) gives meteorological tables for the area in Kenya where his stocks were collected. The mean temperature there ranges from 17° to 21° C, with extremes between about 5° and 28° C, depending on the season. At no time does the daily temperature range stay within one of the extreme high or low ranges required for the induction of pure morphs. Rather, daily temperature means throughout the year stay within the range in which only intermediate morphs are produced in laboratory experiments. Intermediate morphs, however, are rare in nature (Pinhey, 1949; Clark and Dickson, 1957; McLeod, 1968). Thus, although an appropriately high or low rearing temperature may be a sufficient condition for the development of each of the morphs, it would seem that in nature *P. octavia* uses some additional environmental information to remove ambiguity from the temperature signal. Photoperiod is unlikely to play a significant role in nature because the species ranges across equatorial Africa, where the photoperiod stays nearly constant throughout the year.

Temperate Zone Polyphenisms

In temperate regions, photoperiod appears to be the most common cue for the physiological and developmental switches that control seasonal development and polyphenism. Unlike temperature, the changing photoperiod provides an unambiguous predictor of the approach of each season, and many insects have evolved a complex but accurate physiological mechanism for measuring day length or night length (Truman, 1971; Bünning, 1973; Beck, 1980). When a species is capable of alternate developmental responses in different seasons, a critical photoperiod usually determines the developmental switch and a more or less prolonged sensitive stage during which an accumulation of inductive photoperiods must be experienced in order to evoke an effective response. The exact value of the critical photoperiod depends on the latitude at which an insect lives and the lag time between the photoperiod-sensitive stage and the stage in which the developmental response becomes manifest. The critical photoperiod is genetically determined and is an adaptive character of a local population (Danilevski, 1965). In most cases where it has been investigated, temperature exerts a synergistic effect with photoperiod. Cool temperatures usually enhance the effectiveness of short photoperiods; a diurnal temperature cycle in phase with the photoperiod (a so-called thermoperiod—Beck, 1983) has a particularly strong synergistic effect, so that fewer inductive cycles are required to achieve a particular response. Occasionally the synergism is such that a change in temperature alters the critical photoperiod. In *Pieris brassi-*

cae, for instance, a decrease in temperature from 25° to 15° C increases the value of the critical photoperiod by about one hour (Danilevski, 1965); other cases are discussed by Beck (1980) and Saunders (1982). In a few temperate species, temperature alone can serve as the seasonal cue for diapause or the development of alternate seasonal morphs (Tauber et al., 1986).

When temperature is presumed to be the primary environmental stimulus for a particular phenotypic change, it is essential to distinguish between its direct effects on development and its function as a sign stimulus for the induction of the seasonal response. If the altered temperature is directly responsible for an altered phenotype (because it changes the rate of the developmental and physiological processes that generate that phenotype), then the altered phenotype is not a polyphenism but a change in the norm of reaction (Table 5.1). Only when the effective temperature regime or critical period precedes the period during which the process of interest occurs can we be sure that the response is a seasonal adaptation.

In some temperate zone Lepidoptera the development of seasonally polyphenic wing patterns is coupled to pupal diapause. One of the seasonal morphs emerges only from pupae that have undergone diapause; the other, from nondiapausing pupae. Depending on the peak flight season for these alternate forms, they are referred to as the spring versus summer morphs, the summer versus fall morphs, or the cool-weather versus warm-weather morphs. Examples include *Araschnia levana*, *Pieris napi*, *Papilio xuthus*, and *P. zelicaon*.

One of the best understood of these polyphenisms is that of the European map butterfly, *Araschnia levana* (Fig. 6.19; Plate 6A), whose seasonal development has been studied for more than a century (Dorfmeister, 1864;

Weismann, 1875, 1896; Süffert, 1924a). The physiological mediation of the environmental signal and the manner in which polyphenism is linked to diapause in *A. levana* have recently been elucidated by Koch (1985, 1987) and Koch and Bückmann (1985, 1987). In this species both diapause and polyphenism are under strict photoperiodic control. The sensitive period for photoperiodic induction of diapause occurs during the fourth and fifth larval instars (Müller, 1955, 1956). Short-day photoperiods during the sensitive period induce pupal diapause, and the adults that eventually emerge from such pupae all have the spring phenotype (the orange-and-brown *levana* form). Larvae that experience long-day photoperiods during their sensitive period do not undergo pupal diapause; they initiate adult development immediately after pupation and emerge as the summer phenotype (the black-and-white *prorsa* form). Thus the summer phenotype can be considered a kind of default condition, because it is the morph that develops when the life cycle is continuous and uninterrupted. Koch (1985, 1987) has shown that when short-day pupae are parabiosed to long-day pupae immediately after pupation, both animals develop synchronously without diapausing, and both emerge as adults with a normal summer phenotype. Thus morph determination does not occur by photoperiodic induction but depends on some process associated with the developmental standstill that occurs during diapause.

It appears that a developmental switch from the summer to the spring phenotype occurs during the first few days of the pupal stage. Koch showed that larvae reared on short-day photoperiods (12 hours light : 12 hours dark) always produced diapause pupae and that these pupae could be made to initiate adult development at any time by injecting them with the molting hormone, ecdys-

terone. The phenotype of the adult that emerges from such pupae depends critically on the time after pupation at which ecdysterone is injected. Injections done 10 or more days after pupation resulted in adults with a normal spring phenotype, just as would have been produced if the pupae had been allowed to break diapause naturally. By contrast, when ecdysterone was injected on day 3 after pupation or before, adults with a normal summer phenotype emerged, identical to those that would have been produced by nondiapausing long-day animals. The converse experiment was also done. Pupae from larvae reared under a long-day photoperiod (16 hours light : 8 hours dark) can be prevented from developing further by removal of the head and prothorax (thus removing the endocrine centers that normally regulate and produce the hormone). About half of the animals whose head and prothorax were removed 1 day after pupation underwent normal adult development and emerged as normal nondiapause (black-and-white) summer morphs. When the remaining animals that did not initiate development were injected with ecdysterone 14 days later, all developed into the orange spring morph (Koch and Bückmann, 1987).

It is evident from these experiments that in *Araschnia*, determination of the spring morph occurs between 3 and 10 days after pupation. Koch (1985, 1987) and Koch and Bückmann (1987) have shown that ecdysterone injections within this interval can result in a smooth series of intermediate morphologies between the normal spring and summer forms. The later the injection is done, the closer the resemblance to a normal summer morph. The largest steps in this transition occur between days 3 and 5 after pupation. Thus it appears that the transition between the two morphs occurs during the early days of diapause development (Tauber et al.,

1986) and that this transition is gradual. The timing of ecdysterone secretion has also been shown to be the principal determinant of the polyphenism in *Araschnia burejana* (Keino and Endo, 1973). The exact role of ecdysterone in this polyphenic transformation is still unclear. For the time being, it may be assumed that ecdysterone serves merely as a trigger for adult development and that it simply reveals the degree of pattern determination (or transition) achieved at the time at which adult development was initiated.

The great influence that the exact timing of ecdysterone secretion can have on the adult phenotype may explain the confounding effects of temperature on morph determination in *Araschnia,* which long obscured the elucidation of the photoperiodic mechanism. It is possible to partially transform prospective summer forms into spring forms by artificially cooling their larvae and early pupae (Süffert, 1924a; Reinhardt, 1969). This effect can now be readily explained if the lowered temperature served to delay the normal peak of ecdysterone secretion so that it fell within the morph transition period. Temperature effects on form determination appear to be largely laboratory artifacts. Intermediate morphs are rare in nature, and it is clear from the work of Müller (1955, 1956) that in nature, photoperiod is the principal if not the only determinant of polyphenism in *Araschnia.*

A parallel case to that of *Araschnia* may exist in the lycaenid *Lycaena phlaeas.* This species has a seasonal polyphenism that is controlled by a factor from the brain (Endo and Kamata, 1985). When brains from presumptive summer-morph (long-day) pupae are implanted into presumptive spring-morph (short-day) pupae, most of the latter develop into summer morphs. Ecdysterone injections, however, tend to induce the development of a color pattern that resembles the spring morph (Endo and Kamata, 1985). The exact roles of ecdysterone and the presumptive brain factor are unclear, but Endo and Kamata suggest that differences in the relative timing of ecdysterone secretion in long-day and short-day animals may be the primary cause of the phenotypic differences between the seasonal forms of this species. If this is correct, then the brain factor may simply be the prothoracicotropic hormone that normally activates the prothoracic glands to produce ecdysterone.

The seasonal polyphenism of *Papilio xuthus* may be controlled by a mechanism that is somewhat different from that of *Araschnia.* Endo and Funatsu (1985) have shown that when diapausing pupae are parabiosed to nondiapausing pupae immediately after pupation, both individuals usually develop into the summer morph characteristic of the nondiapausing generation. They also showed that when brains derived from larvae, prepupae, or pupae were transplanted into the abdomen of diapausing pupae immediately after pupation, many of the recipients developed into summer morphs. Yet, quite unlike what one would expect based on the finding from *Araschnia,* brains from developing animals continue to induce the summer morph when they are implanted into diapausing pupae as late as 30 days after pupation, and also when they are transplanted into diapausing pupae that have been chilled for two months to break diapause (and that would normally have developed into the spring morph). These results indicate that the brain produces a humoral factor that somehow induces development of the summer pattern. The simplest explanation for these findings is that the brain factor is the prothoracicotropic hormone that is normally responsible for provoking the production of ecdysterone. If this is the case, then *P. xuthus* may differ from *Araschnia* only in the timing of the pattern morph transition relative to the time of pupation and diapause. On the other hand, it is also possible that the brain produces a neuroendocrine factor like the one that is involved in morph determination in *Polygonia* (see below).

The environmental induction of seasonal color pattern polyphenism is not always as straightforward as in the case of *Araschnia levana* or *Papilio xuthus*. In species of the *Pieris napi* complex, temperature and photoperiod interact in a complex way in the environmental control over seasonal polyphenism. The polyphenism of *P. napi* is usually tightly linked to diapause. Nondiapausing pupae give rise to the summer (warm-weather) form, whose wings are unmelanized. Diapausing pupae give rise to the spring (cool-weather) form, whose wings are strongly melanized along the wing veins (Shapiro, 1976, 1978). In some populations it is possible to override diapause-inducing short-day photoperiods by rearing larvae at high temperatures. Thus, by bypassing the photoperiodic signal, warm temperatures induce what appears to be an anomalous cold-weather phenotype. In addition, if long-day nondiapausing pupae are chilled shortly after pupation, they develop into the summer (warm-weather) morph. This result may be due to a phenocopy effect, because it does not appear that development can take place during the 0° to 5° C chilling period. A summary diagram of this control system, which also includes the possibility of obligatory genetic diapause in some montane populations, is given in Figure 6.22. Shapiro (1978) attributes the ecological significance of the phenotypic response to chilling to the fact that this species usually breaks diapause in the middle of winter and that it is therefore naturally exposed to some postdiapause chilling. The phenology of this system is clearly complex, and in that light it is of particular interest to note that an equally complex and nearly perfectly parallel case of environmental induction has also been documented for *Pieris rapae* in Europe (Shapiro, 1978). In regard to the evolution of wing patterns, perhaps the most interesting aspect of this system is the utilization of what appears to be a phenocopy response to induce one of the morphs.

Photoperiodically controlled wing pattern polyphenisms in temperate zone butterflies need not be coupled to pupal diapause. Among the Pieridae the polyphenism of *Colias eurytheme* can be continuously modulated by the photoperiod. Short day lengths give rise to melanic cool-season forms, long day lengths to pale warm-season forms, and intermediate day lengths to animals with intermediate degrees of melanization (Ae, 1957; Hoffmann, 1973, 1974). The photoperiod-sensitive period is during the third and fourth larval instars. *Pontia* (formerly *Pieris*) *protodice* and *P. occidentalis* likewise have a polyphenism that is strictly under photoperiodic control and is also continuously variable between fully warm-season and fully cold-season morphs, depending on the exact day length of the photoperiod regime (Shapiro, 1968, 1973, 1976).

The physiological mechanism behind one such polyphenism that does not involve diapause has been investigated by K. Endo and his co-workers in the oriental anglewing, *Polygonia c-aureum* (Nymphalidae). In this species, long days (16 hours light : 8 hours dark) and high temperatures (28° to 30° C) induce the light phenotype of the summer morph, whereas a short-day photoperiod (8 hours light : 16 hours dark) and low temperatures (19° to 21° C) induce the dark phenotype of the fall morph (Fukuda and Endo, 1966; Hidaka and Takahashi, 1967; Endo, 1984; Endo et al., 1988). The sensitive period is during the third and early fourth lar-

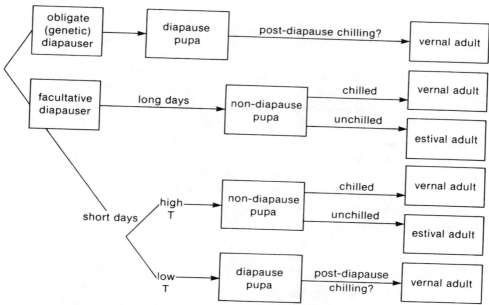

Figure 6.22. Environmental control of seasonal polyphenism in *Pieris napi*. Populations are poly-morphic for diapause, consisting of individuals that are obligate, genetically determined diapausers and individuals that are facultative diapausers responding to day length cues. Individuals with an obligate, genetically determined diapause always emerge as adults of the darker vernal (spring) form. Individuals whose diapause is induced by environmental cues (long days or high temperatures during larval devel-opment) are also seasonally diphenic. Those that do not diapause usually develop into the estival (sum-mer) form, but chilling (cold shock) of the pupa can induce development of the darker spring form. Question marks indicate responses not yet studied in the laboratory. (From Shapiro, 1978)

val instars. Fukuda and Endo (1966) and Endo (1970, 1972, 1984) have shown that this polyphenism is under neuroendocrine control. When the brain is removed from long-day pupae within 28 hours after pupa-tion, the presumptive summer morph is transformed into the fall morph. Simply splitting the brain, thus cutting the nerves between the brain and the corpora cardiaca (the neuroendocrine organs through which brain neurosecretions are released into the hemolymph), likewise causes long-day pupae to develop into adults of the fall phenotype, as does the surgical removal of the medial neurosecretory cells of the brain. When brain ablation is done more than about 34 hours after pupation, the normal presumptive

summer-morph develops. When corpora car-diaca from presumptive summer-form pupae are transplanted into the abdomens of pre-sumptive fall-form pupae, the pupae develop into intermediate forms that resemble the summer form in rough proportion to the number of corpora cardiaca transplanted (Endo, 1984). Apparently the brains of long-day animals produce a neurohormone that in-duces the summer morph. Endo et al. (1988) have succeeded in extracting a water-soluble compound from the brains of long-day pu-pae, probably a polypeptide, that induces the development of the summer phenotype in short-day pupae in a dose-dependent manner. This material is also present in the brains of presumptive fall-morph pupae but in much smaller quantity. Furthermore, substances with identical summer-morph-inducing activity on *Polygonia* can be extracted from the brains of *Papilio xuthus, Lycaena phlaeas,* and *Bombyx mori.*

The identity of the summer-morph-inducing factor is not known. It evidently must act during the first two days after pupation to be effective. It is therefore possible (but not demonstrated) that this factor has prothoracicotropic activity, so that its effect on development could be mediated by the timing of ecdysterone secretion, as in *Araschnia.* In support of this idea, we can note that presumptive fall forms develop more slowly than presumptive summer forms, so a difference in the timing of ecdysterone secretion could be involved. The slower development of presumptive fall morphs is in part due to the lower temperatures at which they are raised (Fukuda and Endo, 1966; Endo, 1972). Lower temperatures could shift the timing of the ecdysterone peak, as is believed to occur in *Araschnia* and *Lycaena,* so that it falls after the morph transition period. This could happen if the temperature acted differentially on the rates of pattern development and ecdysterone secretion; both are delayed by lower temperatures, but one more so than the other, so that they are brought out of phase.

If ecdysterone is not involved, then we must consider the likelihood that the putative neuroendocrine factor affects pattern determination directly. Summer and fall morphs of *Polygonia* differ only in quantitative features of their color pattern (they differ only slightly in the sizes of pattern elements and more considerably in the darkness of the background pigment). It is therefore not difficult to imagine that temporal or quantitative changes in an endocrine regulator of cell metabolism or biosynthesis could provoke the kinds of morphological differences we see. Because the photoperiod-sensitive area is in the brain of insects (Saunders, 1982), and the phenotypic response is systemic, it seems reasonable to expect that photoperiodically induced polyphenisms in general will prove to be mediated by alterations in a neuroendocrine signal. If this proves to be the case, it would bring seasonal color pattern polyphenism in line with the vast majority of color, phase, and caste polyphenisms among the insects, all of which are under endocrine control (Nijhout and Wheeler, 1982).

In the preceding chapters, we have examined the comparative morphology, development, and genetics of color patterns. We have seen that the bewildering amount of information contained in the thousands of color patterns can be organized in several different ways. So far we have considered five systems of order: the nymphalid ground plan, the hierarchy of symmetry systems, wing cell autonomy of patterns, serial homology of pattern elements, and morphoclines. The nymphalid ground plan provides an overall organizing principle that we can use to find the identities of the various dots and stripes that make up any species' color pattern. The elements of the nymphalid ground plan can be interpreted as a hierarchy of symmetry systems. The recognition of wing cells as the compartments for pattern formation, and of the serial homology of pattern elements, allows us to focus our analysis of pattern on a relatively small fraction of the whole, knowing that the overall wing pattern will be made up of an iteration, with variation, of the pattern found in one wing cell. The patterns in a single wing cell can, in turn, be ordered in a set of morphoclines based on those features of pattern element morphology that vary independently from the others.

It is not clear that any of these organizing schemes is more fundamental than the others. They represent different ways of focusing on the problem of how to create order in a massive amount of morphological data. Nor are any of these ordering schemes complete; each addresses different aspects of pattern organization. However, all these methods of ordering pattern can be applied to any species or taxonomic subgroup of the butterflies, which shows that color patterns arise from orderly processes common to all the butterflies. This conclusion is further supported by the finding that pattern elements with dramatically different appearances, such as interven-

Chapter 7

Models and Mechanisms

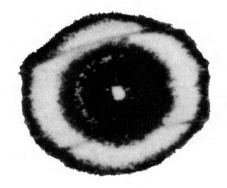

ous stripes and eyespotlike dotting patterns, are genetically similar and developmentally compatible (Fig. 6.7). So we are faced with a highly ordered developmental system that in the course of time has produced an enormous variety of distinctively different pattern morphologies. We have now arrived at a point at which we must give serious thought to deducing the basic processes that underlie the origin and diversity of form in this system.

In this chapter we will put all the preceding information together and see if it is possible to develop a single encompassing model for color pattern development in the Lepidoptera. Specifically, we are looking for a set of processes that can generate the entire diversity of color patterns and out of which emerge all the various forms of order we have observed so far. To do so, it will be necessary first to lay out explicitly what the expectations for performance are for a model of pattern formation. In addition we need to specify the constraints (mechanisms, boundary conditions) within which any model for pattern formation in Lepidoptera must operate.

Expectations of a Model

Some exploratory modeling of wing pattern formation has been done by Nijhout (1978, 1985a), Murray (1981a, 1989), and Bard and French (1984). These authors showed that various diffusion and reaction-diffusion schemes can generate realistic replicas of central symmetry systems, eyespot morphologies, and intervenous stripes. All these efforts, however, were directed at modeling specific pattern shapes in specific species of butterflies or moths. What we would like to develop here, by contrast, is a model that can explain the causal basis of the morphology of any pattern element in any species of butterfly. We are not interested in modeling the

pattern of one species thoroughly, because in general it is possible to devise a nearly unlimited number of models to explain singular instances. The explanation of any singular instance should be an emergent property of a good general model. An indefinitely large range of possible models needs to be constrained to a small number of probable ones.

The following are the requirements for an acceptable model of color pattern formation: (1) It should be consistent with the known physical, chemical, and developmental features of a wing and its color pattern; (2) it should be able to generate all known extant patterns (in addition, it will undoubtedly generate many unknown ones, and we will consider the implications of this as we go along); (3) it should generate this diversity of patterns by simple quantitative changes in the values of its parameters. These three requirements, and particularly the last one, place severe constraints on the acceptability of a model. The rationale behind the first two requirements is obvious. The third requirement emerges from the many observations documented in previous chapters that indicate that almost all patterns can be connected through a graded series of intermediates, and that small genetic changes can result in small as well as dramatic changes in the pattern. Although it is highly unlikely that the last two expectations can be fully met, we ought to try to come as close as possible to this ideal.

General Features of a Model for Color Pattern Formation

Here we list and justify explicitly the mechanisms and boundary conditions that must be used in the construction of a model for color patterns.

1. *Color pattern formation is a two-step process.* Color patterns are specified by inductive sources along the wing cell midline, at the wing veins, and at the wing margin. Experimental perturbation studies show that the positions of the signaling sources on the wing cell midline are established well before they become active in pattern induction. Wing veins mark the boundaries of the compartments for pattern formation and influence the form of the pattern, so whatever determines their position and function must also precede determination of the form of the pattern. The processes of source determination and pattern induction (by those sources) occur sequentially. These are therefore separable processes that may be modeled independently.

2. *Each wing surface is a static monolayer of cells.* The dorsal and ventral wing surfaces are monolayers, and experimental evidence shows that there is no communication between them. During pattern determination the wing surface grows in area through cell division. Pushing by centrally located cells results in some cell rearrangement during growth, as does the final positioning of scale-building cells in some species. These relative cell movements are small (one or two cell diameters, at most) relative to the scale of the pattern (tens to several hundreds of cell diameters). No migration of cells occurs. Relative cell movement can therefore be excluded as a factor in pattern formation.

3. *Cell-to-cell communication is the mechanism of signal transmission.* If cells are static, they can interact only with their immediate neighbors. Their immediate neighbors are those cells in actual physical contact. Direct long-distance communication with non-neighbors through filopodia or epidermal feet is excluded. There is no evidence for the existence of filopodia during the period of pattern determination, and epidermal feet tend to be short relative to the scale of pattern formation. The mechanisms for direct cell-to-cell communication are limited. The cytoplasm of adjacent cells is partially continuous through gap junctions. Thus diffusion of small molecules through gap junctions is a feasible and reasonable mechanism for communication across both short and long distances. Cell-to-cell communication can also occur through the extracellular medium via secretion of molecules that can interact with receptors on target cells (as happens with hormones and neurotransmitters). In the wing the extracellular medium is the hemolymph, which circulates slowly through the lacunae that form the wing veins and percolates among the spindle-shaped epithelial cells. It is unlikely that this circulation serves as a transport mechanism for pattern-inducing signals, because its directionality is not the same in all wing veins, whereas the pattern in all wing cells often is. A final means of transmitting a signal among cells is mechanical, by traction on an extracellular matrix that connects groups of cells. At the time of pattern determination, wing epithelial cells do not extend processes and are not deformed in any way that suggests that mechanical stresses need to be considered as a signal transduction mechanism. Thus diffusion through gap junctions is the only reasonable mechanism of signal transmission in the wing epithelium. This then limits the molecules that can be used for signal transmission to those that can pass through the gap junctions. Insect gap junctions have a molecular weight cutoff at about 1,400 daltons (Caveney and Podgorski, 1975; Caveney, 1980; Caveney and Berden, 1982), so most organic molecules and small polypeptides (10 amino acids or less) are potential signaling molecules.

4. *Special conditions exist only at the wing veins and wing margin.* When pattern determination begins, the lacunar system of the wing imaginal disk (the vein lacunae and the bordering lacuna) constitutes the only differentiated structures in the wing. This lacunar system serves as the only avenue for ingress and egress of material to the wing disk. Thus the lacunae are the only structures in the wing that can serve as initial organizing or inducing centers. Experimental perturbations provide considerable evidence that special properties of the wing veins and the bordering lacuna affect the morphology of the color pattern. The wing veins also serve as compartment boundaries in that they limit the passage of inductive signals between wing cells. They can do this in a passive way and an active way: Gap junctions between wing epithelial cells may be absent in the area of the wing veins (passive), and the veins can act as sinks that can destroy morphogenetic substances (active). These two ways of making boundaries have different effects on the morphology of the color pattern, as we will see below. The veins can also be inactive, or transparent to morphogenetic signals, as happens with the large eyespots of certain Nymphalidae and the central symmetry systems of small moths.

5. *The wing cell is the unit compartment for pattern formation.* Because the overall pattern in most species consists of the repetition of a basic theme from wing cell to wing cell, it is appropriate to model the origin of pattern in a single wing cell rather than in the wing as a whole. Patterns in adjacent wing cells, when not identical, can usually be derived one from the other by displacement and distortion of pattern elements. This indicates that patterns in adjacent wing cells differ from each other only by quantitative variation of one or a few of the generating parameters. These variables may extend in a graded fashion across most of the wing, but as a first approximation they can be modeled as a constant within any one wing cell. A good model should be able to produce the transformation series normally seen in a row of adjoining wing cells.

6. *Pattern morphology is resistant to moderate changes in parameter values.* If pattern diversification comes about primarily through quantitative rather than qualitative variations in the processes of pattern formation, then changes in the absolute or relative values of a model's parameters should have an effect on the pattern. However, a good model should also have a certain resistance to change, just as most developmental processes are buffered against moderate environmental and genetic changes. For instance, the patterns in wing cells whose linear dimensions differ by as much as a factor of 1.5 have a nearly identical morphology. Individuals whose body sizes differ by a similar factor likewise have virtually identical color patterns. Thus the processes that establish the position and the shape of the pattern elements are rather insensitive to the absolute size of the developmental field. A model mechanism should therefore possess a similar degree of size independence. Pattern is also fairly insensitive to temperature during development. Differences in rearing temperature as large as $10°$ to $15°$tsC have no appreciable effect on the color pattern (temperature-sensitive polyphenisms are an exception, but these involve discrete developmental switches in which temperature acts as a token stimulus). Thus, the reactions and processes involved in pattern formation must either have the same Q_{10} or be somehow temperature-compensated. A good (though admittedly arbitrary) general stability requirement, for purposes of evaluating the suitability of a model, is that a 10% to 20% change in the value of any one parameter

should have a relatively small effect on the form of the pattern.

7. *Point sources and line sources should emerge as quantitative variants in the model.* This rather specific requirement stems from the observation that it is possible to construct smooth morphoclines that form a transition between intervenous stripes and nymphalid ground plan elements (see Chapters 3 and 6). That this can be done in several unrelated taxa indicates that continuity between these two patterns is a fundamental property of pattern formation in butterflies.

8. *Within a wing cell, each pattern element develops independently from the others.* This is the most important simplifying assumption we will make, and it is justified by the observation that each pattern element varies independently from the others during development as well as evolution. The presence or absence of an ocellus in a wing cell has no noticeable effect on the form or position of the adjacent parafocal element or central symmetry system band. The genetics of *Heliconius* likewise illustrates that single elements of the color pattern can be switched on and off, or be displaced, by changes at a single genetic locus without affecting the form or position of any of the other elements. Phenotypic correlation studies show moderate to strong correlations among homologous pattern elements in different wing cells but little or no correlation among elements within a wing cell. In general, it should be legitimate to model the development of each pattern element independently.

A Model for Inductive Signals and Their Interpretation

Pattern formation was described above as a two-step process: first, a process that determines the position and characteristics of sources for pattern induction; and second, the induction of the actual pattern by the activity of these sources. To construct a theoretical model for pattern formation, it is best to begin with the construction of a model for the second stage, the induction of pattern morphology, because this is the process about which we have the best information from comparative morphology and experimentation. A model for the first stage, determination of sources, will be developed below, in the section "A Model for Source Formation."

We begin the construction of a model by considering the determination of the two pattern elements whose developmental mechanism is best understood: the border ocelli and the central symmetry system. Both are clearly determined by some sort of activity emanating from a central source. Experimental evidence shows that if the source is destroyed, the current state of the pattern becomes frozen. As noted previously, the dynamics of this determination, at least in the case of border ocelli, are compatible with the hypothesis that a central source produces a diffusible signal substance whose concentration falls off with distance from the source. No measurements have yet been made of the dynamics of central symmetry system determination. The response to cautery is consistent with a diffusion gradient hypothesis, though other mechanisms have not been strictly ruled out.

Small molecules or ions that diffuse from cell to cell through gap junctions provide the most reasonable means of transmitting a signal within an epithelial sheet. Murray (1981a, 1989) and Bard and French (1984) have shown that diffusion can occur at a rate that is adequate to transmit signals over the distances required for pattern formation. Diffusion occurs from an area of high concentration (a source) to an area of low concentration (a sink), and the diffusing molecule may decay with time or react with other molecules and be transformed. In building a model that depends on signal propagation by diffusion, it is necessary to define the properties of the source, the properties of the diffusing substance, and the mechanism by which a concentration gradient of a diffusible molecule (a continuously distributed parameter) can give rise to a discrete pattern element (a local discontinuity).

Assuming a point source for eyespot determination, we must consider whether it is an instantaneous source, a constant-rate source, or a constant-level source. Instantaneous sources, in which a certain amount of material is produced at one instant in time and then diffuses away, are easy to handle mathematically but are unlikely to be involved in eyespot formation, because the cautery evidence shows that the continuous presence of a source is required for continued expansion of the pattern. Constant-rate sources are unlikely to effect eyespot formation, because unless their rate of production is exactly matched to the rate at which material diffuses away (by some type of feedback process), the concentration of material at the source will increase without bound as diffusion gradients become shallower over time. Constant-level sources are therefore biologically most reasonable. They require a simple feedback mechanism that matches the rate of material production to the rate at which it diffuses away so that source cells are maintained at a constant concentration.

Continuous-gradient models for discretely distributed responses require that the gradient be somehow translated into a discontinuous function. This is usually accomplished by the arbitrary assumptions that one or more thresholds exist and that gradient values above a threshold evoke a developmental response different from that evoked by gradient values below threshold. The mechanism whereby a threshold operates is not trivial, however. The simplest and most widely used models for threshold assume that a control molecule, distributed as a concentration gradient, activates an allosteric enzyme by cooperative binding (Fig. 7.1). Because the allosteric response is sigmoidal, the enzyme will remain completely inactive at low concentrations of the control molecule and fully active at high concentrations. The sharpness of the transition depends on the slope of the control gradient and the exponent, or Hill coefficient, of the allosteric response. The Hill coefficient is a function of the number of binding sites on the enzyme, usually between 2 and 4, which yield only a moderately sharp transition zone. An allosteric mechanism for threshold determination is essentially an equilibrium phenomenon, so cells will have no memory of the effect of the signal gradient after it is gone (Lewis et al., 1977), except insofar as the enzyme would allow the accumulation of certain reaction products in "active" cells.

Lewis et al. (1977) have proposed an elaboration of the allosteric model. It produces alternative steady states that are stable and heritable from one cell generation to the next, and the transition from one state to the other is sharp, even with shallow signal gradients (Fig. 7.2). The model assumes that the activity of a gene G is controlled in a linear fashion by a signal substance, S, and in a sigmoidal fashion by its own product, g, which is also degraded at a rate proportional to its concentration. This is a positive feedback system for the production of g. The rate of change of g (dg/dt) is then expressed by the equation

$$\frac{dg}{dt} = k_1 S + \frac{g^2}{1 + g^2} - k_2 g \qquad (7.1)$$

where k_1 and k_2 are reaction constants. The graph of this function is an inclined sigmoi-

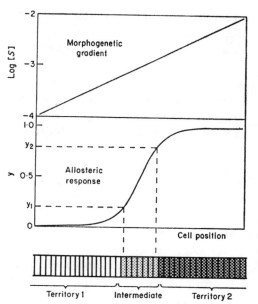

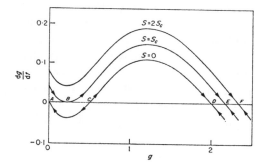

Figure 7.1. A simple allosteric model for a threshold. The upper graph depicts a concentration gradient of a substance, S, which is an activator for an allosteric enzyme that obeys the Hill equation. The degree of saturation, y, of the enzyme for the range of concentrations of S is given in the lower graph. The sigmoid curve results from cooperativity in the allosteric response. The threshold provided by such a mechanism is not sharp. In this case, the transition of the allosteric response requires a change of nearly an order of magnitude in the control variable, S. Thus a steep concentration gradient would be required for this mechanism to give a reasonably sharp threshold. (From Lewis et al., 1977; reprinted with permission of Academic Press Inc., London)

Figure 7.2. The rate of change (dg/dt) of the concentration of a gene product as a function of its own instantaneous concentration, g, for three concentrations of the signal substance, S. The form of these curves is defined by equation 7.1. Arrows on the curves point in the direction of increasing g where $dg/dt < 0$ and of decreasing g where $dg/dt > 0$. Stable steady states occur where the arrows converge on the axis of $dg/dt = 0$. When S = 0, there are two stable points, at A and D. As S increases beyond a critical value, S_c, only one stable point remains, at E or F. Thus, if g starts low, it will tend to remain low as long as $S < S_c$, and g will move to a high steady-state concentration when $S < S_c$. (From Lewis et al., 1977; reprinted with permission of Academic Press Inc., London)

dal curve (Fig. 7.2). When $S = 0$, this function shows that there are two dynamic equilibria for the concentration of g, one at $g = 0$ (point A) and the other at a high concentration of g (point D). The equilibria occur at the concentrations of g where $dg/dt = 0$. If the concentration of g increases slightly, dg/dt becomes negative and the value of g will drop to its equilibrium point; if the concentration of g decreases slightly, dg/dt becomes positive and g will increase to its equilibrium point, where dg/dt is once more zero.

Assuming that the system starts with little or no g, then the dynamics are such that the concentration of g will always be driven to zero. The dynamics are different, however, when the concentration of S is increased. An increase in S causes an overall upward shift of the graph of dg/dt (accompanied by a small increase in the equilibrium concentration of g; Fig. 7.2). At values of S above a critical level, however, there is only one equilibrium point at high concentrations of g (point F). As S rises, g switches abruptly from a dynamically stable steady state at which g is low to a stable steady state at which g is high (upper curve in Fig. 7.2). Figure 7.3 graphs the steady-state concentration of g as a function of S. Thus we have a mechanism whereby a smooth and continuous gradient of a signal substance (S) can be translated into an abrupt change in the activity of a gene (i.e., the formation of g). The exact concentration of S at which this threshold lies depends on the values of the constants in equation 7.1. One of the particularly nice features of this model is that it has "memory." Once g is at its high steady-state concentration, the equilibrium dynamics will cause g to persist at its high steady state even if S subsequently disappears, as can be seen by examining the lower curve in Figure 7.2 once more.

For the purposes of modeling, we will assume, based on the foregoing, that long-distance signals for color pattern formation exist in the form of diffusion gradients. Threshold mechanisms of the kind just described can be used to interpret such gradients, so that genes or enzymes can be activated at a specific and arbitrary concentration of a gradient substance. In the discussions that follow, we will therefore assume the existence of diffusion gradients and thresholds for pattern formation in the wing epithelium.

Basic Patterns

To organize the construction of a specific model, it is useful to deal with the most common patterns first, on the assumption that uncommon and rare patterns are derived from the common ones. The most common shapes among the symmetry system bands, border ocelli, and parafocal elements are illustrated in Figure 7.4. The central symmetry system and border ocelli are the simplest and least diverse, consisting mainly of points, circles, arcs, and lines. The parafocal elements are also mostly arcs, but with some regular elaborations that form the basis of much of the variation in these pattern elements (Fig.

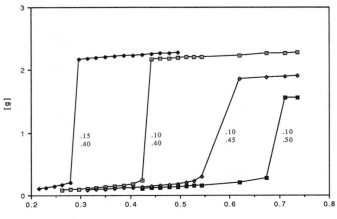

Figure 7.3. Dependence of the steady-state concentration of a gene product, g, on the concentration of the signal substance, S, for various values of the reaction constants in equation 7.1. Both axes are linear. The upper of each pair of numbers refers to k_1; the lower, to k_2. Sharp stable transitions occur across small differences in the concentration of S.

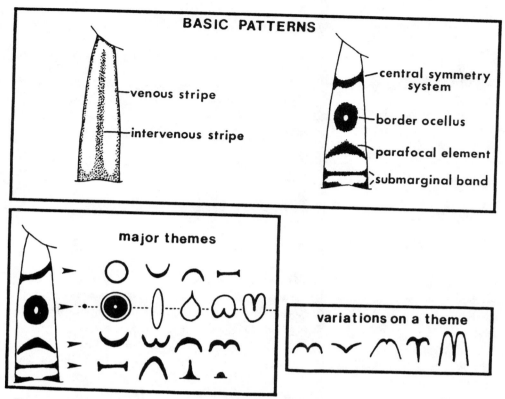

Figure 7.4. The major themes in butterfly wing cell patterns. Intervenous stripe patterns (*top left*) and nymphalid ground plan patterns (*top right*) form two mutually exclusive categories. Although there are patterns that form transitions between them (e.g., Fig. 6.8), the intervenous stripe patterns and nymphalid ground plan patterns are seldom superimposed on one another. Variation in the venous and intervenous stripe patterns consists mainly in variation in the length and width of the stripes. The nymphalid ground plan patterns are diverse. The major themes in the variation of each of the pattern elements are shown (*lower left*), as well as some of the variations on one of the parafocal element themes (*lower right*). (From Nijhout, 1985b)

7.5). Seen from the base of the wing, parafocal elements can be concave or convex arcs and can have a proximally or distally directed peak on the wing cell midline. Parafocal elements can also be expressed as short lines or dashes along the wing cell midline, in essence an intervenous stripe.

We continue the construction of a model by determining the simplest source-sink distributions that could give rise to these shapes. We constrain the possibilities by as-suming that source-sink locations are restricted to the wing veins, wing margin, and wing cell midline, locations for which we have experimental evidence for the existence of organizing centers (Chapter 5). We also assume that only diffusion gradients and thresholds are involved in pattern specification. The bottom panels of Figure 7.6 give examples of the simplest combinations of sources and sinks that can give rise to the patterns above them. Figure 7.7 is a summary of

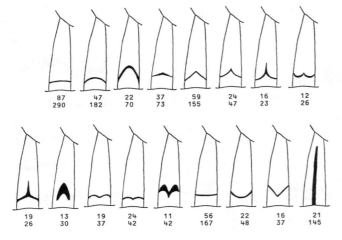

Figure 7.5. The most common pattern morphologies found in the parafocal elements, selected from the survey in Figure 2.21. The numbers shown are as indicated in Figure 2.21. (From Nijhout, 1990)

the source-sink locations that are necessary to produce all of the basic pattern themes seen in the border ocelli and parafocal elements (Nijhout, 1990). The problem now becomes one of determining what additional assumptions have to be made in order to obtain real patterns such as those illustrated in Figures 2.19 and 2.21. The vast majority of real patterns are variations on the themes shown in Figures 7.4 and 7.5. Real parafocal elements differ, among other aspects, in the relative lengths of the legs and central peak of an arc

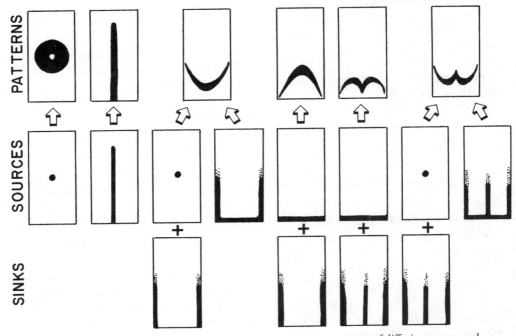

Figure 7.6. Six major pattern themes (*top row*) and the simplest arrangement of diffusion sources and sinks that can give rise to them. (From Nijhout, 1990)

and in the angles between these various parts of an arc. Real border ocelli and central symmetry bands vary as distortions by localized stretching or indentation of circular and arc-shaped patterns. Simple variation in the position or dimension of the sources shown in Figure 7.7 cannot produce the desired diversity of patterns (Nijhout, 1990).

Additional information is clearly required to produce the observed diversity of form. The most conservative assumption for the origin and nature of this additional information is that it is provided by diffusion gradients from discrete sources and, furthermore, that these sources are most likely to be restricted to those locations where sources have been suggested to occur, namely those shown in Figure 5.17. A two-gradient model, the interpretation landscape model of Nijhout (1978, 1981, 1985b), has been used

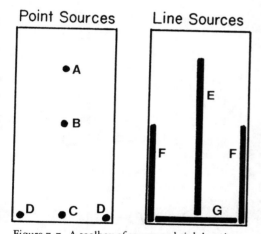

Figure 7.7. A toolbox of source and sink locations that can produce most of the diversity of form found in border ocelli and parafocal elements. By selecting sources indicated by the letters in this figure, an additive two-gradient model can produce most of the diversity of patterns seen in the butterflies. It is assumed that the points and lines shown can be specified as sources or sinks of a diffusible substance and that patterns are specified by a simple threshold of the gradient produced. (From Nijhout, 1990)

to show that pattern diversity can indeed be conceptualized as a series of simple distortions of circular themes. The basic circular pattern is created by a point source producing a cone-shaped diffusion gradient, and the distortions come about through interaction with a more complex second gradient, the interpretation landscape. The interpretation landscape model, however, suffers from two serious deficiencies. First, it requires complex interpretation gradients, sometimes with inflections that would be virtually impossible to produce by diffusion. Second, the gradients required were derived a posteriori, through curve fitting, not from more basic principles as we are trying to do here.

The main reason that the interpretation landscape model required such complex gradients was the assumption it made about the nature of the interaction between the two gradients. The interaction was assumed to be permissive, in which one gradient set the local threshold for the interpretation of the other. Many other types of interactions between two gradients are possible. In the course of nearly a decade of exploring models for pattern formation, I have found only one type of interaction between two gradients that allows a truly broad diversity of realistic patterns to be produced with the restricted source-sink distribution described in Figure 7.7. As it happens, it is also the simplest type of interaction possible, namely a simple addition of the values of the two gradients. In fact, two additive diffusion gradients whose sources are picked from the toolbox of sources in Figure 7.7 can generate virtually the entire diversity of patterns found in nature, as we will see below.

Detailed Model Patterns

We will first examine the manner in which patterns and pattern diversity can be produced from the diffusion source toolbox, leaving for the next section an exploration of how the system could go about actually selecting the sources. Computer simulation of diffusion in two dimensions from constant-level sources provides a convenient means of exploring the shapes of gradients. Figure 7.8 illustrates the shapes of the gradients that are produced by each individual set of sources, assuming that the veins (boundaries of each box) act as barriers to diffusion. These are the gradients that will be combined pairwise to produce actual patterns. An arbitrary number of contours is drawn to define the shape of the gradient. The shape of a contour corresponds to the shape of the pattern that would be produced by a threshold at that height. Evolutionary changes in the pattern are made possible by changing the threshold at which a gradient is read or by changing the parameters that generate the gradients.

There are but a few modes of variation, given a system of two gradients that interact additively. First, the locations of the sources and sinks for the two gradients can change in evolution, and that would alter the shape of the gradient. Second, the relative time of onset of diffusion in the two gradients can change, as can the strength (steady-state concentration) of the sources or sinks for both gradients. Finally, the diffusion coefficient and the rate of breakdown of the diffusing substances can change. Intuitively, one can see that small changes in any one of these parameters should result in small changes in the shape of the pattern.

Even though few modes of variation exist, the number of permutations, given the array of source distributions, is extremely large. Therefore we will examine only a small sample of the possibilities explicitly. Ex-

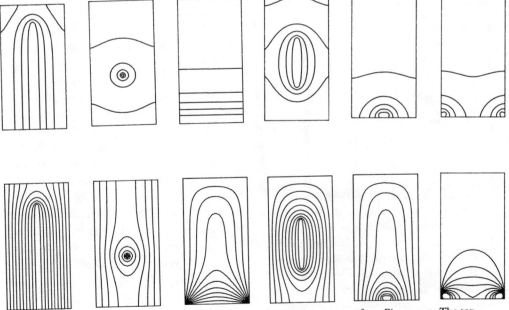

Figure 7.8. Contour plots of the gradients produced by each of the sources from Figure 7.7. The top row illustrates the gradients, assuming no flux at the wing veins. The bottom row illustrates the gradients from the same sources but assumes that the wing veins (the two long sides) act as sinks for the diffusing substance. Each contour represents a potential pattern.

Figure 7.9. Contour plots produced by the additive interaction of two gradients. The interaction of a point source with four other types of sources is shown. Each contour is a threshold and represents a potential pattern.

amples of the patterns generated by a few of the possible gradient interactions are shown in Figure 7.9.

The size and shape of a pattern element and its position within the wing cell are determined by a complex interaction of the strength of the sources, the duration of their activity, the values of the thresholds, and the position of the sources relative to each other. It is, however, possible to draw some general conclusions about the effects of variation in specific parameters of this gradient model on the morphology of the pattern. If wing veins act as sinks, circular patterns become flattened along the long axis of the wing cell, and bands that stretch across the wing cell become bent toward their source (Fig. 7.10). If the vein sink is active for a brief time, this bend is rather sharp (panel A2 in Fig. 7.10);

if the vein sink has been active longer, the bend is wider and less acute (panel A4 in Fig. 7.10). Line sources along the wing cell midline account for the midline peaks and indentations in border ocelli and parafocal elements (Fig. 7.10B,C). The strength (or height) of the source determines the relative size of the peak. Here too, if the source has been active for a relatively brief period, the peaks will be sharp; if the source has been active for a longer time, the peaks will be broad (panels B4 and C4 in Fig. 7.10). It is worth noting that at the peak of a parafocal element there is usually a distinctively colored dash-shaped mark (usually much darker than the rest of the element) that may well correspond to the location of the source responsible for this feature.

All patterns that have peaks or indenta-

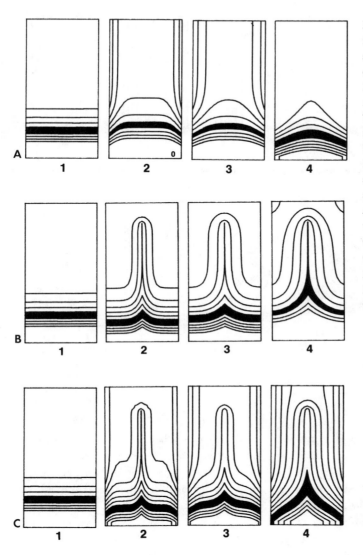

Figure 7.10. Effect that adding venous sinks and midline sources has on the morphology of the pattern produced by a line source along the (bottom) margin of the model wing cell. The leftmost panels in each row show the pattern produced by the marginal source alone. Each row is a time series of the activity of additional venous sinks or intervenous sources. *A*, venous sinks only; *B*, intervenous source only; *C*, both. One set of contours is filled in each panel to show one of the possible patterns that can be produced by this set of gradients.

tions along the wing cell midline require the presence of a line source (seldom a sink) on the midline. A midline source by itself causes a normal intervenous stripe; this stripe can be made to flare open into a wishbone pattern by interaction with a point source at the wing margin (Fig. 7.11). Stretching this point source into a line along the margin widens the flare to the width of the wing cell. As the gradient produced by the marginal line source extends into the wing cell, the flared intervenous stripe shortens to the brace-shaped figure characteristic of many parafocal elements (Fig. 7.11D). Moving a point source along the wing cell midline has only a minor effect on the shape of the surrounding pattern when the slope of the second gradient is fairly shallow; however, when that slope is steep, a shift of source position usually changes the shape of the pattern dramatically.

Morphoclines of gradient shapes can be generated in a variety of ways and can be used to test specific hypotheses about evolutionary relations among patterns. For instance, pattern differences between closely related species should be due to changes in only one or a few of the signaling sources or to a change in thresholds. This can be tested directly by an appropriate change in source or threshold parameters in the model. It is possible to study the probable developmental basis of a pattern morphocline that one has constructed—by modeling the two extremes, interpolating a series of parameter values between the two, and examining whether the patterns generated by such interpolation resemble the morphologies of species that make up the morphocline of real patterns. The model also allows one to explore what would happen to the shape of the pattern when a source is moved or is gradually strengthened or diminished. Such gradual quantitative changes should lead to gradual changes in the pattern, and these might correspond to specific pattern morphoclines observed in a particular taxon. The simplest approach to constructing exploratory morphoclines is to vary the value of one parameter of the model and observe its effects on the shape, size, and position of a pattern. For instance, if one varies only the relative values (strengths) of the sources in the two interacting gradients, then it is possible to construct a series of intermediate patterns that form a transition between any two single-gradient patterns (Fig. 7.12).

To date, Nijhout (1990) has shown that nearly 70% of the patterns illustrated in Figure 2.21 can be generated by the additive two-gradient model, and much of the remainder appears to be feasible as well. As might be expected, the model also produces some patterns that are not seen in nature. This is a reasonable finding, because it is highly unlikely that all possible patterns already exist. In addition, some patterns may be maladaptive. If the model is a correct representation of pattern development, then the novel patterns that it generates represent one aspect of the evolutionary potential of the system: essentially a corner of unoccupied morphospace. Other examples of unoccupied morphospace abound. Eyespots on the wing margin can be easily produced under the assumptions of the model but are not found in nature. Positioning of pattern elements with shapes like those of the parafocal elements proximal to an ocellus is also possible in the model, but such patterns are not found in na-

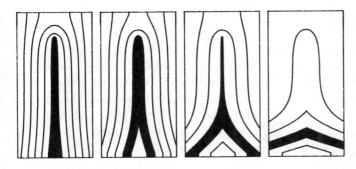

Figure 7.11. Transition from an intervenous stripe to a standard parafocal element. A long midline source is equally active in all four panels, and a progressively stronger source along the bottom margin is added from left to right. The wishbone-shaped intermediates are also found as normal patterns in *Acraea* (e.g., Fig. 2.10D) and other taxa.

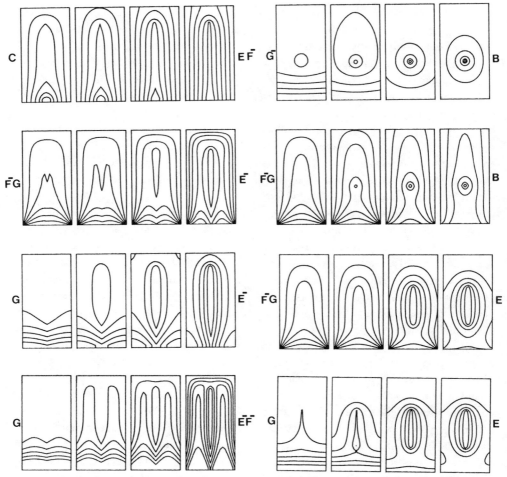

Figure 7.12. Two-gradient interactions. Each set of four panels represents a progression in the relative strength of the sources that produce the two gradients. The two gradients are listed on each side of the series by the letters that represent their source distribution in Figure 7.7. A minus sign to the right of the letter signifies a sink. The contour diagrams for each of these source distributions alone are shown in Figure 7.8.

ture. In these last two examples the model can violate apparently powerful rules of pattern formation. Because these rules do not emerge from the model, they indicate that in this respect the model is incomplete. Various reasons why certain areas of morphospace may remain unoccupied during evolution have been discussed by Maynard Smith et al. (1985) and will be considered in Chapter 8.

There are also certain types of patterns that cannot be generated by this model. Patterns that are asymmetric with respect to the wing cell midline cannot be produced. Such patterns do occur in nature and are probably due to a bilateral asymmetry in one of the gradients. Such an asymmetry could be introduced in an ad hoc fashion, and this would be equivalent to the assumption of an additional gradient orthogonal to the long axis of the wing cell. Graded series of pattern shapes in successive wing cells likewise argue for such global orthogonal gradients, as was noted before. The threshold model we used does not easily account for sparse scaling (see Fig. 7.19) unless additional assumptions are made. Without additional assumptions, the model also cannot account for bands that take on the character of a secondary symmetry system (e.g., Figs. 2.8 and 3.14); this would essentially require that the pattern element itself become a source. The complications that arise in the production of self-symmetrical bands will be considered at the end of this chapter.

A Model for Source Formation

In the preceding discussion we have assumed the existence of a specific source-sink toolbox from which we can pick point and line sources at specific locations. The two-gradient model depends on our being able to take any source or sink at will from this toolbox in the construction of each of the gradients. What we need next is a model for a mechanism that allows us to do just that; we need a model that produces, individually, all the sources of the toolbox and in their proper location.

Nijhout (1990) has shown that the lateral inhibition model of Meinhardt (1982) can do just that, if it is made to function on a domain that resembles the wing cell. The lateral inhibition model is a special case of a reaction-diffusion model. In reaction-diffusion models, two or more substances are assumed to diffuse freely while they react with each other and with their environment. If the reactions are of a specific type, reaction-diffusion systems can give rise to dynamically stable patterns in the distribution

of their components. Meinhardt's lateral inhibition model assumes the existence of two chemicals: an activator, a, whose synthesis is subject to strong positive feedback (autocatalysis), and an inhibitor, i, which suppresses autocatalysis around an area of activator production. This system is described by the following equations:

$$\frac{\delta a}{\delta t} = \frac{ca^2}{i} - k_1 a + D_a \left(\frac{\delta^2 a}{\delta x^2} + \frac{\delta^2 a}{\delta y^2} \right)$$

(7.2)

$$\frac{\delta i}{\delta t} = ca^2 - k_2 i + D_i \left(\frac{\delta^2 i}{\delta x^2} + \frac{\delta^2 i}{\delta y^2} \right)$$

(7.3)

where c, k_1, and k_2 are reaction constants; D_a and D_i are diffusion coefficients for the activator and inhibitor, respectively; and the second derivatives in parentheses represent diffusion in a two-dimensional (x,y) plane.

If the values of the diffusion coefficients and decay rate constants (k_1 and k_2) are chosen so that the inhibitor acts over a longer distance than the activator does, and so that activator and inhibitor production rates are for the most part balanced, it is possible to establish a stable pattern of activator and inhibitor concentration. Edelstein-Keshet (1988) gives an excellent summary of the conditions under which diffusive instability and pattern can arise in reaction-diffusion systems. Meinhardt and Gierer (1974) and Meinhardt (1982) have shown that many pattern formation processes in development can be explained by models of this type. In fact, it is believed that pattern formation in biological and chemical systems requires processes of short-range activation and long-range inhibition, and this requirement appears to be independent of the type of processes or mechanisms that constitute activation and inhibition (Meinhardt, 1982; Edelstein-Keshet, 1988; Murray, 1989).

Nijhout (1990) implemented the above equations in two dimensions by computer simulation. A wing cell was modeled, as we did implicitly in the two-gradient model above, as a simple rectangle in which the two long sides and one of the short sides represent the wing veins, and the other short side the wing margin. If it is assumed that every point in this rectangle is initially at steady state for the above equations ($\delta a/\delta t = \delta i/\delta t = 0$), then this system will maintain constant and homogeneous concentrations of a and i indefinitely. Nijhout (1990) tested the effect of letting the wing veins serve as continuous sources of small amounts of additional activator. As this activator diffuses into the plane of the wing cell, the local steady state is upset, and a complex dynamic of activator and inhibitor reaction-diffusion takes place that has the following general properties. Activator production becomes inhibited along the three sides near the wing veins; after some time, the cells along the wing cell midline become sites of strong activator production. A long midline source of activator production develops that extends from the wing cell margin to some distance from the apex of the model wing cell (Fig. 7.13). At the free end of this line source, activator production increases even further, and an area of inhibition begins to develop around this free end. This causes the line source to withdraw gradually toward the wing margin, leaving one or more point sources of activator production behind (Fig. 7.13). The number of point sources left behind and their exact position along the wing cell midline depend on the relative values of the diffusion coefficients and decay constants of the activator and the inhibitor.

Meinhardt's lateral inhibition model thus produces point sources in some of the exact locations required by the two-gradient model. But it also produces, albeit transiently, midline sources of the type required to make intervenous stripes. The temporal sequences shown in Figure 7.13 bear a remarkable resemblance to the morphoclines that connect intervenous stripe patterns and ground plan element patterns in many taxa (e.g., *Papilio memnon,* Fig. 6.8). If we assume that the positions on the wing surface where local peaks of activator production occur become sources of the morphogenetic signals required in the two-gradient model, then an early switch to signal production would result in a long midline source, whereas a late switch would result in one or more point sources on the midline. This model therefore suggests that the difference between midline sources and point sources is simply due to a difference in developmental timing—a heterochrony.

By varying the relative values of the two diffusion coefficients or of the decay constants in equations 7.2 and 7.3, it is possible to vary the number of stable points that will develop and, to some degree, their position on the wing cell midline. Nijhout (1990) found that only one additional assumption was needed in order for this system to generate the whole diversity of source locations needed for the toolbox, namely that the initial activator source at the wing veins must diminish in strength toward the margin of the wing. Proximo-distal gradients of activity along a wing vein are not an unreasonable assumption; many dependent patterns give evidence of increasing or decreasing pattern-inducing activity along the veins (e.g., Fig. 2.9C). If the initial wing vein sources sloped in a proximo-distal direction, then the following relations were found to obtain. Given a set of parameter values that produce a single, stable

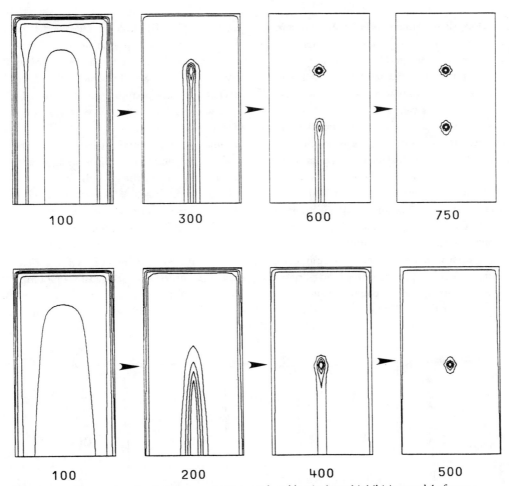

Figure 7.13. Time series of activator concentration produced by the lateral inhibition model of equations 7.2 and 7.3. In each case, a small amount of activator is introduced along three sides. A strong midline source of activator is produced that gradually retreats toward the wing cell margin, leaving one or more stable point sources behind. The number of point sources can be regulated by changing the values of the constants in equations 7.2 and 7.3. (From Nijhout, 1990)

point source, any of three parameter changes will cause this point source to move gradually toward the wing margin: increasing the diffusion coefficient of the inhibitor (or decreasing the D_a/D_i ratio), decreasing the decay constant of the inhibitor (or increasing the k_1/k_2 ratio), or increasing the level of activator production by the wing veins. The source

can, of course, be moved toward the apex of the wing cell by changing the values of one or more of these parameters in the opposite direction. Usually, if the position of the source becomes shifted too much toward the apex of the wing cell, additional sources will develop, evenly spaced between it and the margin. Whether or not such sources develop, their number and distribution depend critically on the slope of the wing vein source and on its strength (Nijhout, 1990).

Figure 7.14 illustrates an array of source distributions that can be produced by varying the parameter values in this model. If we assume that a peak of activator somehow stimulates the production of a diffusible signal molecule, then this model provides us with the mechanism for producing sources and placing them exactly where they are needed (compare Figs. 7.7 and 7.14). What is perhaps most exciting about this model is that it operates on a minimal number of special assumptions, none of which are biologically unreasonable. The model assumes conventional reaction-diffusion and uses only the existing venation system as the initial organiz-

ing principle. By linking the model for source production to the additive two-gradient model, we have achieved a system that can generate a truly broad and realistic diversity of color patterns almost entirely by quantitative changes in the values of fewer than a dozen parameters.

Time is an important parameter in the model. The source determination as well as the threshold component of the model produce the right kinds of patterns only when gradients are set up first and then "read" or "fixed" by a discrete determination event. Assumptions of progressive determination during the diffusional expansion of the various gradients do not produce sensible patterns. Thus it is a prediction of the model that several discrete critical periods exist during

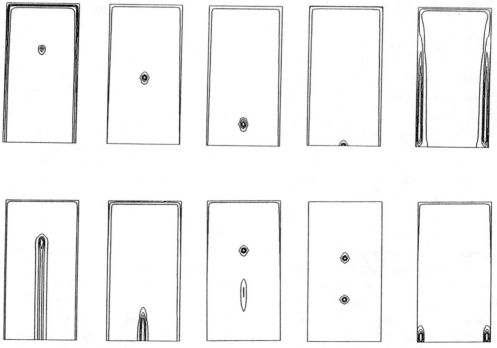

Figure 7.14. An array of activator source distributions that can be produced by altering the values of the constants in equations 7.2 and 7.3 and the rate at which activator is supplied by the veins. (From Nijhout, 1990)

which the status of the gradients is read to determine the position of sources and later the position of the pigment pattern. This also implies that heterochronies afford a mechanism for pattern alteration, and it should be obvious by now that simple quantitative changes in the timing of these critical periods can result in the development of dramatically and qualitatively different color patterns.

The scale invariance requirement is not completely met by this model. Source-to-sink gradients produce patterns that are inherently scale-invariant (Wolpert, 1969, 1971), but reaction-diffusion processes do not (Arcuri and Murray, 1986). Murray (1982) has shown that the range of parameter values for which this type of reaction-diffusion model can generate pattern is narrow. Given a particular set of parameter values, changing the dimensions of the field in which the reaction-diffusion takes place can (but need not always) result in qualitative changes in the position and number of stable sources that are produced. Changes in the values of the decay constants or reaction rates can compensate for changes in field size to produce size invariance (Othmer and Pate, 1980), but this is an ad hoc solution. It would be more elegant if scale invariance was an emergent property of the model. H. Meinhardt (pers. com.) has suggested that the addition of a low background level of excess of inhibitor production throughout the wing cell should make the system much less sensitive to variation in size (see also Meinhardt, 1982).

What is the relation between this model and real life? Surely pattern formation is a great deal more complicated than this model suggests. I believe that this model represents the simplest set of mechanisms that can simultaneously produce realistic-looking patterns and generate the natural diversity found in these patterns. It is unlikely that a model with fewer components will be able to do as well (though it can never be proved that the simplest model has been found). Thus we have a parsimonious model that is probably most useful as a heuristic device to help guide our thinking about the development, genetics, and evolution of patterns. For instance, the reaction-diffusion scheme, the threshold mechanism, and the sources and sinks all involve reaction rates that can be regulated by enzymes. Because enzymes are gene products, it becomes possible to make predictions about the effects of specific mutations on the morphology of the pattern. This, in turn, can lead to predictions about what kinds of evolutionary changes could occur by the accumulation of mutations that have only small and quantitative effects on the properties of gene products. We will attempt to conduct a brief exercise of this type in Chapter 8. In the remainder of this chapter, we will explore several additional aspects of pattern formation that bear in various ways on the simple model we just developed.

Special Cases

Unusual Foci

Arc-Shaped Foci

Throughout the preceding discussion, we have assumed that sources that give rise to ocelli are simple point sources. This assumption is reasonable, because the vast majority of ocelli have a simple dot-shaped focus at their center. This trait is not universal, however, and a survey of pattern diversity in the Nymphalidae disclosed a set of bizarre focal morphologies, particularly in the Brassolinae and Morphinae (Figs. 2.19 and 7.15; Plate 7B). Foci can be elongate, arc-shaped in various ways, and double. It is evident that although ocelli are often roughly concentric around their foci, that is not a necessary con-

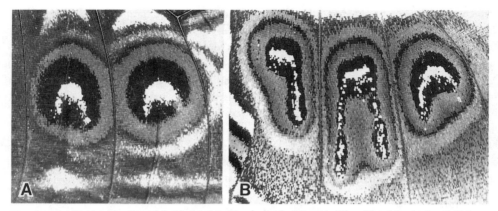

Figure 7.15. Unusual ocelli and foci (white) in *Morpho hecuba* (Nymphalidae: Morphinae). *A*, ventral forewing; *B*, ventral hind wing.

dition: Round ocelli can develop around arc-shaped foci, and vice versa. These morphologies indicate that the shapes of a focus and its surrounding ocellus are at least under partial independent control. How do we reconcile this observation with the experimental finding that development of an ocellus is induced by its focus? Part of the answer to this dilemma emerges from the behavior of the reaction-diffusion model for source determination. Under certain parameter values, the transient source patterns produced by this model resemble the elongated arc-shaped foci of certain morphines and brassolines (Fig. 7.16; compare with Figs. 7.15 and 2.19). Sometime later these arc-shaped sources condense to a single point or a pair of point sources side by side in the wing cell (Fig. 7.16). If the color pattern of a focus is determined sometime before its development ceases, then it would be possible to get an arc-shaped focal pattern followed by a much condensed point source that then induces the ocellus (Nijhout, 1990). Only if the times of focus activation and focus pigment pattern determination are close together will the outline of an eyespot be concentric around the shape of the focus.

Fragmented Foci

Among the Satyrinae are a number of species, particularly in the genera *Lethe* and *Euptychia*, that have well-developed ocelli with an apparently random scatter of small foci within their central field (Fig. 7.17). In *Lethe* these patterns are at one extreme of a morphocline that has perfectly normal-looking dot foci at its opposite extreme (Fig. 7.17). We assume that these scattered dots are foci, because each remains surrounded (irregularly at times) by small areas of black scales. These scattered foci can occupy the entire area of the black central disk of the ocellus, and it looks as if a chaos of point sources has replaced the normal central focus in those specimens. A centrifugal progression is apparent: Fragmented foci occur most commonly near the periphery of the black central disk. The model does not account for such a distribution of foci. At certain parameter values the reaction-diffusion model becomes unstable and produces chaotic peaks of activator, but these are not constrained to a disk-shaped area, as they obviously are in *Lethe*. These fragmented foci are reminiscent of the satellite spots that accompany bands of the central

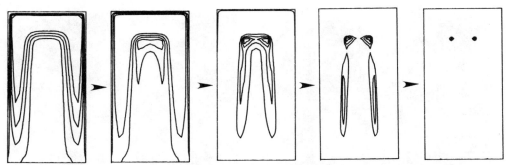

Figure 7.16. Time series of activator concentration patterns from the lateral inhibition model obtained when activator production by the lateral veins diminishes with proximity to the wing margin. The transition from a source resembling the arc ocellus in the hind wing of *Morpho hecuba* (Fig. 7.15B) to a doublet of point sources, common in *Euptychia* (Nymphalidae: Satyrinae), is readily obtained this way.

symmetry system in *Charaxes* and of the fragmentation of these bands in the discal cell of *Caligo* (Fig. 7.18), and we are probably dealing with variants of the same phenomenon in all these cases. Patterns such as these lead to the suppositions that the entire pattern, a band or an ocellus, can behave as a source or organizing center and that an initial central source can activate more-peripheral cells to become sources. If source cells in such a large field gradually turn off in random order, one could get the source distribution patterns shown in Figure 7.17. Clearly the explanation of these patterns will require additional assumptions in the model.

Fuzzy Boundaries, Sparse Patterns, and Heterozygotes

The boundaries between two fields of different colors are seldom sharp. Many scales in the vicinity of a boundary clearly make "mistakes" in their color determination. If we assume that these boundaries reflect the position of a threshold on a continuous gradient, then there must be a stochastic component in a scale cell's ability to sense the gradient or the threshold. Similar stochastic behavior is evident in sparse patterns, which are made up of a peppering of scales of a particular contrasting color, such as those shown in Figure 7.19. Sparse patterns usually correspond to specific pattern elements: They are found as intervenous stripes and as border ocelli. The distribution of colored scales in these patterns appears to be either random or overdispersed and often exhibits a density gradient (Fig. 7.19). Here too, then, we have evidence for a stochastic process at work in determining which scales will be of which color. A third case that leads to the supposition that stochastic processes are at work in pattern determination is found in the color patterns of interspecific or interracial hybrids. The pattern elements in which the two parental races differ are often present in the hybrids as poorly outlined shadow patterns, such as seen in the "sooty" yellow phenotype of hybrids of *Papilio glaucus* (Scriber and Evans, 1986) and in the hybrids of *Heliconius melpomene* (see illustrations in Turner, 1971a; in Sheppard et al., 1985; and in Mallet, 1989). The fuzzy boundaries of these hybrid patterns are not intermediate or graded in color but consist of a mixture of scales of the two colors characteristic of the two parental types.

Wherever we find gradients of color in the wing patterns of butterflies, they are composed of gradations in the proportions of

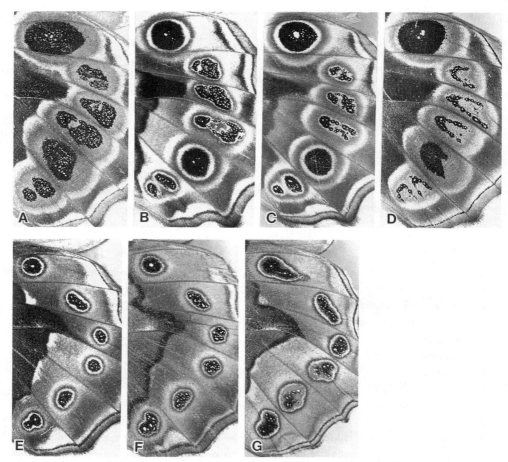

Figure 7.17. Fragmented ocelli in *Lethe* (Nymphalidae: Satyrinae). Pictured are several transformation series from normal ocelli with a single central focus to highly fragmented ocelli with multiple apparent foci. *A, Letha europa; B, L. drypetis; C, L. dyrta; D, L. arete; E* and *F, L. coelestelis; G, L. chandica.*

scales of different colors. In areas of the color pattern where intermediate values for a particular pigment pattern are specified, or where the color assignment is ambiguous, each scale cell evidently makes an all-or-none decision among the set of possible alternatives. In each scale cell, only the genes responsible for one of the possible pigment synthesis pathways become activated. If hybrids are heterozygous at the genes that specify a particular color, then the fuzzy patterns produced in such heterozygotes indicate that

either one or the other allele has become activated in each scale cell and that this occurred in random fashion. Perhaps we can extend this idea of allelic exclusion to explain fuzzy boundaries and sparse patterns as well. The graded scale distributions found in many sparse patterns suggest that the threshold mechanism specifying the boundary between pattern and ground may interact with a mechanism that selects the expression of one gene or the other.

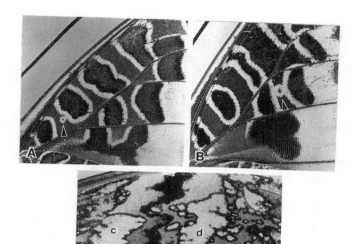

Figure 7.18. Satellite spots (*arrows*) in *Charaxes andara* (*A* and *B*) and fragmented bands in *Caligo beltrao* (*C*).

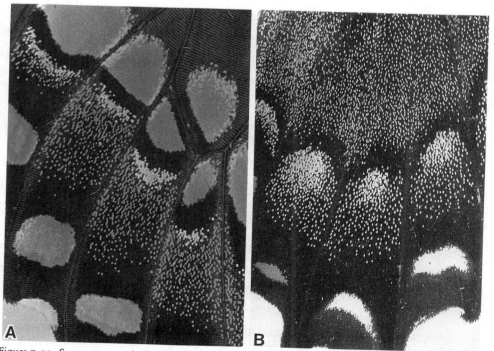

Figure 7.19. Sparse patterns in *Papilio polyxenes* (*A*) and *P. glaucus* (*B*).

The Cases of *Neita, Prepona,* and *Oxeoschistus*

Until now, we have assumed that each pattern element is determined independently from the others. Recall that this assumption emerges from the observation of independent variation and independent evolutionary diversification of pattern elements within a wing cell. In a few cases, however, evidence suggests that adjacent pattern elements in a wing cell may share one or more developmental determinants. In the African genus *Neita* (Satyrinae) and its sister genus, *Coenyropsis,* one of the wing cells of the hind wing bears an unusual fusion between parafocal element and ocellus (Fig. 7.20). The parafocal element (element i) extends as a narrow hairpin loop along the wing cell midline to the location where the focus for the ocellus would have occurred. Some specimens have a bulge at the end of the loop around this presumptive focus, and in yet other specimens this loop has been pinched off to form an oval ring or even a small eyespot. In the latter specimens the parafocal element is reduced and approximates the shape of its homologs in other wing cells. If the looped parafocal element marks one of the contours of a determination gradient, then the outermost rings of an ocellus must be determined by the same threshold as the parafocal element, and the two are therefore homologous. This implies also a similarity or identity in the gradients that specify the outer ocellus and parafocal element.

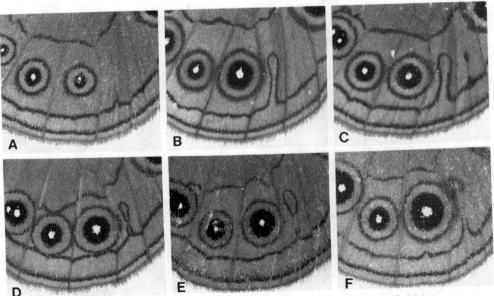

Figure 7.20. Pattern variability in the parafocal elements of the South African butterfly *Neita extensa* (Nymphalidae: Satyrinae). In some wing cells the parafocal element extends as a loop into the region where one would expect the center of an ocellus (*A–C*). This loop can be pinched off to form a small isolated ringlet (*D* and *E*) that may be homologous to the outer ring of a normal ocellus (*F*). Species in the closely related genera *Neocoenyra* and *Coenyropsis* have small but normal-looking ocelli in the location of this ringlet.

The outlines of an ocellus and of a parafocal element appear to be contours of the same gradient. This may help explain why these two pattern elements tend to become fused after temperature shock in many species (Figs. 5.19 and 5.20). If temperature shock either lowers the threshold or raises the level of one or both of the gradients that specify outer ocellus and parafocal element, then a fusion between the two elements, possibly with a slight alteration in shape, is the obvious consequence. Although this hypothesis explains fusion, it does not explain why the fused elements tend to be smaller and narrower than the normal pattern elements. A simple threshold shift would have exactly the opposite effect, and we must therefore conclude that the response to temperature shock, as well as the *Neita* morphocline shown in Figure 7.20, is due to an alteration in the shape of the gradients (in addition to any possible shift in threshold). Thus the *Neita* morphocline represents an orderly series of changes in the development of a gradient for which the pattern marks one of the contours.

The shape of the patterns found in *Neita* in fact corresponds remarkably well to the pattern progression in the reaction-diffusion model (Fig. 7.13), except that in *Neita* the pattern flares out near the margin of the wing. The ocelli and parafocal elements have shared determinants that are revealed by the *Neita* pattern morphocline (as they were by the observation that these two pattern elements often fuse in response to a temperature shock), but they also have unique determinants that are responsible for the inner structure of the ocellus and for the exact shape of the parafocal element. The model can account for the shared determinants as a consequence of the dynamics of source determination: A midline source can break up into several point sources; one of these can become the focus for an ocellus, and another can be-

come part of the determination system for the parafocal element. But to explain the origin of the unique determinants, we have to assume that there must be additional and independent source-determining events for each pattern element. We catch a glimpse here, again, of the possibility that source determination and pigment pattern determination may not always be singular processes and may not always be cleanly separable.

The case of *Neita* provides an introduction to a similar but somewhat more difficult pattern found in the Neotropical genus *Prepona*. The ventral forewing of most species of this genus bears a pattern that looks structurally simple and easy to describe but is difficult to explain within our general model. The pattern is variable; no two individuals are quite alike. Yet there is order, and as so many times before, the patterns can be arranged in a morphological series that suggests the progression of a determination process (Fig. 7.21). The identities of the various stripes that make up the pattern can be deduced from the patterns of primitive members of the genus (sometimes split off as a separate genus, *Archaeoprepona*), in particular those of *P. chalciope* (Fig. 7.22). The central symmetry system (elements **d** and **f**) is easy to identify, and the outermost band of the pattern must therefore belong to the border ocelli system. Presumably this band is homologous to the outer ring of the border ocelli (element **g**; its homology is best visualized by examining pattern element **g** of *Charaxes* in Figs. 3.12 and 3.13, a genus within the same tribe as *Prepona*). The most peculiar feature of this pattern, and apparently unique to *P. chalciope,* is a fusion between this outer ocellar band and the distal band of the central symmetry system near the posterior margin of the wing. This fusion implies a homology between the two bands and, therefore, a topological homology among the fields

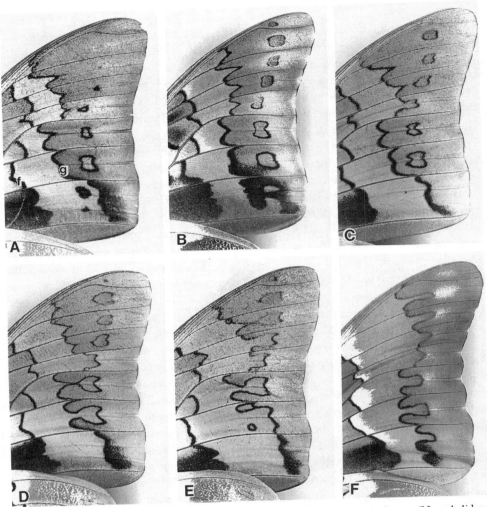

Figure 7.21. Various degrees of fusion between the ocellar ring and element **g** in *Prepona* (Nymphalidae: Charaxinae). *A* and *C–E*, *Prepona omphale; B*, *P. laertes; F*, *P. neoterpe.*

bounded by these bands. The central field of the central symmetry system is thus homologous to the field distal to the outermost band: the border ocellus field. This is the relation that Süffert (1929b) and Henke (1933b) saw in the patterns of many moths (and a few butterflies) and that led them to postulate the hierarchy of symmetry systems (Figs. 2.6–2.8).

The border ocelli of *Prepona* (element **h**) are simple ringlike patterns that generally have

the shape of a heart or an arrowhead (Fig. 7.21). In most cases, these ocelli lie free within the field of the border ocellus, but they are also often connected by a thin line of pigment to pattern element **g**. Occasionally this connection is broad, and the ocellus appears to be nothing more than a long loop of the **g** band. The field inside the ocellus is thus homologous to the fields between elements f and **g**, the "background" between the central symmetry system and border ocellar

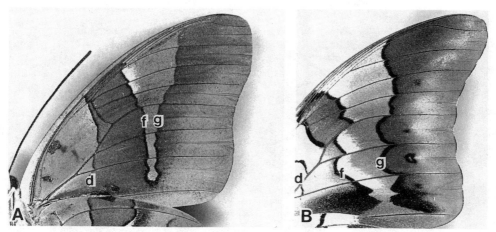

Figure 7.22. Fusion between elements f and g in *Prepona chalciope* (A), and the nearest pattern in *P. brooksiana* (B).

fields. By itself this would be a strange conclusion were it not for the strict parallel to the observations we just made on *Neita*. In *Neita* the central field of an ocellus is homologous to the field distal to the parafocal element, which we also tend to regard as the "background" on which the border ocelli (and central symmetry system) develop.

Thus we have two lines of evidence that the central field of a border ocellus is homologous to the background field outside the ocellar and central symmetry systems. It is necessary to note here that normal-looking ocelli do not occur in the ventral forewing of *Prepona*. They do occur in the ventral hind wing, however, where they look like standard eyespots, but the ventral hind wing pattern appears to be of no help whatever in shedding further light on the pattern relations of the forewing, except to confirm the identities of the various fields. We must therefore look at the patterns themselves for further enlightenment. If we interpret the pattern as a contour of the pattern-determining gradient, and the fields of the central symmetry system and border ocelli as higher than the back-

ground, then the ring-shaped ocelli must be interpreted as dips or valleys in the border ocellar field. The ring ocellus thus seems to have developed around a relative sink, just as appears to be the case in the hind wing of *Precis*. In any event, it is clear that in this case the edge of the border ocellar field (element **g**) and the ocellus itself (element **h**) are not produced by successively higher thresholds on a monotonic gradient, as might be expected if element **g** were merely the homolog of the outer ring of an ocellus.

The patterns of *Neita* and *Prepona* indicate that element **g** and the parafocal element (element **i**) form the proximal and distal boundaries of the border ocellar field (Figs. 2.1 and 2.17) and are therefore homologous bands just like those of the central symmetry system. Homology of bands in the central symmetry system is demonstrated by the fusion of distal and proximal bands to form closed arcs and rings (Fig. 2.5). An analogous fusion between elements **g** and **i** occurs in the Neotropical satyrine *Oxeoschistus pronax* (Fig. 7.23). The "inner" field of the border ocelli system is restricted to an arc parallel to

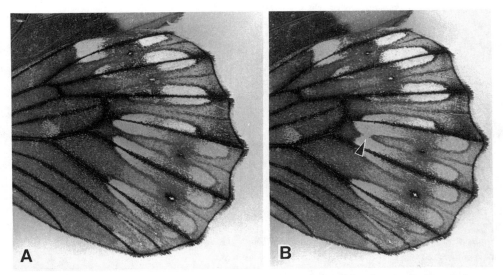

Figure 7.23. Ventral hind wing patterns of *Oxeoschistus pronax* (Nymphalidae: Satyrinae). *A*, typical pattern; *B*, unusual pattern with a break (*arrow*) in the intervenous stripe.

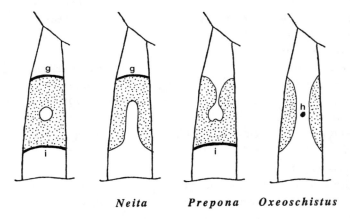

Neita   *Prepona*   *Oxeoschistus*

Figure 7.24. Summary diagram of the various modes of fusion among elements **g** and **i** in *Neita, Prepona,* and *Oxeoschistus.* At left is the ring-shaped pattern common to both *Neita* and *Prepona* and to most of the Nymphalidae. *Oxeoschistus* combines the fusions seen in *Neita* and *Prepona.* Homologous areas in all three taxa are stippled.

the wing veins. The homology of the field in which the focus of the ocellus develops to the "background" field is obvious from the pattern of *Oxeoschistus.* The *Oxeoschistus* pattern can thus be interpreted as the sum of the pattern variants of *Neita* and *Prepona* (Fig. 7.24). Fusions between bands that belong to different symmetry systems have also been observed in *Charaxes* (Nijhout and Wray, 1986), but here the fusions are between the basal symmetry system (element c) and the central symmetry system (element d). Figure 7.25 summarizes the various modes of fusion within and among symmetry systems and shows that such fusions provide yet another way of visualizing the relation between the elements of the nymphalid ground plan (top) and dependent patterns (venous and intervenous stripes, bottom).

If development of an ocellus is indepen-

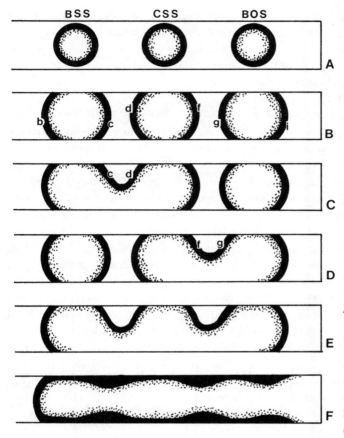

Figure 7.25. The various modes of fusion among the symmetry systems of the nymphalid ground plan. *A*, the simplest condition, in which basal symmetry system (*BSS*), central symmetry system (*CSS*), and border ocelli system (*BOS*) are represented by ring-shaped patterns in each wing cell. *B*, expansion of these rings gives the various elements of the ground plan, labeled by the letter code of Figure 2.17 and Table 2.1. *C*, fusion of elements c and d, common in *Charaxes* (Nijhout and Wray, 1986). *D*, fusion of elements f and g, seen in *Prepona*. *E*, hypothetical fusions that are possible between all symmetry systems. *F*, symmetrical version of *E*, showing what is essentially a venous-intervenous pattern. If each of the symmetry systems arises by virtue of inductive sources at their centers, then the expansion and fusion of those sources along the length of the wing cell produce the pattern in *F*. Transitions between *A*-like patterns and *F*-like patterns have been made often in butterfly pattern evolution, and Figure 7.13 shows how they might be achieved developmentally. Based on present knowledge, neither *A* nor *F*, nor any of the intermediates, can be assumed to be the primitive condition.

dent of the development of elements **g** and **i** (except that the ocellus must develop in the field between **g** and **i**), then how do we explain the loops and rings formed around the presumptive site of the ocellar focus in *Neita* and *Prepona,* respectively? A hypothesis consistent with all the preceding information is the following. Assume, as we have before, that a peak of activator produced by the reaction-diffusion system signals only a singularity that can be switched into either a source or a sink. If a sink is produced, then we get the ring-shaped patterns seen on the dorsal hind wing of *Neita* (and of *Precis*) and the ventral forewing of *Prepona.* But if a source is produced, then we get a normal eyespot as seen on the remainder of the hind wing of *Neita* (and the dorsal forewing of *Precis*) and the ventral hind wing of *Prepona.* This hypothesis has an appealing symmetry, besides helping to explain the puzzling case of *Precis* eyespot determination.

How do we reconcile the fact that elements **g, h,** and **i** vary independently from one another with the fact that they are members of a single symmetry system? A parallel situation actually exists in the central symmetry system. It has been recognized for a long time that the proximal and distal bands of the central symmetry system likewise vary independently of one another (Schwanwitsch, 1924, 1926; Süffert, 1929b; Chapter 3). We can assume, therefore, that the outer bands of both symmetry systems probably share determinants that characterize them as members of a symmetry system, but each band must also come under the influence of unique determinants that act to define its shape and color. The first determinant could be an *Ephestia*-like gradient system that sets an overall background level of morphogen with a symmetrical slope. The second determinant could be the source-sink distributions produced by the reaction-diffusion system. An

interaction between these two gradient systems could then establish a symmetrical arrangement of pattern elements, each with its own characteristic morphology and variability.

Secondary Symmetry Systems

A significant problem with a simple diffusion gradient model for central symmetry system determination is that in some species the bands of the central symmetry system are themselves symmetry systems. These are the secondary symmetry systems of Süffert (Fig. 2.8) and the complex bands of *Charaxes* (Fig. 3.14). In a diffusion gradient model of the kind we have been dealing with so far, the bands of a symmetry system correspond to contours of the gradient. The system is symmetrical because the gradient is symmetrical. But because each band is a contour on a locally monotonic (and therefore asymmetrical) gradient, it is clearly impossible for the band itself to be made up of a symmetrical arrangement of pigments. How then do we account for the origin of a secondary symmetry system within a band?

The simplest hypothesis is that in secondary symmetry systems the band itself has become a source of pattern determination. This hypothesis suggests that instead of turning on a pigment synthetic pathway at a given threshold in the gradient, what is turned on instead is the synthesis of a new diffusible substance. The actual pigment pattern will then be determined by thresholds on this new, secondary gradient. We have already considered the possibility that pattern elements of the nymphalid ground plan act as centers for the outward spread of the pigment pattern in deriving the color patterns of *Heliconius* (Chapter 3). Pattern variation within a species in *Heliconius* consists mostly of changes in the size of pattern elements. Round pattern elements, such as element **d** in *Idea* and *Heliconius,* can contract to a single point and expand to cover most of the wing cell. If this expansion is due to a change in threshold, then the gradient that gives rise to this pattern must have a hump, or local maximum, located at the center of expansion.

One way to produce such a maximum is by having a source at that point. Another way of getting a local hump in a concentration gradient is by rapidly destroying the gradient substance peripherally. In a wing cell this could happen if the wing veins became sinks in the presence of a high background concentration of the gradient material. Soon the concentration near the wing veins would drop, and a hump of high concentration would be left along the wing cell midline. Although this scenario might account for the apparent expansion and contraction of spots in *Heliconius,* it cannot account for the complex bands of *Charaxes,* because these bands run all the way to the wing veins. If the wing veins were sinks, the bands would always be rounded and have a portion of their contour parallel to the wing vein. A few of the complex bands in *Charaxes* are indeed closed and rounded off in this way (Fig. 3.14), but most are not. We are thus left with the hypothesis that there must be a source at the center of a complex band. If the source is small, a rounded figure will result that will look just like the rounded symmetry system in *Mantaria* (Fig. 2.5B). If the source is larger and runs from wing vein to wing vein, then an open secondary symmetry system will develop around it, which in *Charaxes* we recognize as a complex band.

Perhaps the hierarchy of symmetry systems (central symmetry system, flanked by basal and ocellar symmetry systems, with every band developed into a secondary symmetry system) evolved by a succession of steps like those hypothesized above. If this interpretation of pattern evolution is correct, it leads to the following generalization: Each contour on a gradient can be interpreted in two ways, as the prospective site of pigment synthesis and as the prospective site for a new source for pattern determination. The vast majority of butterflies do not have complex bands or secondary symmetry systems, so in most cases the pattern elements of the nymphalid ground plan are expressed directly as pigment patterns. But in a few cases, such as *Charaxes, Idea,* and *Heliconius,* some of the pattern elements become sources, and a qualitatively different type of pattern results.

Color patterns are images of a slice in time. They represent a frozen moment in two dynamic processes: development and evolution. The developmental processes that produce gradients, thresholds, and ultimately patterns change continually during ontogeny. At a given point in time, their state becomes fixed into a differentiated pattern of pigments. The parameters that control the dynamics and interactions of the processes change during evolution, and we observe the results of those altered parameters as a change in pattern. In this chapter we will try to reinforce the image we have built up so far of patterns as the product of a dynamic and interactive system, and we will examine the consequences of such dynamics for the study of morphological evolution. First we will look at some specific problems in color pattern evolution and suggest some of the ways in which one could interpret and study them. Then we will look at the problem of morphological evolution in a more general way. We will examine how changes in genes and development alter form, and we will find that genotype affects phenotype in a most indirect and nonintuitive way. In conclusion we will examine the consequences of the characteristics of this developmental system for interpreting and studying the evolution of processes and the patterns they produce.

Patterns as Adaptations

Most color patterns have a function. Those on the wings of butterflies serve many different functions. As butterflies rest with their wings closed, the ventral surface of the hind wing is exposed to the environment, and we find that most ventral hind wings have finely detailed patterns with a camouflaging effect. The ventral forewing, which is covered by the hind wing except at its tip, usually bears a bold

Chapter 8

Evolution of a Process

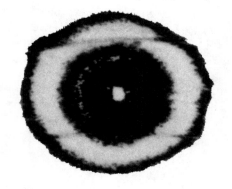

but poorly detailed pattern except at the tip; the pattern on the forewing tip is usually similar in color and detail to that of the hind wing. Hence, the exposed ventral surfaces of forewing and hind wing blend together to present an integrated pattern (Fig. 8.1). A close match between patterns on forewing and hind wing when wings are held in their resting position also occurs on the dorsal side of many species with a distinct striped pattern (Fig. 8.2), a phenomenon known as Oudemans' principle (Oudemans, 1903; Graham, 1950; Sibatani, 1987b).

Although many ventral patterns serve to camouflage, most do not mimic a specific background. Rather, a filigree of dark and light earth colors is used to produce general-purpose patterns that can blend into many of the substrates on which the butterfly might alight, such as bark, leaf litter, or soil. If this is the correct interpretation for the functional significance of such ventral color patterns, then camouflage as an adaptation requires only that the pattern be finely detailed and of a particular blend of colors; but there is no requirement that the elements of the pattern be of a specific shape. One of the consequences of this conclusion is that the detailed

A

B

Figure 8.1. Illustration of Oudemans' principle on the ventral wings of (*A*) *Morpho catenarius* (Nymphalidae: Morphinae) and (*B*) *Mantaria maculata* (Nymphalidae: Satyrinae). Wings on the left are mounted in the conventional way; wings on the right are mounted as they would be held at rest. The hind wing overlaps all but the tip and anterior margin of the forewing, which bears a pattern that fairly resembles that of the ventral hind wing. The portion of the forewing that is covered at rest often bears bright marks (as in *B*) that can be flashed, presumably to startle a potential predator.

Figure 8.2. Illustration of Oudemans' principle on the dorsal wings of (*A*) *Cyrestis camillus* (Nymphalidae: Limenitinae) and (*B*) *Papilio glaucus* (Papilionidae). The specimens at the top are mounted in the conventional way; those at the bottom are mounted with their wings as they would be held at rest. The stripes on forewing and hind wing form a continuous pattern when the wings are held at rest. The function of this overall coordination of the pattern is unknown.

shapes of the pattern elements have no adaptive function. But it is obvious that even the smallest pattern elements have fairly constant and species-specific shapes. We are thus faced with the dilemma of having to interpret the functional significance of the constancy and diversity of those shapes.

We can begin to develop such an interpretation by noting that the details of the ventral color pattern are often of a much finer scale than the details of the background they must blend into. This is probably why the arrangement of pattern elements is seldom random or evenly spread across the wing surface. Pattern elements are often arranged in rows or clusters and form coherent patches of color and texture, and it is the scale of these arrays of pattern elements that approximates

the scale of visual complexity (or visual noise) present in the animal's environment. It is difficult to see how the precise shape of, for instance, a parafocal element could contribute to the function of the overall pattern. Whether a parafocal element is W- or M-shaped (Fig. 7.5) scarcely has an effect on the appearance of the pattern, and it is difficult to conceive of an agent of natural selection that operates at that level of resolution. The infinitely varied details of the ventral patterns are unlikely to be the products of specific selection by external agents. It seems more likely that the butterflies have hit upon a developmental mechanism that can produce a pattern with a large variety of textures and colors, and that it is the general texture of the pattern that is under selection. The precise structural details of that texture are dictated by the developmental mechanisms of pattern formation.

Occasionally, as in the case of eyespots, there is (or has been) selection in favor of a specific morphology in one of the pattern elements. Such selection and morphological specialization then burden further evolution of that pattern element (Riedl, 1978), so that its evolutionary descendants tend to look like small ringlets and dots, even though such miniature ocelli have no specific function (other than to add to the general noisy texture of the wing pattern) in the species in which they occur. Put in the simplest terms, this hypothesis holds that the details of shape of many pattern elements often are the way they are because they are developmentally easy to make that way. The pattern that develops in the region of the border ocelli looks like a series of rings because developmental mechanisms tend to produce point sources for pattern induction in that region of the wing. As we saw in Chapter 7, the distribution of organizing centers for pattern formation is relatively simple and largely tied to preexisting structural features of the wing: the wing veins and the wing margin. The developmental mechanism that uses these organizing centers makes it possible to produce the filigreed patterns that are required for the textural effect. These patterns will have finely detailed, specific shapes, but those shapes are an incidental consequence of the developmental mechanism.

The regularity of many patterns, however, indicates that special mechanisms also exist that organize the larger structure of the wing pattern in many species. Oudemans' principle, pierellization (Fig. 3.2), and many other examples of the perfect alignment of nonhomologous pattern elements such as we see in *Kallima* (Fig. 3.7) are all manifestations of successful evolutionary attempts to bring elements of different and distant origins together into an integrated whole. In all these cases the pattern that has evolved gives the impression of visual continuity of lines and curves. In *Kallima* the selective pressure is easy to see, namely faithful leaf-vein mimicry. But in *Pierella* it is difficult to imagine what selective agent could possibly favor the perfect alignment of those thin lines of pattern. All we can say, in retrospect, is that the alignment is evidently important, or it would have been broken up as a consequence of genetic drift.

From a developmental perspective, the correlated development of homologous pattern elements in different wing cells could come about by the elimination of veins as barriers or sinks, so that homogeneous gradients can be set up across large areas of the wing. This could be accomplished easily by allowing gap junctions to persist between cells at the venous lacunae or by eliminating catabolism of the gradient substances, respectively. If this is what happens in pierellization, then we must also assume that the compartment boundary at vein M_3 remains in effect, because there is a sharp dislocation of the pattern across this vein. Alternatively, if the wing cells are not in communication with each other, it is possible to have homologous pattern elements perfectly lined up in adjacent wing cells if the processes of pattern formation proceed identically in both. In practice it may be possible to distinguish between these two alternatives by looking for small dislocations of homologs in adjoining wing cells in a large sample of animals. If two adjacent wing cells are not in communication, occasional misalignment of a stripe on either side of a venous boundary should be expected; but when adjacent wing cells form a single compartment, there should be no dislocation at all where a band crosses the intervening vein.

It is much more difficult to imagine what kind of developmental mechanism would ensure the perfect alignment of nonhomologous pattern elements. As we have seen, the distal band of the central symmetry system (element **f**) can become aligned with the proximal band of the central symmetry system (element **d**) as well as with the proximal band of the border ocelli system (element **g**), depending on the species. Distal and proximal bands are developmentally independent. If these bands indeed develop as contours on gradients with opposite slopes, then we cannot assume that the alignment is due to the loss of a venous boundary, because that would cause the gradients to blend and cancel each other, and one would not expect to get any pattern at all. Besides, the "transition" patterns found in some species of *Pierella* and in some specimens of *Kallima* show that we are dealing with a true dislocation of gradients along a compartment boundary. The critical question is whether some kind of developmental integration occurs between the compartments on either side of the dislocation to ensure alignment, or whether the alignment is due to independent convergence on particular absolute positions of pattern elements in each compartment, for instance, through a strong directional selection to move sources or contours to the appropriate locations. An analysis of correlation and covariation of elements on either side of the alignment boundary could be used to obtain evidence for integration.

Integration of forewing and hind wing patterns (Fig. 8.2) presents a parallel case. Such alignment clearly cannot be due to direct communication between the wings during development and must therefore be due to independent convergence of the two patterns to particular positions on each of the wings. Studies of the genetic correlations between aligned and nonaligned patterns on forewings and hind wings could provide evidence for convergent evolution of pattern position. Comparative study of the genetic variation of within-wing and between-wing pattern alignment within single species should provide a powerful tool for the analysis of pattern evolution at this level of organization.

Dorsal patterns have evolved independently from ventral patterns in the vast majority of species and are adapted for a wide variety of functions (Fig. 8.3; Plate 8). Sexual signaling, crypsis, thermoregulation, and aposematism and its mimicry are some of the better-known functions for dorsal patterns. In contrast to most ventral patterns, dorsal patterns tend to be bold and bright and have relatively little fine detail. The elements of the nymphalid ground plan are often not easily discernible in the dorsal pattern because of loss or fusion of pattern elements. However, the developmental mechanisms that generate the elements of the nymphalid ground plan (or a system of dependent patterns) clearly operate on the dorsal surface. Thus, the development and evolution of dorsal patterns must be explainable by the same mechanisms that operate on the ventral side. Integration of pattern by the alignment of homologous and nonhomologous pattern elements is as common on the dorsal as on the ventral surfaces and presents the same problems of interpretation.

Dorsal patterns that have a clear signaling function, such as those involved in aposematism and mimicry, and the many presumptive sexual signals of male butterflies, are almost invariably composed of large patches or bars of color without an easily definable shape and often without a sharp outline. The scale of these patterns is presumably some function of the level of resolution at which the recipient of the signal, a potential predator or mate, can detect and recognize the signal. Al-

Figure 8.3. Independent evolution of dorsal (*top*) and ventral (*bottom*) wing patterns. *A*, *Charaxes castor* (Nymphalidae: Charaxinae); *B*, *Baeotus baeotus* (Nymphalidae: Limenitinae).

though birds may be able to discern fine details of the pattern when a butterfly sits still, a butterfly on the wing presents a blur of color, and a finely detailed pattern would be useless as a signal. As for mate recognition, it is unlikely that the compound eye of butterflies can resolve any but the grossest features of the wing pattern of a potential mate or suitor. Thus selective pressures on the dorsal surface must often be in favor of patterns that are scaled to the resolution of an observer with a relatively poor spatial acuity.

The evolution of Müllerian and Batesian mimicry has occurred mostly on the dorsal surface (though many mimics have ventral patterns that are nearly identical to, but slightly vaguer than, the dorsal ones). Accurate mimicry is facilitated by the simplicity of the developmental-genetic toolbox that is used for pattern formation. In fact, it is difficult to distinguish pattern mimicry from convergent evolution, because often the same pattern elements are used in identical ways by both model and mimic. Because the required pattern alterations are relatively straightforward and within the repertoire of variability of many species, it is clear that mimicry does not present any unusually difficult developmental or evolutionary problems. That is not to say, however, that the evolution of mimicry is necessarily simple or easy. Prior divergence of two species results in the accumulation of many uniquely derived genetic and developmental traits in each, and these may well make it difficult to alter the pattern of one or the other in a particular direction. This, however, is a problem of directional selection in general, of which mimicry is but one type. The freedom with which butterfly species are able to shift, realign, hide, unmask, and alter the color and shape of their pattern elements suggests that in principle there is no pattern to which that of another species could not converge. In practice, there are limits, set by the genetics and development of pattern formation, that constrain the short-term evolutionary potential of many patterns and that may make the initial steps of convergence toward particular kinds of patterns (mimicking or other) unlikely to occur except through the accumulation of a large number of fortuitous mutations.

Whether or not a particular evolutionary path is open to a given species is determined by the kind of heritable variation present in its color pattern. Gradualistic evolution may be the norm for butterfly color patterns as it appears to be almost everywhere else, yet there are a remarkably large number of cases in which evolution has been erratic and in which there appears to have been saltation from one phenotype to another. *Heliconius* can clearly jump (and has done so frequently) among dramatically different phenotypes without graded intermediates and without prior variation in the "right" direction. In many taxa, singular species have patterns that deviate significantly from the norm of the taxon. Often this deviation is due to the loss or addition of one or two pattern elements or to their severe displacement or distortion (a few examples will make the point: in *Oxeoschistus*, *O. pronax* differs greatly from the pattern norm of the genus, in *Charaxes*, *C. analava* and *C. zingha;* in *Callicore*, *C. sorana;* in *Idea, I. blanchardii;* in *Precis, P. westermanii*). The adaptive value of these odd patterns is not obvious, so a gradualistic response to selection is difficult to imagine, and if they arose by drift, one would expect to find some intermediates and an increase in

variance in the outliers. Instead, we find singular aberrant species or small clusters of closely related species that produce their own variations on an aberrant theme (for instance, the *chlorochroa* group in *Eunica* and the *velutina* group in *Hamadryas*), and this provokes the strong suspicion that some pattern evolution may have gone in rather large and probably discontinuous steps. In the sections that follow, we will explore the general characteristics of gradualistic and discontinuous evolution of color patterns and discover possible reasons why erratic saltation may have occurred at various times during color pattern evolution.

The Evolution of Patterning Mechanisms

The many things that we have learned about pattern formation and pattern diversity in the previous chapters reveal color patterns as the end products of a fairly complex set of interacting processes. Changes in genes and environment can alter these processes and thereby lead to a change in pattern. If we want to understand the evolution of patterns, we need to understand both the nature and the evolution of the processes that generate them. This perspective on morphological evolution—the perception of a pattern as a snapshot of a dynamic process—has significant implications for how we study comparative morphology and for how we interpret the relation between genotype and phenotype. Many of the principles that are dealt with in the present chapter will apply to morphological evolution and the analysis of form in general, although we will continue to be specifically concerned with the evolution of color patterns. We will begin our discussion with a brief review of the basics of pattern formation and then explore a number of ways of dealing, conceptually and practically, with the evolution of the processes that give rise to the patterns.

The basic wing pattern is an array of three symmetry systems: basal symmetry system, central symmetry system, and border ocelli system. The various modes of fusion between the bands of these symmetry systems (Fig. 7.25) reveal that they are homologous or, perhaps more accurately, developmentally identical. The two bands that define a symmetry system can be fused to form a locally closed pattern. This we have assumed to be the simplest pattern element, an ocellus. The expansion of such an ocellus into a symmetry system provides an intuitively obvious demonstration of the homology of the two bands of a symmetry system. Adjoining bands that belong to different symmetry systems can also fuse (Figs. 7.22 and 7.25). Although this phenomenon is fairly rare, it nevertheless shows that the symmetry systems themselves are homologous. Thus, we may assume with a considerable degree of certainty that primitively all (six) bands of the three symmetry systems are homologous. This means that the three systems arise by identical processes, presumably around identical diffusion gradients.

This fundamental similarity of symmetry systems stands in sharp contrast to the results from comparative morphology and genetics, which show that each pattern element in each wing cell has its own characteristic and species-specific morphology and can change independently from other such pattern elements. Thus, although symmetry systems reveal a highly integrated and organized patterning system, genetics and comparative morphology reveal a mosaic system in which details of morphology are under a very locally restricted control. Changes at single genetic loci can turn individual pattern elements on and off or can change their position and color.

Although most genes (in *Heliconius,* at least) appear to affect entire ranks of serial homologs (thus the entire band of one of the symmetry systems), a few genes affect individual pattern elements in single wing cells, and their mere existence suggests that there may be more of them. After all, the wing pattern of almost every species is a testimony to the independent evolutionary potential of each wing cell.

How did such a system, with high-level order (in the symmetry systems) and local variability (in the individual pattern elements), evolve? Clearly what is needed is two separate developmental systems: one to set up the symmetry systems, and another to act locally on each band to specify its presence or absence, its form, and its precise position. Could it be that the two-step determination process—determination of sources followed by determination of pattern (for which we developed a model in Chapter 7)—is the remnant or, perhaps more appropriately, the derived version of these two developmental systems? It is easy to see that the source determination process could correspond to, or have evolved from, the original symmetry-determining system and that the pattern determination processes in each wing cell could correspond to the local form-determining system. Although we may never be able to discover the true evolutionary origin of this pattern formation mechanism, the scenario just presented at least forms an attractive working hypothesis whose implications for the evolution of patterns we can explore.

The fundamental similarity of symmetry systems provides a simple explanation for the lack of overlap between the elements of two adjacent systems. If the bands of each symmetry system are fundamentally similar and arise by identical processes (presumably involving identical gradients), then the elements of adjacent systems may fuse to form a single compound structure, but it is impossible to get an element from one system to intersect that of another, or to place an element of one system within the field of another. The local regulation of the shape of pattern elements implies that each pattern element is individually subject to selection and drift.

Having a two-stage pattern specification system means that two semi-independent modes of variation exist. There can be variation in source determination and variation in pattern determination. These two modes of variation interact, of course, because the process of source determination restricts sources to particular shapes and locations, and that constrains the types of patterns that can be subsequently induced by those sources. This interaction helps explain why different taxa among the butterflies adhere to the nymphalid ground plan with different degrees of faithfulness. A fully formed pattern resembling the nymphalid ground plan, with all its elements easily recognizable, requires the source determination process to have gone far toward completion. If source determination proceeds to a lesser degree, then there will be a tendency for sources to be elongated and for patterns to be concentrated along the wing cell midline. The relation of such patterns to the symmetry systems and to the nymphalid ground plan becomes increasingly difficult to visualize as patterns tend more and more toward intervenous stripes.

Simple symmetry systems that dominate the wing pattern, such as are found in many moths, are rare among the butterflies. If we assume that the simple geometrid symmetry patterns represent the primitive condition, then it seems fair to conclude that the butterflies as a group use a derived version of the original lepidopteran pattern formation system. For purposes of discussion in the remainder of this chapter, we will assume that

this derivative system closely resembles the model developed in Chapter 7.

From Genotype to Phenotype

Genes code for proteins. Proteins, in turn, interact with each other and with their cellular environment. This interaction gives rise to many of the biological properties of cells and tissues. Gene products therefore affect processes. Many gene products are enzymes that regulate the rates of biochemical reactions, others serve as signal receptors and signal transducers, and yet others provide the physical structure that supports or compartmentalizes the activities within and around a cell. Somehow, this fundamentally simple and restricted repertoire of gene products must be reconciled with the observation that changes in genes (mutations) often result in specific and stable changes in form, far down the chain of events. This correlation between changes in the genotype and changes in the phenotype leads us to assume that genes somehow control form. But the relation between genotype and phenotype, even in this relatively simple pigment patterning system, is far from direct.

Most of the reactions and interactions in the model we developed in Chapter 7 can be assumed to require enzymes and other proteins. Several different enzymes are required for the synthesis of most pigments. The threshold reaction (Figs. 7.2 and 7.3) is essentially an enzyme kinetics reaction. The reaction-diffusion system likewise requires enzymes to catalyze the synthesis, breakdown, and interaction of the reactants. The diffusion systems that set up the final two-gradient interaction likewise require enzymes for the synthesis, breakdown, and reaction of the gradient substances. It is not difficult to see that changes in any of these enzymes—

not just their inactivation but also small changes in the rate constants for the reactions they catalyze—could lead to changes in the positions of sources, the slopes of gradients, or the levels of thresholds.

The model provides an explicit system of interactions in which it is possible to trace the steps required to make a specific pattern and in which we can come to understand how genetic and environmental change alters pattern. A given pattern is not specified by any one of the variables in the model, nor is any one variable more, or less, responsible than any other for the characteristics of any given feature of the pattern. Most or all of the interacting variables in the model are necessary to produce a pattern. But we can isolate specific effects by observing how genetic change causes a change in the pattern. For instance, the *Fs* (forewing shutter) locus of *Heliconius melpomene* affects the position of pattern element h, the border ocellus homolog. Homozygosity for one allele, Fs^p, results in a more proximal placement of element h than homozygosity for the other allele, Fs^c. In the model, placement of a source is affected by the rate of breakdown of the inhibitor in the lateral inhibition reaction. All other things being equal, a higher rate of breakdown results in a more proximal source position. Thus the *Fs* gene could code for the enzyme that catabolizes the inhibitor, and the two alleles could simply code for enzymes with slightly different efficiencies. But source position is also affected by the relative rates of diffusion of the activator and the inhibitor, by the rate of activator breakdown, and by the rate at which activator is introduced into the wing at the veins. Changes in any one of these parameters will cause a change in the position of a source, whereas changes in more than one parameter can synergize or cancel this effect. Although the *Fs* gene may well code for a catabolic enzyme, it is clear that

this gene does not have exclusive control over the position of element h.

Along the same lines, we can analyze the possible mechanism by which locus *F* (fused) of *H. melpomene* affects the pattern. The dominant allele of this locus enhances the pattern of the wing veins so that the forewing band appears to be broken up into a series of spots. This effect could come about by an increase in the rate at which veins produce a gradient substance. But a similar effect could be produced by a decrease in the rate of breakdown of the gradient material or by a decrease in the threshold mechanism by which the local concentration of the gradient is interpreted. The dominance relation of the alleles suggests that the phenotypic effect is due to an increase in some function, not a decrease, so the first interpretation just given is more likely to be correct than the other two. This example once more shows that no single gene exercises exclusive control over a feature, even though a specific genetic change may always produce a specific change in pattern when all other developmental parameters are held constant.

In addition to using the model for interpreting the effects of a known gene, we can also use it to examine the consequences of hypothetical changes in the genotype. One example will suffice to illustrate the point. We could ask, What will happen to a pattern when a threshold is raised slightly? Such a change in a threshold could come about by a mutation in the gene that codes for the allosteric enzyme responsible for the sigmoidal shape of the reaction curve in Figure 7.2. Because the threshold defines one of the contours of a gradient as the pattern, raising the threshold causes a different contour to be selected. Examination of Figures 7.10 and 7.12, which have several contours drawn in, shows that a change in threshold will change both the shape and the position of the pattern. This tells us immediately that a change in the shape of a pattern element without a change in its position cannot be accomplished by a change in thresholds alone. If changes in thresholds are involved at all, then there must also be some compensatory change (like a decrease in the value of the gradient) that ensures that the pattern element does not move. Using this thought experiment as a prediction, one could then try to test a threshold hypothesis by measuring whether a genetic or evolutionary change in the shape of a given pattern element is accompanied by a shift in its position. One potential consequence of raising a threshold is the elimination of a portion of the pattern. The underlying gradient would still be there, but it would not be expressed if no portion of it is high enough to intersect the threshold. The loss of pattern elements has been a common feature of color pattern evolution. However, if the loss of a pattern element is a simple threshold effect, then this evolutionary step could easily be reversed by the introduction of alleles that lower the threshold or by mutations that raise the level of the gradient. Such mutations would give the impression of causing an atavism: the reappearance of a previously lost character.

These examples given show that small changes in the activity of a single gene can have complex effects on the phenotype, a form of pleiotropy. They also show that changes in several different kinds of genes can have either similar or exactly complementary effects on the phenotype. The interaction among the presumptive gene products within the model makes it clear that pleiotropy and epistasis (the nonadditive interaction of genetic effects so that the phenotypic consequences of changes at two or more loci simultaneously cannot be predicted on the basis of knowledge of the effects of changes at each locus by itself) should be the rule in the rela-

tionship between genes and form. Strict additivity of the phenotypic effects of genes required by theories of quantitative inheritance (Falconer, 1981; Wright, 1982) is possible, of course, but should not be expected to be the rule. The developmental model thus provides a realistic means for exploring the nature of genetic mechanisms that affect form and lead to functional integration of parts.

How Do Patterns Change during Evolution?

In Chapter 4 we developed the idea of a multidimensional morphospace of independent variables. If all morphological variables could be represented in such a morphospace, then the pattern of each species would be represented as a point in this space. Furthermore, one can visualize the morphological evolution of any species as a trajectory through this multidimensional space, and we can ask the question, What would such a trajectory look like? To answer that question, we need to examine the general characteristics of morphological evolution, and the relation between that morphospace and the pattern formation model of Chapter 7.

Two types of change are possible in morphological evolution: gradual and discontinuous. Gradual, or orthogenetic, change is easy to explain as a direct extension of microevolutionary theory (Futuyma, 1986). The accumulation, through natural selection or drift, of mutations with small effects leads to a gradual alteration of the characters affected by the genes in question. Discontinuous morphological evolution can have two very different causes. The first cause is the occurrence of mutations with large effects on the phenotype. Such mutations are generally believed to be rare, though they seem to have been the rule in pattern evolution in *Helicon-*

ius. The second way to get discontinuous change is through nonlinearities and bifurcations in the processes that lead to pattern, so that small continuous changes in one parameter lead to an abrupt change in another when a critical value is passed (Murray, 1981b, 1989; Oster and Alberch, 1982). Several examples of such bifurcations emerge from the model. In the lateral inhibition mechanism for source determination, the linear progress of time leads to a breakup of a long line source into one or more point sources. The time at which a point source pinches off and inhibits the line source in its vicinity is an example of a bifurcation that leads to the sudden appearance of a source with a different morphology from the one that existed before. The threshold mechanism is another example of a bifurcation; it causes an abrupt switch in gene activity at a critical point along the smoothly varying control parameter. Small changes of the control parameter near its critical point can cause the threshold to shift and even to disappear abruptly, with a concomitant abrupt change in the pattern. The trajectory of morphological evolution can thus be smoothly continuous through morphospace, and it can be discontinuous or saltatory. Presumably both modes of evolution occur at different times and in different regions of morphospace.

Each of the axes of our morphospace was chosen to represent variation in a single aspect of the pattern: the size of a spot, its position, the aspect ratio of an arc, a simple shape transformation. The underlying assumption in constructing such morphological axes is that if it is possible to break down the variation and diversity in the shape of a pattern element into all the components that vary independently from one species to another, then the variation represented on each axis is likely to be controlled independently

from the others. Presumably, independent variation means independent developmental mechanism and independent potential for evolutionary change. Considering the highly interactive nature of development expressed in the model, it is fair to ask how independent the axes of our morphospace really are. As we just noted, even a simple change in threshold that could be accomplished by altering a single genetic determinant can alter both shape and position of a pattern element (but only if the gradient is asymmetric). Yet these two features would be depicted as independent (perpendicular) axes of variation in morphospace because, a priori, it is reasonable to assume that shape and position are independent. In cases in which variation in shape is linked to changes in position (a developmental constraint on variation), the trajectory of morphological evolution cannot be parallel to either the shape axis or the position axis; the trajectory will be along some diagonal with simultaneous change in both features. We must conclude therefore that in some cases the axes of morphospace may not correspond to variation in independent genetic or developmental factors.

Many axes of morphospace unambiguously represent independent variation. The size of a pattern element that is produced by a symmetric gradient (an ocellus, for instance) can vary without altering its position or any aspect of its shape. Thus the size axis in this case could represent variation in a single genetic or developmental determinant. But there is a problem here too, because it is clear that the size of a pattern element is not determined by a single factor. Variation in size can be due to variation in threshold, source strength, relative timing of determination, and various environmental variables. Thus a size axis represents not an axis of variation in a single genetic or developmental determinant but only an independent axis of morphological variation. In any given species or population, variation along that axis is probably due to variation in a single determinant, but morphological evolution along that axis could be due to the accumulation of several different kinds of genetic or developmental changes.

Thus the evolutionary trajectory through pattern morphospace is sometimes constrained by the developmental mechanisms of pattern formation. Some types of change are easy to make and can be accomplished by many different mechanisms; other changes are difficult or impossible to make unless several processes change simultaneously. There may be "preferred zones" in morphospace that represent patterns that can be arrived at through a number of different developmental pathways, just as there are undoubtedly "forbidden zones" in morphospace that represent patterns that cannot be produced by any amount of alteration of parameters in the model. The evolutionary trajectory through morphospace is probably smooth and continuous most of the time in most species. We suspect this because it is almost always possible to construct a smooth series of morphoclines from the patterns of closely related species, and because hybrids between races of a species that differ in pattern often have patterns that are intermediates between those of the parents (for instance, the case of *Papilio memnon*, Fig. 6.7). The existence of smooth morphoclines between species and of intermediate morphologies in hybrids suggests that the genes responsible for these differences have additive effects on the development of the pattern and that pattern evolution in these cases follows a gradualistic track. The actual track of the trajectory is, of course, dictated by the action of natural selection and drift. But when a trajectory approaches a bifurcation in the process that generates pattern, it will disappear and immediately reappear elsewhere in morphospace.

The usefulness of thinking about morphological evolution as a trajectory through morphospace is limited by the complex relation between the developmental process and the resultant form. It is impossible to know from morphological data alone whether or not it is possible to get from one place to another. As a corollary, it is also impossible to deduce how extant patterns came to be where they are and the trajectory by which two patterns diverged from a common ancestor. Although it may usually be perfectly safe to assume a linear interpolation of shapes between ancestor and descendant (Bookstein, 1978, 1980), the point is that the morphology itself can give us no clue as to when it is indeed safe to assume so and when it is not. Yet we would like to be able to reconstruct the path that links ancestors to descendants as well as the paths by which species diverge from a common ancestor. That is after all the purpose of many studies in evolution: the reconstruction of the past.

Perhaps the resolution of this dilemma lies in understanding the evolution of developmental processes. Suppose we represent each parameter of the reaction-diffusion and two-gradient models and their respective threshold mechanisms (diffusion coefficients, rate constants, and so on), and time, by independent axes whose dimensions represent the entire range of variation in that parameter that result in pattern (for instance, a threshold value that is an order of magnitude higher than the peak concentration of any substance would be well outside the range of useful variation for our purposes). If these axes are arranged as an orthogonal coordinate system, then we have an analogue of morphospace;

we'll call it developmental space. Each point in this multidimensional developmental space corresponds to a unique combination of parameter values that, when substituted in the model, result in a specific morphology or pattern. Thus each point in developmental space corresponds to a point in morphospace. But the points in developmental space do not map evenly and homogeneously into morphospace. Several developmental points will more than likely map onto the same point in morphospace, and when all points in developmental space have been mapped, many empty regions will remain in morphospace.

At the population and species level, there will be a certain amount of variability in the value of each developmental parameter because different individuals bear alleles of genes that code for slightly more or less active enzymes. Each species therefore becomes represented by a cloud of points that occupies a small volume of a particular size and shape in developmental space and in morphospace. Evolution proceeds by the accumulation of mutations, which widen the species' volume in each of these multidimensional spaces, and by selection and drift, which narrow the distributions and shift their mean position in space. The link between development and evolution lies in the fact that the origin of diversity occurs in developmental space, whereas natural selection occurs in morphospace.

Perhaps the most important point to note in conclusion is that all the data presented in previous chapters (concerning development, comparative morphology, genetics, and so on) speak with a single voice: It is possible (and straightforward) to have both gradualistic change and saltational change within this system. In fact, numerous kinds of saltational changes in phenotype emerge from gradual changes in genotype, according to the model. In addition, the ability of different kinds of

parameter changes to produce similar phenotypic effects argues strongly that even if one kind of parameter change is constrained from causing a particular morphology, some other kind of parameter change could do so. Almost all changes can be compensated for by a different parameter in the model. Thus, although there may be a short-term constraint on the evolution of certain morphologies, this system, if we have interpreted everything correctly, is virtually free of constraint in the long term.

How Can We Study Pattern Evolution?

The path of morphological evolution could be reconstructed if it were possible to reconstruct the path of evolution in developmental space (we assume, of course, as we have done implicitly throughout this chapter, that we have no fossil record or other tangible evidence of the actual morphology of ancestors). In theory, it should be possible to use molecular genetic methods to identify the genes that account for each of the parameters in the developmental model and to compare their sequences and the relative biological activities of their products in different species. Differences in the accumulated mutations could then be used to reconstruct the enzymes of the ancestor. When the parameter values for those enzymes are substituted in the model, it should produce the pattern of the ancestor. Whether all this could be done in practice is another matter.

A more practical and immediately useful approach to the study of pattern evolution is obviously desirable. In practice, most studies of morphological evolution start with the assumption that evolution in the group under consideration has been gradualistic, or orthogenetic. Ancestors and descendants are assumed to be connected by all grades of inter-mediates. Even when direct evidence for such intermediates is unavailable (which is usually the case), it is assumed, usually implicitly, that they must have existed. A particularly common approach in studies of morphologi-cal evolution is to compare the form of two extant species and to assume that some kind of transformation between them represents evolution. This approach to morphological evolution derives its strong emotional appeal from the visually exciting Cartesian transfor-mations of fish and dinosaurs published by D'Arcy Thompson (1917). Most current morphometric methods for measuring mor-phology and for statistically analyzing the differences between forms (e.g., Bookstein, 1978; Benson et al., 1982; Bookstein et al., 1985) are inspired by the work of Thompson and extend his method in elegant ways. Comparing two extant forms, however, no matter how carefully and ingeniously done, can tell us little about their evolution unless one of the two has diverged but little from their common ancestor. If both forms have diverged significantly from a common ances-tor, then an analysis that assumes that one is derived from the other, or that one is a trans-formation of the other, would be misguided. Analysis of the morphometric differences be-tween extant forms is useful for taxonomic purposes, however, because it provides a way of quantifying morphology. Such a quantita-tive description of form, in turn, can provide an objective way of distinguishing among forms and classifying them according to their degree of similarity.

The differences between extant species are indeed due to evolution, but to evolution from a common ancestor that may not have looked like either of the two descendants. If we wish to study and understand morphologi-cal evolution, we must find a way of obtain-ing or reconstructing ancestors. This can be done by means of the logical methods of phy-logenetic systematics, worked out by Hennig (1965, 1966). Phylogenetic systematics at-tempts to identify monophyletic groups of species by first identifying primitive and de-rived traits among homologous structures and then using shared derived characters to identify clades (monophyletic groups) that share a common ancestor (Wiley, 1981). The characteristics of an ancestor can then be de-duced from the distribution of primitive and derived traits among its descendants. Carte-sian transformation methods (and their de-scendants) can be applied to the study of morphological evolution only when the char-acters to be compared are known to be prim-itive and derived versions of each other that diverged gradualistically, and when develop-mental data demonstrate that smooth and continuous processes give rise to each. Al-though this may often be true for morphol-ogy in general and color patterns in particu-lar, exceptions certainly exist, as we have just seen. Obviously it is desirable to develop an analytical approach to the study of pattern evolution that does not depend on assump-tions of gradualism and continuity but can accommodate both gradualism and saltation.

To develop a practical way of studying pat-tern evolution, one could proceed as follows. First it is necessary to select a monophyletic taxon, with a number of morphologically similar and a few morphologically dissimilar species. We could, for instance, pick any of the taxa whose morphospace we explored in Chapter 4. We could then try to deduce the model parameters necessary to generate each of these patterns. This is not necessarily an easy task, but one could begin by trying to model the pattern of one species precisely and then see if small changes in parameters can give rise to the patterns of related species. Each species thus becomes identified by a list of parameter values that encode its pattern in a developmentally meaningful way. These parameter values will allow us to visualize the distribution of species in developmental space. We can then use phylogenetic systematic methods on the parameter value data to reconstruct the parameter values for an ancestor (without regard to morphology), and we can use morphological data from pattern morphoclines to reconstruct the morphology of the same ancestor. The interesting question then is, Do the parameter values of the ancestor, when substituted into the developmental model, produce the pattern predicted from strictly morphological considerations? If evolution is gradual, as it likely to have been in the derivation of two fairly similar-looking sister patterns from their immediate ancestor, then the answer to this question should be yes. If a cluster of similar-looking patterns can be analyzed in this way, then it should be possible to use the model as an exploratory device to see whether progressive alterations of parameter values result in a gradualistic or saltatory change to the more dissimilar species of the clade.

The foregoing scenario assumes, of course, that the model is largely a correct representation of pattern formation in butterflies and that it is possible to identify the primitive and derived states of all characters correctly. This kind of phylogenetic approach, in contrast to a two-taxon morphometric comparison, is the only way in which the evolution of developmental processes and the evolution of developmental patterns can be critically addressed. Furthermore, other questions of similarity—such as between pattern and process, between pattern elements at different levels of homology, and between evolutionarily convergent phenotypes in mimicry—are most effectively addressed by such a phylogenetic approach, and it is the only method that is likely to give deep insight into morphological evolution in general.

As a rule, attempts at phylogenetic systematics have yielded multiple, equally parsimonious trees, and two authors seldom agree on the details of a phylogeny. The references accompanying the phylogenies below should be consulted for caveats and alternative schemes.

Appendix A

Classification and Systematics of the Butterflies

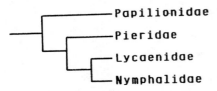

Figure A.1. Phylogeny of the families of butterflies. (After Ackery, 1984)

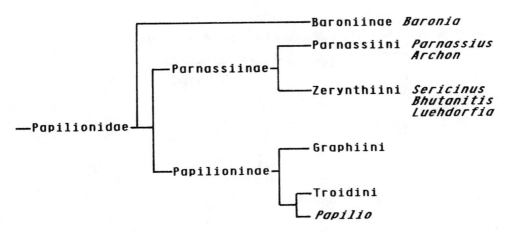

Figure A.2. Phylogeny of the tribes of Papilionidae, with principal genera in the Parnassiini. The genus *Papilio,* with more than one third of the species of Papilionidae, forms a yet unsatisfactorily resolved taxon. (After Miller, 1987)

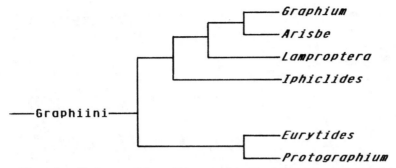

Figure A.3. Phylogeny of the major genera of the Graphiini. (After Miller, 1987)

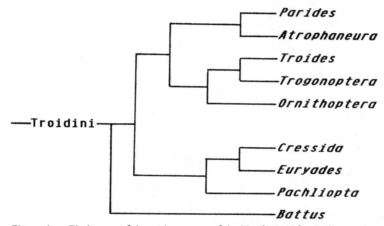

Figure A.4. Phylogeny of the major genera of the Troidini. (After Miller, 1987, and Munroe and Ehrlich, 1960)

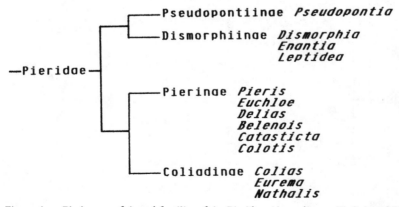

Figure A.5. Phylogeny of the subfamilies of the Pieridae. (According to Ehrlich and Ehrlich, 1967)

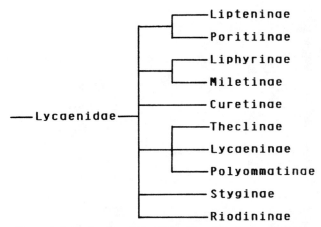

Figure A.6. Phylogeny of the subfamilies of the Lycaenidae. (According to Ackery, 1984)

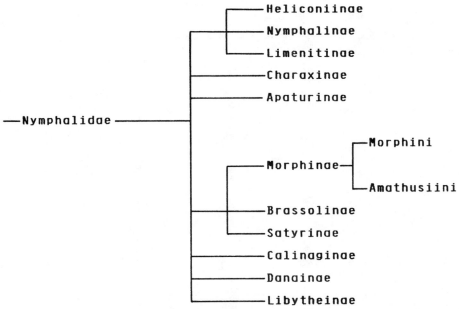

Figure A.7. Phylogeny of the subfamilies of the Nymphalidae. The phylogenetic relations among the subfamilies and tribes remain largely unresolved. (After Ackery, 1984, and based on the classification of Harvey, in Appendix B, which presents a tribe- and genus-level classification of the Nymphalidae)

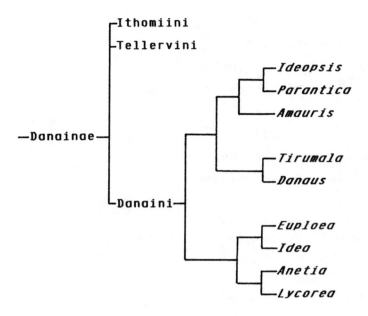

Figure A.8. Phylogeny of the tribes and principal genera of the Danainae. (After Ackery, 1984, and Ackery and Vane-Wright, 1984)

Higher Classification of the Nymphalidae

The Nymphalidae contains about 6,452 species (Shields, 1989) and is perhaps the most biologically well-studied family of Lepidoptera. The systematic relationships within the Nymphalidae, however, are poorly understood. This is particularly true of the subfamily Nymphalinae sensu Ehrlich (1958), especially at the generic and tribal levels. To date, no classification has been proposed that treats the genera on a worldwide basis. I propose a new higher classification of the Nymphalidae (especially the genera of Nymphalinae sensu Ehrlich) that recognizes, to the extent currently possible, monophyletic higher taxa. I base my classification on a synthesis of systematic studies and character information available in the literature and on my own unpublished research. One result of this new classification is that the taxonomic rank and content of some higher taxa are changed. For example, genera of Ehrlich's Nymphalinae are placed in four subfamilies (Heliconiinae, Nymphalinae, Limenitidinae, and Apaturinae), and other familiar groups (Acraeinae, Heliconiinae, and Melitaeinae of authors) are dropped in rank. These changes are necessary to show more clearly the phylogenetic relationships between taxa. (To avoid confusion, future references to higher taxa will refer to the groups as defined here, unless otherwise noted.)

Ackery (1988), in an excellent review of the larval host plants of nymphalid butterflies, presented a classification that differed from that of both Ehrlich (1958) and Ackery (1984). The main differences between the classification proposed here and that of Ackery (1988), excluding the fact that genera without known hosts were usually omitted from his review, can be summarized as follows: The Heliconiinae, with four tribes, includes his Acraeinae, Heliconiinae, and Argynninae; the Nymphalinae, with three tribes, includes his Nymphalinae (after removal of his Coloburini) and Melitaeinae; the Limenitidinae includes his Limenitinae plus Coloburini, with considerable differences in tribal arrangement; and the Morphinae includes his Morphinae and Amathusiinae.

My classification owes a great debt to the work of Wilhelm Müller (1886). Müller reared more than 60 species of South American Nymphalidae

Donald J. Harvey
Department of Entomology
National Museum of Natural History
Smithsonian Institution

and carefully studied and described the behavior and morphology of their immature stages. He then produced a hierarchical classification based on comparative larval morphology. (Curiously, although recognizing natural groups, he never proposed taxonomic changes in accord with his results.) The wealth of character information in Müller (1886) has remained largely untapped, and his often brilliant insights on nymphalid morphology and relationships have been ignored by virtually all subsequent authors. I hope that this present work will, to some extent, make up for more than a century of neglect.

This work is divided into three sections. The first, Table B.1, reproduces Müller's (1886) classification, with the nomenclature of included genera brought up to date. The second, Table B.2, presents my classification of the Nymphalidae. Each taxonomic group in Table B.2 is referenced to the third section—notes that provide access to taxonomic literature (see also references in Ackery, 1984), address problems of nomenclature, and often discuss character state distributions supporting monophyly of that taxon. Note 1 introduces a character system of mature larvae that is useful in determining relationships.

For the family-group names in this classification, I employ the endings "-inae," "-ini," and "-iti" for subfamilies, tribes, and subtribes, respectively. I use the oldest available family-group name for each category in the hierarchy, based on the list compiled by Bridges (1985).

Some comment on the choice of generic names included in Table B.2 is in order, because the number of available generic names (see Hemming, 1967) greatly exceeds the number used. I follow the most recent taxonomic treatment for groups that have been revised on a worldwide basis (see "Notes"). In groups for which no such works are available, my choice of generic names has been of necessity somewhat arbitrary. Inclusion of a generic name does not necessarily reflect my advocacy of its use—I have tried to err on the side of "splitting" to increase the number of names that are placed systematically.

Although I believe that this classification is an improvement over previous ones, many problems remain. The relationships of the genera of Heliconiini are unresolved. There is no evidence for the monophyly of the Limenitidinae, although at least two of its component tribes (i.e., Limenitidini and Biblidini) appear natural. The Satyrinae and its component groups also need further study. Finally, the relationships among most of the subfamilies are still obscure. (Ehrlich, 1958, and Scott, 1985, proposed phylogenies of the Nymphalidae, but the contents of their subfamilies differ from those used here.)

Adult specimens used in this study are in the collections of the Allyn Museum of Entomology, Florida State Museum, Sarasota (AME); Carnegie Museum of Natural History, Pittsburgh (CMNH); and the National Museum of Natural History, Smithsonian Institution, Washington, D.C. (USNM). Larval specimens examined (including representatives of more than 70 nymphalid genera) are from the collections of AME, CMNH, and USNM and the private collections of Roy Kendall, John Rawlins, and the author.

I thank Fred Nijhout for the opportunity to publish this classification in his book, Lee and Jacqueline Miller for unlimited access to the collections of the Allyn Museum in 1980 and 1981 and during subsequent visits, John Rawlins for long discussions on systematics, and Gerardo Lamas for help with literature. Drafts of the manuscript profited from comments by Tim Friedlander, Lee and Jacqueline Miller, John Rawlins, Bob Robbins, and Adrienne Venables.

Table B.1

Müller's (1886) system of the Nymphalidae

I. [Not named]
  A. Acraeinae: *Acraea* [*Actinote*]
  B. Heliconiinae: *Heliconius* [*Heliconius, Laparus*], *Eueides, Colaenis* [*Philaethria, Dryas*], *Dione* [*Agraulis*]
  C. [Not named]: *Argynnis* [*Argynnis, Issoria*], *Cethosia*
II. [Not named]
  A. Vanessinae
    a. *Hypanartia*
    b. *Pyrameis* [*Cynthia*], *Vanessa, Grapta* [*Polygonia*]
  B. Diademinae
    a. *Phyciodes* [*Eresia*], *Melitaea* [*Eurodryas*]
    b. *Victorina* [*Siproeta*], *Anartia, Junonia, Doleschallia, Precis, Hypolimnas*
III. [Not named]
  A. *Gynaecia* [*Colobura*], *Smyrna*
  B. *Ageronia* [*Hamadryas*], *Ectima*
  C. Epicaliinae
    a. *Myscelia, Catonephele, Eunica*
    b. *Temenis, Pyrrhogyra, Callicore, Haematera, Catagramma* [*Callicore*]
  D. *Dynamine*
  E. *Didonis* [*Biblis*]
  F. Adelphinae: *Athyma, Adelpha, Limenitis, Neptis*
IV. [Not named]
  A. *Prepona, Agrias, Siderone*
  B. *Anaea, Hypna, Protogonius* [*Consul*]
  C. *Nymphalis* [*Charaxes*]
[Uncertain "*zweifelhafter*" placement, = Apaturinae]:
  Apatura [*Doxocopa*], *Thaleropis*
Brassolinae:
  Opsiphanes, Dynastor, Caligo, Narope, Brassolis
Morphinae:
  Morpho, Amathusia, Discophora
Satyrinae:
  Pedaliodes, Taygetis, Euptychia, Antirrhea
Danainae:
  Danaus, Dircenna, Ceratinia, Ithomia, Thyridia [*Methona*], *Mechanitis*

Note: Generic names used by Müller are followed by modern equivalents, in brackets, if they differ. No numbers were employed by Müller beyond his group IV.

Table B.2
Revised classification of the Nymphalidae

NYMPHALIDAE[1]
- I. HELICONIINAE[2]
  - A. Pardopsini[3]
    - *Pardopsis*
  - B. Acraeini[4]*
    - *Acraea*
    - *Actinote*
  - C. Cethosiini[5]*
    - *Cethosia*
  - D. Heliconiini[6]
    - 1. Heliconiiti
      - *Philaethria*
      - *Podotricha*
      - *Dione*
      - *Agraulis*
      - *Dryadula*
      - *Dryas*
      - *Eueides*
      - *Neruda*
      - *Laparus*
      - *Heliconius*
    - Subtribe uncertain
      - *Vindula*
      - *Cirrochroa*
      - *Paduca*
      - *Terinos*
      - *Lachnoptera*
      - *Smerina*
      - *Cupha*
      - *Vagrans*
      - *Phalanta*
      - *Euptoieta*
    - 2. Argynniti
      - *Argyreus*
      - *Damora*
      - *Childrena*
      - *Pandoriana*
      - *Argynnis*
      - *Fabriciana*
      - *Speyeria*
      - *Mesoacidalia*
      - *Brenthis*
      - *Issoria*
      - *Argyronome*
      - *Nephargynnis*
    - 3. Boloriiti
      - *Boloria*
      - *Clossiana*
      - *Proclossiana*
    - Subtribe uncertain
      - *Yramea*
      - *Kuekenthaliella*
      - *Prokuekenthaliella*

- II. NYMPHALINAE[7]
  - A. Nymphalini[8]
    - *Vanessa*
    - *Cynthia*
    - *Bassaris*
    - *Aglais*
    - *Inachis*
    - *Nymphalis*
    - *Polygonia*
    - *Kaniska*
    - *Antanartia*
    - *Hypanartia*
    - *Symbrenthia*
    - *Mynes*
    - *Araschnia*
  - B. Kallimini[9]
    - *Anartia*
    - *Metamorpha*
    - *Siproeta*
    - *Napeocles*
    - *Precis*
    - *Junonia*
    - *Hypolimnas*
    - *Salamis*
    - *Yoma*
    - *Kallima*
    - *Kallimoides*
    - *Doleschallia*
    - *Amnosia*
    - *Vanessula*
    - *Catacroptera*
    - *Rhinopalpa*
  - C. Melitaeini[10]
    - 1. Euphydryiti
      - *Euphydryas*
      - *Hypodryas*
      - *Occidryas*
      - *Eurodryas*
    - 2. Melitaeiti
      - *Mellicta*
      - *Melitaea*
      - *Poladryas*
      - *Didymaeformia*
      - *Chlosyne*
      - *Thessalia*
      - *Texola*
      - *Dymasia*
      - *Microtia*
      - *Gnathotriche*
      - *Gnathotrusia*
      - *Higginsius*
      - *Antillea*
    - 3. Phycioditi
      - *Phyciodes*

  - *Phystis*
  - *Anthanassa*
  - *Dagon*
  - *Telenassa*
  - *Ortilia*
  - *Tisonia*
  - *Tegosa*
  - *Eresia*
  - *Castilia*
  - *Janatella*
  - *Mazia*
  - Subtribe uncertain
    - *Atlantea*
- III. LIMENITIDINAE[11]
  - A. Coloburini[12]
    - *Historis*
    - *Coea*
    - *Pycina*
    - *Baeotus*
    - *Smyrna*
    - *Colobura*
    - *Tigridia*
  - B. Biblidini[13]
    - 1. Bibliditi
      - *Biblis*
    - 2. Euryteliti
      - *Mesoxantha*
      - *Ariadne*
      - *Eurytela*
      - *Neptidopsis*
      - *Laringa*
      - *Byblia*
      - *Mestra*
      - *Archimestra*
      - *Vila*
    - 3. Epicaliiti
      - *Myscelia*
      - *Catonephele*
      - *Nessaea*
      - *Cybdelis*
      - *Eunica*
      - *Libythina*
      - *Sallya*
    - 4. Ageroniiti
      - *Hamadryas*
      - *Ectima*
      - *Panacea*
      - *Batesia*
    - 5. Epiphiliti
      - *Asterope*
      - *Pyrrhogyra*
      - *Temenis*
      - *Epiphile*
      - *Lucinia*

  - *Bolboneura*
  - *Nica*
  - *Peria*
  - 6. Dynaminiti
    - *Dynamine*
  - 7. Catagrammiti
    - *Diaethria*
    - *Perisama*
    - *Antigonis*
    - *Cyclogramma*
    - *Haematera*
    - *Paulogramma*
    - *Callicore*
    - *Catacore*
  - C. Limenitidini[14]
    - 1. Limenitiditi
      - *Limenitis*
      - *Parasarpa*
      - *Moduza*
      - *Sumalia*
      - *Auzakia*
      - *Adelpha*
      - *Basilarchia*
      - *Ladoga*
      - *Athyma*
      - *Pandita*
      - *Parathyma*
      - *Tacoraea*
      - *Cymothoe*
      - *Harma*
      - *Pseudacraea*
      - *Pseudoneptis*
    - 2. Neptiti
      - *Pantoporia*
      - *Lasippa*
      - *Neptis*
      - *Phaedyma*
      - *Aldania*
    - 3. Partheniti
      - *Parthenos*
      - *Lebadea*
    - 4. Euthaliiti
      - *Tanaecia*
      - *Euthalia*
      - *Neurosigma*
      - *Abrota*
      - *Dophla*
      - *Lexias*
      - *Bassarona*
      - *Catuna*
      - *Hamanumida*
      - *Aterica*
      - *Cynandra*
      - *Euryphura*
      - *Pseudargynnis*

*See "Addendum" in "Notes to Table B.2."

Diestogyna
Euphaedra
Euriphene
Bebearia
D. Cyrestidini[15]
Marpesia
Cyrestis
Chersonesia
Dichorragia
Incertae sedis[16]
Pseudergolis
Stibochiona
IV. CHARAXINAE[17]
A. Charaxini
Polyura
Murwareda
Charaxes
Haridra
Zingha
Stonehamia
Eriboea
B. Euxanthini
Euxanthe
Godartia
Hypomelaena
C. Pallini
Palla
D. Prothoini
Agatasa
Prothoe
E. Preponini
Anaeomorpha
Noreppa
Archaeoprepona
Prepona
Agrias
F. Anaeini
1. Zaretiditi
Coenophlebia
Zaretis
Siderone
2. Anaeiti
Hypna
Anaea
Polygrapha
Consul
Cymatogramma
Fountainea
Memphis
V. APATURINAE[18]
Apatura
Apaturina
Apaturopsis
Asterocampa
Chitoria

Dilipa
Doxocopa
Euapatura
Eulaceura
Euripus
Helcyra
Herona
Hestina
Hestinalis
Mimathyma
Rohana
Sasakia
Sephisa
Timelaea
Thaleropis
VI. MORPHINAE[19]
A. Morphini[20]
1. Antirrheiti
Caerois
Antirrhea
2. Morphiti
Morpho
B. Amathusiini[21]
Faunis
Aemona
Xanthotaenia
Taenaris
Morphotenaris
Stichophthalma
Amathusia
Amathuxidia
Zeuxidia
Thaumantis
Thauria
Discophora
Enispe
VII. BRASSOLINAE[22]
Brassolis
Orobrassolis
Catoblepia
Dynastor
Opoptera
Mimoblepia
Opsiphanes
Mielkella
Penetes
Selenophanes
Caligo
Eryphanis
Dasyophthalma
Narope
Aponarope
VIII. SATYRINAE[23]
A. Haeterini
B. Biini

1. Biiti
2. Melanititi
C. Elymniini
1. Lethiti
2. Zetheriti
3. Elymniiti
4. Mycalesiti
D. Eritini
E. Ragadiini
F. Satyrini
1. Hypocystiti
2. Ypthimiti
3. Euptychiiti
4. Coenonymphiti
5. Manioliti
6. Erebiiti
7. Diriti
8. Pronophiliti
9. Satyriti
10. Melanargiiti
IX. CALINAGINAE[24]
Calinaga
X. DANAINAE[25, 26]
A. Danaini
1. Amauriti
Parantica
Ideopsis
Amauris
2. Danaiti
Tirumala
Danaus
B. Euploeini
1. Euploeiti
Euploea
Idea
Protoploea
2. Ituniti
Lycorea
Anetia
XI. TELLERVINAE[27]
Tellervo
XII. ITHOMIINAE[28]
A. Tithoreini
Elzunia
Tithorea
B. (New tribe)
Aeria
C. (New tribe)
Roswellia
Athesis
Patricia
D. Methonini
Methona
E. (New tribe)
Placidula

F. Melinaeini
Athyrtis
Olyras
Eutresis
Melinaea
G. Mechanitini
Paititia
Thyridia
Sais
Scada
Forbestra
Mechanitis
H. Oleriini
Hyposcada
(New genus)
Oleria
(New genus)
I. Napeogenini
Epityches
Rhodussa
Aremfoxia
Napeogenes
Garsauritis
Hyalyris
Hypothyris
J. Ithomiini
Miraleria
Pagyris
Ithomia
K. Dircennini
Callithomia
Velamysta
(New genus)
Dircenna
Hyalenna
L. (New tribe)
Ceratinia
Ceratiscada
Prittwitzia
Episcada
Pteronymia
M. Godyridini
Godyris
Dygoris
Pseudoscada
Hypomenitis
Greta
Hypoleria
Mcclungia
Veladyris
Heterosais
XIII. LIBYTHEINAE[29]
Libythea
Libytheana

Notes to Table B.2

1. Nymphalidae

In this note, I briefly describe a character system of mature larval morphology—filiform setae—that has proven useful in defining groups within the Nymphalidae. Filiform setae, when present on a given body segment of mature larvae, can be distinguished by their morphology and position. Morphologically, filiform setae are usually extremely thin relative to other setae, and they arise from sockets that are modified so that the setae are extremely flexible. Presumably, filiform setae respond to sound (Markl and Tautz, 1975; Tautz, 1977). The position of these setae on a given segment, when they are present, is relatively constant among nymphalid subfamilies. A full segmental complement of filiform setae is present in larvae of Nymphalini and Melitaeini and is distributed as follows (a pair of setae—one seta on each side—will be referred to in the singular). On the prothorax (T1), there is one seta just anterior to the spiracle and a second one in a slightly more dorsal position; these will be referred to as the anterior seta, T1(A), and the posterior seta, T1(P), respectively. On the mesothorax and the metathorax (T2 and T3), filiform setae are laterodorsal. On abdominal segments 1 through 8 (A1–A8), filiform setae occur just ventral and posterior to the spiracle. The filiform seta on A9 is slightly more dorsal than on preceding segments. To my knowledge, these setae have been mentioned in Nymphalidae only by Garcia-Barros (1989), who illustrates their position on the mature larva of *Hipparchia fidia* (Satyrinae), where they occur on T1(P), T2, T3, A5, A6, and A9. (Hereafter, series of segments describing the distribution of filiform setae will use the letters *T* and *A* only on first mention.)

Filiform setae occur in the same positions on larvae of other lepidopteran families. The noctuid moth *Mamestra brassicae* has four filiform setae on the thorax, corresponding to T1(A,P)–3, which respond to sound (Markl and Tautz, 1975; Tautz, 1977). A single seta, T1(P), occurs on the lateral margin of the prothoracic shield of Lycaenidae and Riodinidae (Lawrence and Downey, 1966 [as "a"]; Downey and Allyn, 1979, 1984; Wright, 1983

[as "XD2"]; Ballmer and Pratt, 1989; Harvey and Gilbert, n.d.; Harvey et al., n.d.). In *Parnassius* (Papilionidae: Parnassiinae), filiform setae occur on T1(P),2 and A5–7; and in *Papilio, Pterourus,* and *Heraclides* (Papilioninae: Papilionini) and in *Eurytides* (Papilioninae: Leptocircini), they occur on T1(P),2 and A4–7 (Harvey, unpub. data). Toliver (1987b) noted "primary seta . . . weakly present (L1? on A3–7)" on mature larvae of *Papilio* (*Heraclides*) *cresphontes*. These correspond to what I call filiform setae, although I did not find them on A3 in this species. In the Pieridae, I was unable to distinguish filiform setae in most species but found that *Colias cesonia* had setae on T1(A,P),2 and A10.

Within the Nymphalidae, I have examined mature larvae of representatives of all subfamilies recognized here (except Calinaginae and Tellervinae) for the distribution of filiform setae. (Nymphalid larvae differ from other papilionoid families in the presence of a filiform seta on A9.) I comment on these setae in notes 2, 7–10, 23, and 25, when their presence or absence is restricted to a certain group or when their position (i.e., arising from the sclerotized base of a scolus rather than from the surface of the larval body) is unique.

I interpret filiform setae on different segments as being serial homologs, even though their position differs on the thorax and the abdomen. The topological shift between T3 and A1 is probably related to the different structure of these segments: T3 has true legs and lacks spiracles, whereas A1 lacks true legs but has spiracles. The unique morphology of filiform setae indicates that they share an identical, complex developmental pathway, which is much stronger evidence for homology than position (see Roth, 1984).

It is possible to trace filiform setae back through larval ontogeny from the last to the first instar. In the Lycaenidae and Riodinidae, the single filiform seta is morphologically different from other setae of the same instar and always occurs in the same relative position on the prothoracic shield (Downey and Allyn, 1979; Wright, 1983; Ballmer and Pratt, 1989; Harvey, unpub. data). In the Nymphalidae, filiform setae are less differentiated in earlier instars and may or may not be morphologically differentiated in the first instar, but their position on a given segment does not change. For these reasons, I argue that filiform setae are serially homologous primary setae (sensu Stehr, 1987) that are retained throughout larval development. The filiform setae of Nymphalidae on T2,3 and A1–8 correspond to primary seta 4 sensu Müller (1886). Because these setae are homologs, Müller's concept of serial homology for these segments is corroborated. Later authors treating nymphalid chaetotaxy (e.g., Fleming, 1960; Kitching, 1984) followed Hinton's (1946) system of setal nomenclature (see also Stehr, 1987), in which this seta is treated as SD1 (or SD2) on T2,3 and as L1 on A1–8. Thus, Hinton's system, when applied to these taxa, implies incorrect setal homologies. The presence of two filiform setae on T1 in noctuids, papilionids, and most nymphalids suggests that, in the dorsal portion of this segment, two sets of homologous setal elements are present in the first and later instars.

2. Heliconiinae

I use the Heliconiinae in a considerably more inclusive sense than previous authors. It corresponds to Müller's group I, with the inclusion of many additional genera. Although Ehrlich (1958), Ehrlich and Raven (1965), and Ackery (1984, 1988), among others, have noted affinities among the Acraeinae, Heliconiinae, and Argynninae, none of these authors has credited Müller (1886) with the original insight or has cited character evidence to support this relationship, aside from patterns of host plant use (i.e., association of most taxa with the related families Violaceae, Flacourtiaceae, and Passifloraceae—see Ehrlich and Raven, 1965; Ackery, 1988).

Müller (1886) discovered a key larval character: the presence of lateral scoli (sensu Beebe et al., 1960, = Sst of Müller) arising from between segments T1 and T2 and between segments T2 and T3. In other nymphalids with lateral scoli on the thorax (i.e., Nymphalinae and some Limenitidinae), the lateral scoli are located at the midpoints of segments T2 and T3 (Müller, 1886; Harvey, unpub. data). The lateral thoracic scoli of heliconiines are thus not homologous to those of other

nymphalids and are a synapomorphy for the subfamily. Müller scored this character for the genera of group I listed in Table B.1. According to larval descriptions and illustrations in the literature, this character state also occurs in *Pardopsis*, *Acraea*, and *Actinote* (Van Son, 1963; and see note 4); *Philaethria*, *Dione*, *Agraulis*, *Dryadula*, *Eueides*, *Laparus*, and *Heliconius* (Beebe et al., 1960); *Cethosia* and *Vagrans* (Johnston and Johnston, 1980); *Vindula* (D'Abrera, 1977); *Euptoieta* (Toliver, 1987a); *Brenthis*, *Argynnis*, *Damora*, *Argyreus*, *Argyronome*, *Fabriciana*, and *Mesoacidalia* (Shirôzu and Hara, 1960–62); and *Speyeria* and *Clossiana* (Harvey, unpub. data).

A second unique character of larvae of Heliconiinae that I have examined (*Actinote*, *Agraulis*, *Heliconius*, *Euptoieta*, *Argyreus*, *Fabriciana*, *Speyeria*, and *Clossiana*) is the position of filiform setae on A2–8: They arise from the sclerotized base of, or sometimes a small distance up, the sublateral scoli (sensu Beebe et al., 1960). The filiform setae on these segments arise from the body in all other nymphalids I examined, including those groups with scoli on these segments (i.e., Nymphalinae and most Limenitidinae).

A third feature of larval morphology that characterizes the Heliconiinae genera cited in previous paragraph is the position of the scolus on A9: It is dorsal to the filiform seta. A scolus on A9 is also present in Nymphalinae (see notes 7–10), but it is ventral and posterior to the filiform seta. The scoli on A9 of Heliconiinae and Nymphalinae arise from different positions on the larva and are thus not homologous. Other than in these two subfamilies, scoli are absent on the dorsal and subdorsal regions of A9, with the exception of *Smyrna*, in which the scolus differs by being branched (Müller, 1886).

3. Pardopsini

The monotypic African tribe Pardopsini, consisting of the single species *Pardopsis punctatissima*, was considered to be most closely related to the Acraeini by Ehrlich (1958), Van Son (1963), and Pierre (1987). Its larvae differ from those of all other Heliconiinae in that they have one row of abdominal scoli above the line of spiracles (Van Son,

1963), whereas other known Heliconiinae have two rows (see references for larval descriptions in note 2).

4. Acraeini

The group Acraeini corresponds to Müller's group IA, Acraeini sensu Ehrlich (1958), and Acraeinae sensu Pierre (1987). Pierre (1985, 1986, 1987) treated the group as a single genus, *Acraea,* consisting of two sister lineages: one (subgenus *Actinote*) containing the traditional Neotropical *Actinote* and some African *Acraea,* and the second (subgenus *Acraea*) containing the remaining African *Acraea* and *Bematistes.* I include *Actinote* as a genus for convenience.

Pierre (1987) considered the following characters to be synapomorphies of the Acraeini: female abdomen with *"gland sous-papillaire"* (= anal pouch of Van Son, 1963) well developed and with lateral glands; male tarsal claws asymmetric (see also Ehrlich, 1958); and vinculum of male genitalia massive and longer than high. (See "Addendum," below.)

5. Cethosiini

The tribe Cethosiini contains a single genus, *Cethosia,* which is restricted to the Old World tropics. Brown (1981) suggested that it be included in the Heliconiiti, but it lacks the dorsal abdominal gland found in all Heliconiini (see note 6). I examined the female abdomens of four species of *Cethosia* and found that all have an extremely well-developed *"gland sous-papillaire"* (see Pierre, 1986, 1987; and note 4). The presence of this unusual structure suggests that *Cethosia* might be the sister group of Acraeini. (See "Addendum," below.)

6. Heliconiini

The tribe Heliconiini includes taxa placed in the subfamilies Heliconiinae and Argynninae by Ackery (1988). It is characterized by the presence of an eversible gland on the dorsum of the adult female abdomen between segments 7 and 8 (= dorsal scent gland of Emsley, 1963). This gland was apparently first noticed by F. Müller (1877,

1878), who called it a stink gland because of the nauseating odor it produces. The gland has since been found in all Heliconiiti examined, including representatives of all genera (see Eltringham, 1925; Emsley, 1963, 1965; Brown, 1972, 1973). Glands of the same structure and location were also found in *Speyeria* by Clark (1926) and Eltringham (1928) and in *Boloria* (*Clossiana*) by Clark (1926). I examined the type-species (see Hemming, 1967) of all genera of Heliconiinae listed in Table B.2 for these glands (with the exception of Heliconiiti, in which they are well documented) and found them present in all Heliconiini but not in Pardopsini, Acraeini, and Cethosiini.

F. Müller (1878) also discovered a second structure associated with this gland, and he called the structures "stink clubs" (= abdominal processes of Emsley, 1963). Stink clubs are sclerotized processes that bear scales distally, fit into the dorsal abdominal gland, and are exposed when the dorsal abdominal gland is everted. They are present in all Heliconiiti except *Dryadula* (see Müller, 1878; Eltringham, 1925; Emsley, 1963, 1965; Brown, 1972, 1973). While examining the Heliconiinae for the dorsal abdominal gland (see previous paragraph), I also found stink clubs in *Cirrochroa, Paduca, Terinos, Lachnoptera,* and *Smerina* (in which they are reduced to a tiny nub).

The relationships within the Heliconiini are not resolved. I tentatively recognize three subtribes, along with a series of genera whose relationships and subtribal placement are uncertain.

The Heliconiiti is an exclusively New World group that corresponds to Müller's group IB, the Heliconiinae of Emsley (1963), and the Heliconiini of Brown (1981—but rejecting his possible inclusion of *Cethosia* and *Vindula*). A phylogeny of the group was presented by Brown (1981), although most of the nodes were not supported by published character information.

Emsley (1963) listed four characters as defining the Heliconiiti: a simple, basally recurved humeral vein on the hind wing, androconia on the male hind wing, a dorsal abdominal gland in females, and abdominal processes (= stink clubs) in the female abdomen (except *Dryadula*). Not all of these characters are unique to the Heliconiiti, however. Conspicuous androconial patches occur on the hind wings of *Terinos* and *Lachnoptera* (Harvey, unpub. data), and the dorsal scent gland occurs in all Heliconiini. Furthermore, stink clubs also occur in other genera (see above). A basally recurved humeral vein on the hind wing thus appears to be the only derived character that unites this subtribe.

The unplaced series of genera *Vindula* through *Euptoieta* were included in the Argynninae by Ackery (1988). With the exception of the New World *Euptoieta*, this group is restricted to the Old World tropics. The presence of stink clubs in *Cirrochroa* through *Smerina* suggests that these genera are most closely related to the Heliconiiti.

The Argynniti and Boloriiti roughly correspond to the Argynninae of Warren and co-workers (Warren, 1944, 1955; dos Passos and Grey, 1945; Warren et al., 1946), after the removal of *Euptoieta*. The division into two subtribes follows Shirôzu and Saigusa (1973), who recognized the subtribes Argynnina and Boloriina. (I differ in retaining *Speyeria* and *Mesoacidalia* as distinct genera.) I treat *Yramea, Kuekenthaliella*, and *Prokuekenthaliella* as Heliconiini of uncertain subtribal position.

One character of first-instar chaetotaxy—the presence of secondary setae (sensu Stehr, 1987)—in *Speyeria* and *Clossiana* (Scott and Mattoon, 1982; Harvey, unpub. data) may characterize the Argynniti + Boloriiti. Secondary setae are absent in first-instar larvae of Pardopsini and Acraeini (Müller, 1886; Van Son, 1963), Heliconiiti (Fleming, 1960), and *Euptoieta* (Harvey, unpub. data). Many additional genera need to be examined for this character, however.

7. Nymphalinae

The subfamily Nymphalinae corresponds to Müller's group II and includes genera listed by Ackery (1988) in his Nymphalini (distributed in two tribes here) and Melitaeinae, plus *Amnosia* from his Pseudergolini. I divide this group into three tribes: Nymphalini, Kallimini, and Melitaeini.

The dorsolateral scoli on A9 of larvae of Nymphalinae are in a more ventral position than those of Heliconiinae larvae (see note 2). The distribution of filiform setae (see note 1) on mature larvae differs among the three tribes. In Nymphalini and Melitaeini (but not Kallimini), filiform setae are also present on A1,2, a character not found in other Nymphalidae. In addition, in Kallimini and Melitaeini (but not Nymphalini), the filiform seta on A9 is present on the sclerotized base of the scolus, a derived character not found in other Nymphalidae (see notes 8 through 10 for genera examined). The distributions of these two derived characters conflict.

The tribes Nymphalini and Kallimini can be separated by two adult characters. I examined the type-species (see Hemming, 1967) of each genus in these tribes and found that the eyes are hairy (with conspicuous setae) in Nymphalini but are naked (without setae) in Kallimini. In addition, the palpi in Nymphalini (except *Mynes*) have stiff, projecting bristlelike scales, but the palpi in Kallimini are smooth, lacking such scales (in *Catacroptera* and *Vanessula*, however, projecting flat scales are present).

Müller's inclusion of both Kallimini and Melitaeini in group IIB (his Diademinae) suggests a possible sister-group relation. This idea is strengthened further by host plant relations. Ackery (1988) noted a striking parallel between his Melitaeinae (Melitaeini here) and the Acanthaceae series of his Nymphalini (genera included here in the Kallimini): Both include taxa dependent on iridoid glycosides as larval feeding stimulants (Bowers, 1983, 1984). Iridoid-containing hosts are virtually absent outside these groups. If they are confirmed as a monophyletic group, the location of the filiform seta of A9 on the base of the scolus would be a synapomorphy of Kallimini + Melitaeini.

8. Nymphalini

The tribe Nymphalini corresponds to Müller's group IIA (his Vanessinae). See note 7 for characters of adults that separate it from Kallimini.

Known larvae of Nymphalini can be distinguished from those of other Nymphalidae by the distribution of filiform setae on mature larvae (see notes 1 and 7). The following genera were examined: *Vanessa, Cynthia, Bassaris, Aglais, Inachis, Nymphalis, Polygonia, Kaniska,* and *Hypanartia*.

9. Kallimini

The tribe Kallimini (Hypolimnini of authors, in part) corresponds to Müller's group IIBb. See note 7 for characters of adults that separate it from Nymphalini.

Known larvae of Kallimini can be distinguished from those of other Nymphalidae by the distribution of filiform setae on mature larvae (see notes 1 and 7). The following genera were examined: *Anartia, Siproeta, Junonia, Precis,* and *Hypolimnas*.

10. Melitaeini

The tribe Melitaeni corresponds to Müller's group IIBa and the Melitaeinae sensu Higgins (1981). I follow Higgins's arrangement of genera, but I downrank his three tribes to subtribes. Higgins was unsure of the relationship of the West Indian *Atlantea* to other melitaeines, so I place this genus at the end of the group, with subtribal position uncertain. Higgins (1981) provided keys to his tribes but did not provide any character that was diagnostic for the melitaeines as a group. The male genitalia of melitaeines are unique among the Nymphalidae in having a notched saccus (Harvey and J. E. Rawlins, unpub. data).

Known larvae of Melitaeini can be distinguished from those of other Nymphalidae by the distribution of filiform setae on mature larvae (see notes 1 and 7). The following genera were examined: *Euphydryas, Occidryas, Mellicta, Poladryas, Chlosyne, Thessalia, Texola, Dymasia, Phyciodes, Anthanassa,* and *Eresia*. In *Texola, Dymasia,* and *Microtia* the scolus on A9 is represented by a sclerotized plate.

11. Limenitidinae

The subfamily Limenitidinae corresponds to Müller's group III, after the addition of Cyrestidini (not examined by Müller, 1886). Müller noted that this group was not natural and that no character

defined the group. It is used here for convenience only.

One character of larval morphology, the presence of highly branched scoli, characterizes the Coloburini, the Limenitidini, and most of the Biblidini (Müller, 1886) but not the Cyrestidini (see note 15). In Heliconiinae and Nymphalinae the scoli are unbranched (Müller, 1886).

Müller (1886) noted an unusual character of larval behavior found only in this group (Coloburini, Biblidini, and Limenitidini) and the Charaxinae: Early-instar larvae build frass chains extending from the edge of the leaf and rest on them when not feeding. This behavior also occurs in *Marpesia petreus* (Cyrestidini) (Harvey, pers. obs.). Documentation of this behavior is poor in Old World genera, and its systematic significance is uncertain. I am unaware of any reports of its occurrence outside the Limenitidinae and Charaxinae (see also Müller, 1886).

12. Coloburini

The Coloburini is a small group of seven genera confined to the Neotropics (see Seitz, 1914; Miller and Brown, 1981) and corresponds to Müller's group IIIA. This tribe has not been characterized, but the genera seem to divide into two groups: *Historis* through *Baeotus,* and *Smyrna* through *Tigridia* (T. P. Friedlander, pers. com.). Although placed as a tribe by Ackery (1988) in his Nymphalinae (along with his Nymphalini, = Nymphalini + Kallimini here), coloburine larvae known to me (*Historis, Colobura, Smyrna,* and *Tigridia*) lack filiform setae on segments A1,2, and those on A9 are not on the base of the scolus (see note 7).

13. Biblidini

The tribe Biblidini (Eurytelinae of authors) corresponds to Müller's groups IIIB–E. The generic content and subtribal relationships follow Jenkins (1983, 1984, 1985a,b, 1986, 1987, 1989, and pers. com.), downranking his subfamily to tribe and his tribes to subtribes and elevating his subtribe Ageroniini from his Epicaliini. The Biblidini includes (but is not limited to) the genera listed by Ackery (1988) in his Biblini, Ageroniini, and Epicaliini. Jenkins is preparing a cladistic analysis of this group that will provide justification for the subdivisions employed here.

The tribe is characterized by the presence of a modified eighth sternite of the male abdomen, termed the hypandrium, which is unique in the Nymphalidae (see Dillon, 1948; Munroe, 1949; papers by Jenkins cited above; Harvey and J. E. Rawlins, unpub. data).

14. Limenitidini

The tribe Limenitidini is equivalent to the Limenitini of Chermock (1950), who treated the genera of this tribe on a worldwide basis. According to Chermock, members of this tribe (except *Neptis*) can be distinguished from all other nymphalids by the following character of forewing venation: "The first anal vein is preserved as a short spur at the base of the cubitus."

This tribe may be also be characterized by egg morphology (Eliot, 1978). Known eggs in this group have polygonal cells, with short processes arising from their vertexes; see illustrations of Limenitiditi and Neptiti (Shirôzu and Hara, 1960–62) and of Euthaliiti (Johnston and Johnston, 1980). Eggs of Partheniti are apparently unknown.

For subtribal divisions, I follow Eliot (1969, 1978; see also Ackery, 1988). Eliot's (1969) Neptini is downranked to subtribe, as is Eliot's (1978) Parthenini. The Euthalini of Eliot (1978)—as modified by Ackery (1988), with the addition of some genera listed in Chermock (1950)—is downranked to subtribe, as is the Limenitidini (sans Neptini) of Eliot (1978), with modifications by Ackery (1988). The considerable confusion in use of generic names in the Limenitidini results in part from the plethora of available names (more than 130, most authored by Moore; see Chermock, 1950) and from disagreement among authors over how they should be applied.

15. Cyrestidini

The Cyrestidini (Marpesiini of authors) traditionally includes the genera *Marpesia* (New World) and *Chersonesia* and *Cyrestis* (Old World), which are grouped together on the basis of hosts in the Moraceae and smooth larvae with elongate middorsal scoli (see Ackery, 1988, and included references). In addition, Eliot (1978) noted that eggs (presumably those of *Chersonesia* and *Cyrestis*) have an unusual "trapdoor" through which the larva emerges, a character also found in eggs of *Marpesia petreus* (pers. obs.). I add *Dichorragia* to Cyrestidini on the suggestion of T. P. Friedlander (pers. com.), who notes that its larva (see Shirôzu and Hara, 1960–62; Johnston and Johnston, 1980) is also smooth (but lacks well-developed middorsal scoli) and has similar head capsule morphology. Its larvae differ from those of other cyrestidines in feeding on Meliosmaceae and Anacardiaceae (Ackery, 1988).

16. Incertae Sedis

Eliot (1978) used the subfamily Pseudergolinae for the genera *Pseudergolis, Dichorragia, Stibochiona,* and *Amnosia,* characterizing it by the presence of a developed claw on the female foretarsus. Although females of most Nymphalidae lack foretarsal claws, the claws are also present in Calinaginae and Libytheinae (Ehrlich, 1958) and in some Ithomiinae (Fox, 1956). I treat Eliot's subfamily as a group of uncertain status and subfamilial placement, transfer *Amnosia* to Kallimini and *Dichorragia* to Cyrestidini (see note 15). Although this group is doubtless polyphyletic, I have no information on characters that would shed light on the relationships of these genera to other nymphalids.

17. Charaxinae

The Charaxinae corresponds to Müller's group IV. The classification of this group basically follows Rydon (1971), downranking his family to subfamily, and his subfamilies to tribes. I also follow Ackery (1988), who, incorporating ideas of Comstock (1961), downranked Rydon's Zaretidinae and Anaeinae to subtribes of Anaeini.

18. Apaturinae

The genera of the subfamily Apaturinae are listed alphabetically and were supplied by Friedlander, who revised the genus *Asterocampa* (Friedlander, 1988) and is undertaking a phylogenetic analysis of the Apaturinae. Several of the genera show affinities with others: *Euripus, Hestina, Hestinalis,* and *Sasakia; Eulaceura* and *Herona; Asterocampa* and *Chitoria;* and *Dilipa* and *Thaleropis* (Friedlander, letter to author, 1989). Additional information on this group may be found by consulting Stichel (1938) and Le Moult (1950).

19. Morphinae

Opinions differ as to the content of the subfamily Morphinae. Ehrlich (1958) included taxa placed here in Morphiti, Amathusiini, and Brassolinae. Ackery (1988) treated Morphini, Amathusiini, and Brassolinae as distinct subfamilies but considered them a monophyletic group on the basis of a character of the larval head capsule from DeVries et al. (1985): "vertex horns with narrow base, transition to head abrupt." This character, however, was also reported by DeVries et al. (1985) to occur in the Charaxinae. Unfortunately, the researchers did not illustrate this character, and their photographs of larvae indicate the lack of horns in head capsules of Morphini. I examined head capsules of several species of *Morpho* and found that vertex horns are absent. I follow Müller (see Table B.1) in including the Amathusiini in Morphinae and excluding Brassolinae.

20. Morphini

Classification of the tribe Morphini follows De-Vries et al. (1985), who recognized two subtribes after adding *Caerois* and *Antirrhea* from the Satyridae sensu Miller (1968).

21. Amathusiini

I follow Kirchberg (1942) in the generic content of the tribe Amathusiini (which he treated as a subfamily). Kirchberg recognized two series of genera: *Faunis* through *Stichophthalma,* and *Amathusia* through *Enispe.* Other authors recognized additional genera: *Hyantis* and *Morphopsis* were used by D'Abrera (1977), and *Melanocyma* was separated from *Faunis* by Eliot (1978).

22. Brassolinae

As used here, the Brassolinae corresponds to the Brassolinae of Miller (1968), after its removal

from his Satyridae. To the genera listed by Miller are added four genera subsequently described by Casagrande (1982). On the basis of similarities in male scent organs, Vane-Wright (1972) suggested that the satryine genus *Bia* (Biiti) (see Ehrlich, 1958, and Miller, 1968) may be most closely related to the brassolines. Here, *Bia* is left in the Satyrinae.

23. Satyrinae

The classification of the Satyrinae is modified from Miller (1968), who treated the group as a family. Miller's Brassolinae is removed as a separate subfamily (see note 22). From his Biinae, his tribe Antirrhini is removed (as Antirrheiti) to Morphini (see note 20). The remaining classification follows Miller's scheme, downranking his subfamilies and tribes to tribes and subtribes, respectively. This classification is identical to that used by Ackery (1988). For the systematic placement of the numerous genera in this group, see Miller (1968).

No single derived character defines the Satyrinae as used here, and Miller (pers. com.) considers that his phenetic classification will be modified as phylogenetic studies on the group progress. It has been suggested recently that the Satyrinae is paraphyletic in terms of the Charaxinae (DeVries et al., 1985).

A unique condition of filiform setae (see note 1) occurs in some Satyrinae: Seta T1(A), present in all other nymphalids examined, is absent in *Hipparchia* (Satyrini: Satyriti) (Garcia-Barros, 1989) and in *Pierella* (Haeterini), in *Cissia, Taygetis, Megisto, Paramacera,* and *Chloreuptychia* (Satyrini: Euptychiiti), and in *Calisto* (Satyrini: Pronophiliti) (Harvey, unpub. data). Additional groups of satyrines need to be checked for this character, particularly those from the Old World tropics.

Although not mentioned by Miller (1968), the genus *Penthema* has genitalia typical of Satyrinae (Harvey and J. E. Rawlins, unpub. data), but its position within the subfamily is uncertain.

24. Calinaginae

Ehrlich (1958) placed a single genus, *Calinaga,* in the subfamily Calinaginae.

25. Danainae + Tellervinae + Ithomiinae

Ackery and Vane-Wright (1984) considered the subfamilies Danainae, Tellervinae, and Ithomiinae to form a probable monophyletic group. Although no unique apomorphic character of adult morphology characterizes it, Ackery and Vane-Wright considered the grouping justifiable on biological grounds, particularly the association of these taxa with pyrrolizidine alkaloids.

Müller (1886) placed the Danainae and Ithomiinae in one group (his Danainae) but also referred to both groups as subfamilies. One unique character, of first-instar larval chaetotaxy, unites the Danainae (*Danaus*) and Ithomiinae (*Dircenna, Ceratinia, Ithomia, Methona,* and *Mechanitis*). It was discovered by Müller (1886) but has been overlooked by all subsequent authors except for a passing reference by Gilbert and Ehrlich (1970). Müller noted the presence of an extra primary seta, which he called 1a, located directly below his seta 1 (D1 of authors). This extra primary seta also occurs in *Tithorea tarricina* (Ithomiinae) (Harvey, unpub. data). A scanning electron micrograph the first-instar larva of *Tellervo* (Ackery, 1987) is unfortunately not clear enough to score this character. Kitching (1984) described the first-instar chaetotaxy of *Danaus gilippus* and compared it with that of *Heliconius melpomene* (Fleming, 1960). Apparently unaware of Müller's (1886) discovery, Kitching (1984) proposed several incorrect setal homologies between *Danaus* and *Heliconius.* Comparison of the chaetotaxies of the first abdominal segment shows that Kitching's seta D2 is homologous to seta 1a of Müller (there is no homolog on *Heliconius*—Müller, 1886), and Kitching's SD2 is homologous to seta 2 of Müller and D2 of Fleming.

A further tentative character that supports monophyly of this group is the absence of filiform setae (see note 1) on T2,3 of mature larvae of *Danaus* (Danainae) and *Tithorea* (Ithomiinae) (Harvey, unpub. data). In all other Nymphalidae, filiform setae are present on both segments.

It will be interesting to see the state of these larval characters in *Tellervo,* because the relationships among the three subfamilies are unresolved (Ackery and Vane-Wright, 1984).

26. Danainae

The classification of the group Danainae follows the phylogenetic analysis of Ackery and Vane-Wright (1984). They characterized the Danainae by the presence of paired, sheathed and eversible hair pencils in the adult male and of strongly clubbed, spinose, four-segmented protarsi in the adult female. However, a four-segmented female protarsus also occurs in some Ithomiinae (Fox, 1956), some Satyrinae (Miller, 1968), and the genus *Eueides* (Heliconiiti) (Emsley, 1963; Brown, 1981). The Danainae are unique in having unscaled antennae (Jordan, 1898; Ehrlich, 1958).

27. Tellervinae

Tellervo, the single genus included in the Tellervinae, was recently revised by Ackery (1987).

28. Ithomiinae

The classification of the Ithomiinae follows that presented by Brown (1987), after removal of *Tellervo.* This group is characterized by the presence of a costal fringe of long androconial scales on the hind wing of males (Fox, 1956; Ehrlich, 1958).

29. Libytheinae

The small group Libytheinae has been treated as a family distinct from the Nymphalidae (Ehrlich, 1958; Scott, 1985) or as a subfamily of the Nymphalidae (Kristensen, 1976; Ackery, 1984, 1988). The Libytheinae share with the remaining Nymphalidae a tricarinate antenna (Jordan, 1898), a condition unique in the Lepidoptera, and together they form a monophyletic group (Kristensen, 1976).

Addendum

Recent observations on characters of male genitalia of Nymphalidae (Harvey and J. E. Rawlins, unpub. data) require a change in the present classification. The shape of the vinculum and saccus of *Cethosia* species is the same as in *Acraea* (see note 4) and differs from that of all other Heliconiinae. This finding confirms the sister group relationship of these genera that is suggested in note 5. I place *Cethosia* in the Acraeini and abandon Cethosiini.

Bibliography

Ackery, P. R.

1984. Systematic and faunistic studies on butterflies. In *The biology of butterflies,* ed. R. I. Vane-Wright and P. R. Ackery, 9–21. Academic Press, London.

1987. The danaid genus *Tellervo* (Lepidoptera, Nymphalidae)—a cladistic approach. *Zoological Journal of the Linnean Society* 89:203–274.

1988. Hostplants and classification: A review of nymphalid butterflies. *Biological Journal of the Linnean Society* 33:95–203.

Ackery, P. R., and R. I. Vane-Wright

1984. *Milkweed butterflies: Their cladistics and biology.* British Museum (Natural History), London.

Ballmer, G. R., and G. F. Pratt

1989. A survey of the last instar larvae of the Lycaenidae (Lepidoptera) of California. *Journal for Research on the Lepidoptera* 27:1–81.

Beebe, W., J. Crane, and H. Fleming

1960. A comparison of the eggs, larvae, and pupae in fourteen species of heliconiine butterflies from Trinidad, W.I. *Zoologica* 45:111–154.

Bowers, M. D.

1983. The role of iridoid glycosides in hostplant specificity of checkerspot butterflies. *Journal of Chemical Ecology* 9:475–494.

1984. Iridoid glycosides and host-plant specificity in larvae of the buckeye butterfly, *Junonia coenia* (Nymphalidae). *Journal of Chemical Ecology* 10:1567–1577.

Bridges, C. A.

1985. Lepidoptera: Rhopalocera. In *Synonymic list of family-group names,* part 1 of *Notes on family- and genus-group names,* 1–5. Privately printed.

Brown, K. S.

1972. The heliconians of Brazil (Lepidoptera: Nymphalidae), 3: Ecology and biology of *Heliconius nattereri,* a key primitive species near extinction, and comments on the evolutionary development of *Heliconius* and *Eueides. Zoologica* 57:41–69.

1973. The heliconians of Brazil (Lepidoptera: Nymphalidae), 5: Three new subspecies from Mato Grosso and Rondônia. *Bulletin of the Allyn Museum* 13:1–19.

1981. The biology of *Heliconius* and related genera. *Annual Review of Entomology* 26:427–456.

1987. Chemistry at the Solanaceae/Ithomiinae interface. *Annals of the Missouri Botanical Gardens* 74:359–397.

Casagrande, M. M.

1982. Quatro gêneros novos do Brassolinae (Lepidoptera, Nymphalidae). *Revista Brasileira de Entomologia* 26:355–356.

Chermock, R. L.

1950. A generic revision of the Limenitini of the world. *American Midland Naturalist* 43:513–569.

Clark, A. H.

1926. Notes on the odors of some New England butterflies. *Psyche* 33:1–5.

Comstock, W. P.

1961. *Butterflies of the American tropics: The genus* Anaea *Lepidoptera Nymphalidae.* American Museum of Natural History, New York.

D'Abrera, B.

1977. *Butterflies of the Australian region.* 2d ed. Lansdowne Press, Melbourne.

DeVries, P. J., I. J. Kitching, and R. I. Vane-Wright

1985. The systematic position of *Antirrhea* and *Caerois,* with comments on the classification of the Nymphalidae (Lepidoptera). *Systematic Entomology* 10:11–32.

Dillon, L. S.

1948. *The genus* Catagramma *and allies.* Part 1 of *The tribe Catagrammini (Lepidoptera: Nymphalidae).* Reading Public Museum and Art Gallery, Scientific Publications, no. 8. Reading, Pa.

dos Passos, C. F., and L. P. Grey

1945. A genitalic survey of the Argynninae (Lepidoptera, Nymphalidae). *American Museum Novitates* 1296:1–29.

Downey, J. C., and A. C. Allyn
 1979. Morphology and biology of the imma-
    ture stages of *Leptotes cassius theonus* (Lu-
    cas) (Lepid.: Lycaenidae). *Bulletin of the
    Allyn Museum* 14:1–47.
 1984. Chaetotaxy of the first instar larva of
    Hemiargus ceraunus antibubastus (Hbn.)
    (Lycaenidae). *Bulletin of the Allyn Mu-
    seum* 90:1–4.
Ehrlich, P. R.
 1958. The comparative morphology, phylog-
    eny, and higher classification of the
    butterflies. *University of Kansas Science
    Bulletin* 39:305–370.
Ehrlich, P. R., and P. H. Raven
 1965. Butterflies and plants: A study in co-
    evolution. *Evolution* 18:586–608.
Eliot, J. N.
 1969. An analysis of the Eurasian and Austra-
    lian Neptini. *Bulletin of the British Mu-
    seum (Natural History), Entomology Sup-
    plement* 15:1–155.
Eliot, J. N., ed.
 1978. *The butterflies of the Malay Peninsula*, by
    A. S. Corbet and H. M. Pendlebury.
    Rev. (3d) ed. Malay Nature Society,
    Kuala Lumpur.
Eltringham, H.
 1925. On the abdominal glands in *Heliconius*
    (Lepidoptera). *Transactions of the Ento-
    mological Society of London*, 269–275.
 1928. On the abdominal glands in certain
    North American argynnids (Lep.,
    Nymphalidae). *Transactions of the Ento-
    mological Society of London*, 97–99.
Emsley, M. G.
 1963. A morphological study of imagine Hel-
    iconiinae with a consideration of the
    evolutionary relationships within the
    group. *Zoologica* 48:85–130.
 1965. Speciation in *Heliconius* (Lep., Nym-
    phalidae): Morphology and geographic
    distribution. *Zoologica* 50:191–254.
Fleming, H.
 1960. The first instar larvae of Heliconiinae
    (butterflies) of Trinidad, W.I. *Zoologica*
    45:91–110.

Fox, R. M.
 1956. A monograph of the Ithomiinae. Part
    1. *Bulletin of the American Museum of
    Natural History* 111:1–76.
Friedlander, T. P.
 1988. Taxonomy, phylogeny, and biogeogra-
    phy of *Asterocampa* Röber 1916 (Lepi-
    doptera, Nymphalidae, Apaturinae).
    Journal for Research on the Lepidoptera
    25:215–338.
Garcia-Barros, E.
 1989. Morfología larvaria de *Hipparchia* (*Pseu-
    dotergumia*) *fidia* (L., 1767) (Lepidop-
    tera, Nymphalidae). *Nouvelle Revue
    d'Entomologie*, n.s., 6:71–83.
Gilbert, L. E., and P. R. Ehrlich
 1970. The affinities of the Ithomiinae and Sa-
    tyrinae (Nymphalidae). *Journal of the
    Lepidopterists' Society* 24:297–300.
Harvey, D. J., and L. E. Gilbert
 n.d. Ant association, larval and pupal mor-
    phology of the Neotropical riodinid
    butterfly, *Pandemos palaeste. Journal of
    Natural History*. In press.
Harvey, D. J., J. Mallet, and J. Longino
 n.d. The biology and morphology of imma-
    ture *Nymphidium cachrus* (Riodinidae).
    Journal of the Lepidopterists' Society. In
    press.
Hemming, F.
 1967. The generic names of butterflies and
    their type-species (Lepidoptera, Rhopa-
    locera). *Bulletin of the British Museum
    (Natural History), Entomology Supplement*
    9:1–509.
Higgins, L. G.
 1981. A revision of *Phyciodes* and related gen-
    era, with a review of the classification of
    the Melitaeinae. *Bulletin of the British
    Museum (Natural History), Entomology*
    43:77–243.
Hinton, H. E.
 1946. On the homology and nomenclature of
    the setae of lepidopterous larvae, with
    notes on the phylogeny of the Lepidop-
    tera. *Transactions of the Royal Entomolog-
    ical Society of London* 97:1–37.

Jenkins, D. W.

1983. Neotropical Nymphalidae, 1: Revision of *Hamadryas*. *Bulletin of the Allyn Museum* 81:1–146.

1984. Neotropical Nymphalidae, 2: Revision of *Myscelia*. *Bulletin of the Allyn Museum* 87:1–64.

1985a. Neotropical Nymphalidae, 3: Revision of *Catonephele*. *Bulletin of the Allyn Museum* 92:1–65.

1985b. Neotropical Nymphalidae, 4: Revision of *Ectima*. *Bulletin of the Allyn Museum* 95:1–31.

1986. Neotropical Nymphalidae, 5: Revision of *Epiphile*. *Bulletin of the Allyn Museum* 101:1–70.

1987. Neotropical Nymphalidae, 6: Revision of *Asterope* (= *Callithea* auct.). *Bulletin of the Allyn Museum* 114:1–66.

1989. Neotropical Nymphalidae, 7: Revision of *Nessaea*. *Bulletin of the Allyn Museum* 125:1–38.

Johnston, G., and B. Johnston

1980. *This is Hong Kong butterflies.* Hong Kong.

Jordan, K.

1898. Contributions to the morphology of Lepidoptera. *Novitates Zoologicae* 5:374–415.

Kirchberg, E.

1942. Genitalmorphologie und natürliche Verwandtschaft der Amathusiinae (Lep. Nymphal.) und ihre Beziehungen zür geographischen Verbreitung der Subfamilie. *Mitteilungen der Münchener Entomologischen Gesellschaft* 32:44–87.

Kitching, I. J.

1984. The use of larval chaetotaxy in butterfly systematics, with special reference to the Danaini (Lepidoptera: Nymphalidae). *Systematic Entomology* 9:49–61.

Kristensen, N. P.

1976. Remarks on the family-level phylogeny of butterflies (Insecta, Lepidoptera, Rhopalocera). *Zeitschrift für Zoologische Systematik und Evolutionsforschung* 14:25–33.

Lawrence, D. A., and J. C. Downey

1966. Morphology of the immature stages of *Everes comyntas* Godart (Lycaenidae). *Journal for Research on the Lepidoptera* 5:61–96.

Le Moult, E.

1950. *Revision de la classification des Apaturinae de l'Ancien Monde suivi d'une monographie de plusieurs genres.* Miscellaenea Entomologica (Supplement), vol. 1.

Markl, H., and J. Tautz

1975. The sensitivity of hair receptors in caterpillars of *Barathra brassicae* L. (Lepidoptera, Noctuidae) to particle movement in a sound field. *Journal of Comparative Physiology* 99:79–87.

Miller, L. D.

1968. *The higher classification, phylogeny, and zoogeography of the Satyridae.* Memoirs of the American Entomological Society, vol. 24. Philadelphia.

Miller, L. D., and F. M. Brown

1981. *A catalogue/checklist of the butterflies of America north of Mexico.* Lepidopterists' Society Memoir no. 2.

Müller, F.

1877. The maracujá [passionflower] butterflies. Translated and reprinted (1912) in *Butterfly hunting in many lands,* by G. B. Longstaff, 651–654. Longmans, Green, and Co., London.

1878. The stink-clubs of the female "maracujá butterflies." Translated and reprinted (1912) in *Butterfly hunting in many lands,* by G. B. Longstaff, 664–666. Longmans, Green, and Co., London.

Müller, W.

1886. Südamerikanische Nymphalidenraupen: Versuch eines natürlichen Systems der Nymphaliden. *Zoologische Jahrbücher* 1:417–678.

Munroe, E. G.

1949. A new genus of Nymphalidae and its affinities (Lepidoptera, Rhopalocera). *Journal of the New York Entomological Society* 57:67–88.

Pierre, J.

1985. Morphologie comparée de l'appareil génital mâle des Acraeinae (Lepidoptera, Nymphalidae). *Annales de la Société Entomologique de France,* n.s., 21:381–391.

1986. Morphologie comparée de l'appareil génital femelle des Acraeinae (Lepidoptera, Nymphalidae). *Annales de la Société Entomologique de France,* n.s., 22:53–65.

1987. Systématique cladistique chez les *Acraea* (Lepidoptera, Nymphalidae). *Annales de la Société Entomologique de France,* n.s., 23:11–27.

Roth, V. L.

1984. On homology. *Biological Journal of the Linnean Society* 22:13–29.

Rydon, A. H. B.

1971. The systematics of the Charaxidae. [5 parts.] *Entomologist's Record and Journal of Variation* 83:219–233, 283–287, 310–316, 336–341, 384–388.

Scott, J. A.

1985. The phylogeny of the butterflies (Papilionoidea and Hesperioidea). *Journal for Research on the Lepidoptera* 23:241–281.

Scott, J. A., and S. O. Mattoon

1982. Early stages of *Speyeria nokomis* (Nymphalidae). *Journal for Research on the Lepidoptera* 20:12–15.

Seitz, A.

1914. Gruppe: Gynaeciidae. In *Gross-Schmetterlinge der Erde,* vol. 5, ed. A. Seitz, 470–473. Alfred Kernan Verlag, Stuttgart, West Germany.

Shields, O.

1989. World numbers of butterflies. *Journal of the Lepidopterists' Society* 43:178–183.

Shirôzu, T., and A. Hara

1960–1962. *Early stages of Japanese butterflies in colour.* [In Japanese.] 2 vols. Osaka.

Shirôzu, T., and T. Saigusa

1973. A generic classification of the genus *Argynnis* and its allied genera (Lepidoptera: Nymphalidae). *Sieboldia* 4:99–104.

Stehr, F.

1987. Order Lepidoptera. In *Immature insects,* vol. 1, ed. F. Stehr, 288–305. Kendall/Hunt, Dubuque, Iowa.

Stichel, H.

1938. Subfam. Apaturinae Kirby. *Lepidopterorum Catalogus* 86:156–374.

Tautz, J.

1977. Reception of medium vibration by thoracal hairs of caterpillars of *Barathra brassicae* L. (Lepidoptera, Noctuidae), 1: Mechanical properties of the receptor hairs. *Journal of Comparative Physiology* 118:13–31.

Toliver, M.

1987a. Nymphalidae. In *Immature insects,* vol. 1, ed. F. Stehr, 448–453. Kendall/Hunt, Dubuque, Iowa.

1987b. Papilionidae. In *Immature insects,* vol. 1, ed. F. Stehr, 438–441. Kendall/Hunt, Dubuque, Iowa.

Van Son, G.

1963. *Nymphalidae: Acraeinae.* Part 3 of *The butterflies of southern Africa.* Transvaal Museum, Pretoria.

Vane-Wright, R. I.

1972. Pre-courtship activity and a new scent organ in butterflies. *Nature* 239:338–339.

Warren, B. C. S.

1944. Review of the classification of the Argynnidi, with a systematic revision of the genus *Boloria. Transactions of the Royal Entomological Society of London* 94:1–101.

1955. A review of the classification of the subfamily Argynninae (Lepidoptera: Nymphalidae), 2: Definition of the Asiatic genera. *Transactions of the Royal Entomological Society of London* 107:381–391.

Warren, B. C. S., C. F. dos Passos, and L. P. Grey

1946. Supplementary notes on the classification of the Argynninae (Lepidoptera: Nymphalidae). *Proceedings of the Royal*

Entomological Society of London, ser. B, 15:71–73.

Wright, D.
1983. Life history and morphology of the immature stages of the bog copper butterfly *Lycaena epixanthe* (Bsd. & Le C.) (Lepidoptera: Lycaenidae). *Journal for Research on the Lepidoptera* 22:47–100.

Appendix C

Genera Surveyed for Figures 2.19 and 2.21

Abrota	1
Acraea	107
Acrophthalmia	1
Acropolis	1
Actinote	45
Adelpha	31
Admiratio	1
Aeropetes	1
Aglais	2
Agrias	16
Amathusia	1
Amnosia	1
Anaea	40
Anartia	4
Anetia	3
Antanartia	3
Anthanassa	22
Antillea	2
Antirrhea	27
Apatura	7
Apaturina	3
Apaturopsis	1
Aphantopus	1
Aphysoneura	2
Araschnia	3
Archimestra	1
Argynnina	2
Argynnis	2
Argyreus	1
Argyronympha	1
Argyrophenga	1
Argyrophorus	1
Ariadne	4
Asterocampa	7
Asterope/Sallya	9
Aterica	3
Athyma	6
Atlantea	1
Baeotus	2
Batesia	1

Bebearia	36
Bematistes	19
Bia	1
Bicyclus	36
Blepolensis	2
Bolboneura	1
Bolorial/Clossiana	24
Brassolis	10
Brenthis	2
Brintesia	1
Byblia	2
Caerois	5
Caligo	9
Caligopsis	1
Calinaga	3
Calisto	1
Callarge	1
Callerebia	4
Callicore	34
Callithea	5
Cassionympha	1
Castilia	10
Catacroptera	1

Catagramma	1	*Dymasia*	1	*Harma*	1
Catonephele	3	*Dynamine*	7	*Harmilla*	1
Catoplebia	10	*Dynastor*	3	*Harsiesis*	3
Catuna	6	*Ectima*	3	*Helcyra*	1
Cercyonis	6	*Elymnias*	8	*Heliconius*	43
Cethosia	8	*Elymniopsis*	3	*Henotesia*	29
Charaxes	54	*Enispe*	1	*Herona*	2
Chazara	3	*Epiphile*	11	*Hestina*	2
Childrena	1	*Erebia*	4	*Heteronympha*	5
Chlosyne	20	*Erebiola*	1	*Heteropsis*	2
Cirrochroa	4	*Eresia*	24	*Higginsius*	2
Cithaerias	3	*Eretris*	2	*Hipparchia*	5
Coelites	2	*Ergolis*	3	*Historis*	2
Coenonympha	2	*Erites*	4	*Houlbertia*	8
Coenophlebia	1	*Erycinidia*	2	*Hyalodia*	1
Coenyra	3	*Eryphanis*	6	*Hyantis*	1
Coenyropsis	3	*Etcheverinus*	2	*Hypanartia*	4
Colobura	1	*Ethope*	1	*Hypocysta*	3
Consul	1	*Eueides*	5	*Hypolimnas*	15
Corades	4	*Eulaceura*	1	*Hyponephele*	2
Cosmosatyrus	9	*Eunica*	35	*Issoria*	3
Crenidomimas	1	*Euphaedra*	37	*Janatella*	3
Cupha	4	*Euphydryas/*	10	*Junea*	7
Cybdelis	2	*Melitaea*		*Kallima*	5
Cyclogramma	1	*Euptera*	5	*Kirinia*	1
Cymothoe	69	*Euptoieta*	2	*Kumothales*	1
Cyrestis	5	*Euptychia*	94	*Lachnoptera*	2
Daedalma	1	*Euriphene*	37	*Lamprolenis*	1
Dagon	3	*Euripus*	2	*Laparus*	1
Damora	1	*Euryphaedra*	1	*Laringa*	1
Dasyophthalma	2	*Euryphura*	14	*Lasiommata*	1
Diaethria	9	*Eurytela*	4	*Lasiophila*	3
Dichorragia	2	*Euthalia*	2	*Lebadea*	2
Didonis	2	*Euthaliopsis*	1	*Lethe*	22
Dilipa	1	*Euxanthe*	5	*Lexias*	1
Dingana	2	*Fabriciana*	4	*Limenitis*	10
Dione/Agraulis	3	*Faunis*	3	*Lopinga*	1
Dira	4	*Geitoneura*	4	*Lucinia*	1
Dodonidia	1	*Gnathotriche*	4	*Lymanopoda*	3
Doleschallia	7	*Gnophodes*	3	*Maniola*	3
Doxocopa	7	*Haetera*	1	*Mantaria*	1
Drucina	1	*Hallelesis*	3	*Marpesia*	11
Dryadula	1	*Hamadryas*	8	*Mashuna*	2
Dulcedo	1	*Hamanumida*	1	*Masoura*	5

Mazia	1	*Pantoporia*	5	*Satyrus*	29
Melampias	1	*Paralethe*	5	*Selenophanes*	4
Melanargia	17	*Paramacera*	1	*Sephisa*	1
Melanitis	8	*Pararge*	2	*Siproeta*	2
Meneris	1	*Pardopsis*	1	*Smerina*	1
Mesoacidalia	1	*Parthenos*	6	*Smyrna*	2
Mesoxantha	1	*Patala*	1	*Speyeria*	24
Mestra	1	*Pedaliodes*	19	*Splendeuptychia*	1
Metamorpha	1	*Penetes*	1	*Steroma*	1
Morpho	29	*Penthema*	2	*Stibochiona*	2
Morphopsis	3	*Percnodaimon*	1	*Stichophthalma*	3
Morphotenaris	1	*Peria*	1	*Strabena*	10
Mycalesis	15	*Perisama*	19	*Stygionympha*	5
Mygona	1	*Phaedyma*	5	*Symbrenthia*	1
Mynes	10	*Phalanta*	4	*Taenaris*	18
Myscelia	7	*Philaethria*	2	*Tanaecia*	12
Narope	6	*Phyciodes*	77	*Tarsocera*	7
Neita	7	*Physcaeneura*	4	*Tatinga*	1
Nemetis	1	*Phystis*	2	*Taygetis*	21
Neocoenyra	8	*Pierella*	6	*Tegosa*	7
Neomaenas	2	*Pieridopsis*	2	*Telenassa*	5
Neominois	2	*Platypthima*	4	*Temenis*	2
Neope	10	*Podotricha*	2	*Terinos*	3
Neorina	3	*Polygonia*	7	*Texola*	1
Neptidopsis	2	*Polyura*	5	*Thaleropis*	1
Neptis	42	*Precis/Junonia*	15	*Thauria*	1
Nesoxenica	1	*Prepona*	21	*Thessalia*	1
Nessaea	7	*Prothoe*	2	*Thiemeia*	1
Neurosigma	1	*Pseudacraea*	25	*Tigridia*	1
Nica	2	*Pseudargynnis*	2	*Timelaea*	2
Nymphalis	4	*Pseudathyma*	7	*Tisiphone*	4
Oeneis	3	*Pseudomaniola*	2	*Torynesis*	4
Opoptera	9	*Pseudonympha*	15	*Vagrans*	1
Opsiphanes	15	*Ptychandra*	1	*Vanessa*	4
Oreixenica	5	*Punargenteus*	1	*Vila*	1
Oressinoma	1	*Pycina*	1	*Vindula*	6
Orsotriaena	1	*Pyrrhogyra*	5	*Xanthotaenia*	1
Ortilia	6	*Ragadia*	1	*Yoma*	2
Oxeoschistus	4	*Rhinopalpa*	1	*Ypthima*	9
Palla	4	*Salamis*	12	*Ypthimomorpha*	1
Pampasatyrus	1	*Sasakia*	2	*Yramea*	5
Panacea	4	*Satyrodes*	1	*Zaretis*	1
				Zethera	4

Ackery, P. R.

1984. Systematic and faunistic studies on butterflies. In *The biology of butterflies,* ed.
R. I. Vane-Wright and P. R. Ackery,
9–21. Academic Press, London.

Ackery, P. R., and R. I. Vane-Wright

1984. *Milkweed butterflies: Their cladistics and
biology.* Cornell University Press, Ithaca, N.Y.

Ae, S. A.

1957. Effects of photoperiod on *Colias eurytheme. Lepidopterists' News* 11:207–214.

Allyn, A. C., and J. Downey

1976. Diffraction structures in the wing scales
of *Callophrys* (*Mitoura*) *siva siva* (Lycaenidae). *Bulletin of the Allyn Museum* 40:
1–6.

Arcuri, P., and J. D. Murray

1986. Pattern sensitivity to boundary and initial conditions in reaction-diffusion
models. *Journal of Mathematical Biology*
24:141–165.

Ashby, W. R.

1966. *An introduction to cybernetics.* Wiley, New
York.

Bard, J. B. L.

1977. A unity underlying the different zebra
striping patterns. *Journal of Zoology*
183:527–539.

1981. A model for generating aspects of zebra
and other mammalian coat patterns.
Journal of Theoretical Biology 93:363–
385.

Bard, J. B. L., and V. French

1984. Butterfly wing patterns: How good a
determining mechanism is the simple
diffusion of a single morphogen? *Journal
of Embryology and Experimental Morphology* 84:255–274.

Bateson, W.

1894. *Materials for the study of variation.* Mac-
millan, New York.

Baust, J. G.

1967. Preliminary studies on the isolation of
pterins from the wings of heliconiid
butterflies. *Zoologica* 52:15–20.

Bibliography

Beck, S. D.
1980. *Insect photoperiodism.* 2d ed. Academic Press, New York.
1983. Insect thermoperiodism. *Annual Review of Entomology* 28:91–108.

Behrends, J.
1935. Über die Entwicklung des Lakunen-, Ader-, und Tracheensystems während die Puppenrube im Flügel der Mehlmotte *Ephestia kühniella* Zeller. *Zeitschrift für Morphologie und Ökologie der Tiere* 30:573–596.

Benson R. H., R. E. Chapman, and A. F. Siegel
1982. On the measurement of morphology and change. *Paleobiology* 8:328–339.

Bookstein, F. L.
1978. *The measurement of biological shape and shape change.* Springer-Verlag, Berlin.
1980. When one form is between two others: An application of biorthogonal analysis. *American Zoologist* 20:627–641.

Bookstein, F. L., B. Chernoff, R. L. Elder, J. M. Humphries, G. R. Smith, and R. E. Strauss
1985. *Morphometrics in evolutionary biology.* Academy of Natural Sciences of Philadelphia, Philadelphia.

Brakefield, P. M.
1984. The ecological genetics of quantitative characters in *Maniola jurtina* and other butterflies. *Symposia of the Royal Entomological Society of London* 11:167–190.

Brakefield, P. M., and T. B. Larsen
1984. The evolutionary significance of dry and wet season forms in some tropical butterflies. *Biological Journal of the Linnean Society* 22:1–12.

Brakefield, P. M., and J. Van Noordwijk
1985. The genetics of spot pattern characters in the meadow brown butterfly *Maniola jurtina* (Lepidoptera: Satyrinae). *Heredity* 54:275–284.

Brändle, K.
1965. Die Beeinflussbarkeit der Flügelmusterdetermination bei *Plodia interpunctella* während und nach der Ausbreitungsphase. *Zoologische Jahrbücher, Anatomie* 82:243–298.

Braun, W.
1936. Über das Zellteilungsmuster im Puppenflügelepithel der Mehlmotte *Ephestia kühniella* Z. in seiner Beziehung zur Ausbildung des Zeichnungsmusters. *Wilhelm Roux' Archiv für Entwicklungs Mechanik der Organismen* 135:494–520.

Brower, L. P., and J. V. Z. Brower
1972. Parallelism, convergence, divergence, and the new concept of advergence in the evolution of mimicry. *Transactions of the Connecticut Academy of Arts and Sciences* 44:57–67.

Brown, K. S.
1981. The biology of *Heliconius* and related genera. *Annual Review of Entomology* 26:427–456.

Brown, K. S., and W. W. Benson
1974. Adaptive polymorphism associated with multiple Müllerian mimicry in *Heliconius numata* (Lepid. Nymph.). *Biotropica* 6:205–228.

Brown, K. S., P. M. Sheppard, and J. R. G. Turner
1974. Quarternary refugia in tropical America: Evidence from race formation in *Heliconius* butterflies. *Proceedings of the Royal Society, London,* ser. B, 187:369–378.

Bünning, E.
1973. *The physiological clock.* Springer-Verlag, New York.

Burgeff, H., and L. Schneider
1979. Elektronenmikroskopische Untersuchungen zur Korrelation zwischen Farbe und Struktur bei Flügelschuppen des Widderchens *Zygaena ephialtes* (Lepidoptera: Zygaenidae). *Entomologia Generalis* 5:135–142.

Caveney, S.
1980. Cell communication and pattern formation in insects. In *Insect biology in the future,* ed. M. Locke and D. S. Smith, 565–582. Academic Press, New York.

Caveney, S., and R. Berden
1982. Selectivity in junctional coupling between cells of insect tissues. In *Insect*

ultrastructure, ed. R. C. King and H Akai, 434–465. Plenum, New York.

Caveney, S., and C. Podgorski

1975. Intercellular communication in a positional field. Ultrastructural correlates and tracer analysis of communication between insect epidermal cells. *Tissue and Cell* 7:559–574.

Charlesworth, D., and B. Charlesworth

1975. Theoretical genetics of Batesian mimicry, 1: Single-locus models. *Journal of Theoretical Biology* 55:282–303.

1976a. Theoretical genetics of Batesian mimicry, 2: Evolution of supergenes. *Journal of Theoretical Biology* 55:305–324.

1976b. Theoretical genetics of Batesian mimicry, 3: Evolution of dominance. *Journal of Theoretical Biology* 55:325–337.

Cheverud, J. M.

1984. Quantitative genetics and developmental constraints. *Journal of Theoretical Biology* 110:155–171.

1988. A comparison of genetic and phenotypic correlations. *Evolution* 42:958–968.

Choussy, M., and M. Barbier

1973. Pigments biliaires des lépidoptères: Identification de la phorcabiline I et de la sarpedobiline chez diverses espèces. *Biochemical Systematics* 1:199–201.

Clark, G. C., and C. G. C. Dickson

1957. Life history of *Precis octavia*. *Journal of the Entomological Society of South Africa* 20:257–259.

Clarke, C. A., and P. M. Sheppard

1955. A preliminary report on the genetics of the *machaon* group of swallowtail butterflies. *Evolution* 9:182–203.

1959. The genetics of some mimetic forms of *Papilio dardanus,* Brown, and *Papilio glaucus,* Linn. *Journal of Genetics* 56:237–259.

1960a. The evolution of dominance under disruptive selection. *Heredity* 14:73–87.

1960b. The evolution of mimicry in the butterfly *Papilio dardanus. Heredity* 14:163–173.

1962. The genetics of the mimetic butterfly

Papilio glaucus. Ecology 43:159–161.

1963. Interactions between major genes and polygenes in the determination of the mimetic pattern of *Papilio dardanus. Evolution* 17:404–413.

1971. Further studies on the genetics of the mimetic butterfly *Papilio memnon* L. *Philosophical Transactions of the Royal Society, London,* ser. B, 263:35–70.

1972. The genetics of the mimetic butterfly *Papilio polytes* L. *Philosophical Transactions of the Royal Society, London,* ser. B, 263:431–458.

1973. The genetics of four new forms of the mimetic butterfly *Papilio memnon* L. *Proceedings of the Royal Society, London,* ser. B, 184:1–14.

1975. The genetics of the mimetic butterfly *Hypolimnas bolina* (L.). *Philosophical Transactions of the Royal Society, London,* ser. B, 272:229–265.

Clarke, C. A., P. M. Sheppard, and I. W. B. Thornton

1968. The genetics of the mimetic butterfly *Papilio memnon* L. *Philosophical Transactions of the Royal Society, London,* ser. B, 254:37–89.

Clarke, C. A., P. M. Sheppard, and U. Mittwoch

1976. Heterochromatin polymorphism and colour pattern in the tiger swallowtail butterfly *Papilio glaucus* L. *Nature* 263:585–587.

Comstock, J. A.

1927. *Butterflies of California.* Comstock, Los Angeles.

Comstock, J. H.

1918. *The wings of insects.* Comstock, Ithaca, N.Y.

Danilevski, A. S.

1965. *Photoperiodism and seasonal development of insects.* Oliver and Boyd, London.

Descimon, H.

1965. Ultrastructure et pigmentation des écailles des lépidoptères. *Journal de Microscopie* 4:130.

1973. Les ptérines des Pieridae et leur biosynthèse, 4: Métabolisme de la ptérine, de la xanthoptérine, et de leurs dérivés hydrogénés chez *Colias croceus. Biochimie* 55:907–917.

1975a. Biology of pigmentation in Pieridae butterflies. In *Chemistry and biology of pteridines,* ed. W. Pfleiderer, 805–840. DeGruyter, Berlin.

1975b. The pterins of the Pieridae and their biosynthesis, 5: Metabolism of D-erythro-neopterin and its 7,8-dihydro derivative in *Colias croceus.* In *Chemistry and biology of pteridines,* ed. W. Pfleiderer, 841–847. DeGruyter, Berlin.

1977. Les ptérines des insectes. *Publications du Laboratoire Zoologique du Ecole Normale Supérieur* 8:43–130.

Dohrmann, C. E., and H. F. Nijhout

1990. Development of the wing margin in *Precis coenia* (Lepidoptera: Nymphalidae). *Journal for Research on the Lepidoptera.* In press.

Dorfmeister, G.

1864. Über die Entwicklung verscheidener während der Entwicklungsperioden angewendeter Wärmegrade auf die Färbung und Zeichnung der Schmetterlinge. *Mitteilungen der Naturwissenschaftliche Verein, Steiermark,* 99–108.

Douglas, M. M., and J. W. Grula

1978. Thermoregulatory adaptations allowing ecological range expansion in the pierid butterfly *Nathalis iole* Boisduval. *Evolution* 32:776–783.

Downey, J. C., and A. C. Allyn

1975. Wing-scale morphology and nomenclature. *Bulletin of the Allyn Museum* 31:1–30.

Duncan, I. M.

1986. The bithorax complex. *Annual Review of Genetics* 21:285–319.

Eassa, Y. E. E.

1953. The development of imaginal buds in the head of *Pieris brassicae* Linn. (Lepidoptera). *Transactions of the Royal Entomological Society of London* 104:39–50.

Edelstein-Keshet, L.

1988. *Mathematical models in biology.* Random House, New York.

Ehrlich, P. R., and A. H. Ehrlich

1967. The phenetic relationships of the butterflies, 1: Adult taxonomy and the nonspecificity hypothesis. *Systematic Zoology* 16:301–317.

Emsley, M. G.

1965. The geographical distribution of the colour-pattern components of *Heliconius erato* and *H. melpomene* with genetical evidence for the systematic relationships between the species. *Zoologica* 49:245–286.

Endo, K.

1970. Relation between ovarian maturation and activity of the corpora allata in seasonal forms of the butterfly, *Polygonia c-aureum* L. *Development Growth and Differentiation* 11:297–304.

1972. Activation of the corpora allata in relation to ovarian maturation in the seasonal forms of the butterfly, *Polygonia c-aureum* L. *Development Growth and Differentiation* 14:263–274.

1984. Neuroendocrine regulation of the development of seasonal forms of the Asian comma butterfly, *Polygonia c-aureum* L. *Development Growth and Differentiation* 26:217–222.

Endo, K., and S. Funatsu

1985. Hormonal control of seasonal morph determination in the swallowtail butterfly, *Papilio xuthus* L. (Lepidoptera: Papilionidae). *Journal of Insect Physiology* 31:669–674.

Endo, K., and Y. Kamata

1985. Hormonal control of seasonal-morph determination in the small copper butterfly, *Lycaena phlaeas daimio* Seitz. *Journal of Insect Physiology* 31:701–706.

Endo, K., T. Masaki, and K. Kumagai

1988. Neuroendocrine regulation of the development of seasonal morphs in the Asian comma butterfly, *Polygonia c-aureum* L.: Difference in activity of summer-morph-producing hormone from brain-extracts of the long-day and short-day pupae. *Zoological Science* 5:145–152.

Falconer, D. S.
  1981. *Introduction to quantitative genetics.* 2d
          ed. Longman, London.
Feldotto, W.
  1933. Sensibele Perioden des Flügelmusters
          bei *Ephestia kühniella* Z. *Wilhelm Roux'
          Archiv für Entwicklungs Mechanik der Or-
          ganismen* 128:299–341.
Feltwell, J., and L. R. G. Valadon
  1970. Plant pigments identified in the com-
          mon blue butterfly. *Nature* 225:969.
Fischer, E.
  1895. *Transmutation der Schmetterlinge infolge
          Temperaturveränderungen.*   Friedlander
          und Sohn, Berlin.
  1907. Zur Physiologie der Aberrationen- und
          Varietäten-Bildung der Schmetterlinge.
          *Archiv für Rassen- und Gesellschafts-
          Biologie* 4:761–793.
Ford, E. B.
  1936. The genetics of *Papilio dardanus* Brown
          (Lep.). *Transactions of the Royal Entomo-
          logical Society of London* 85:435–466.
  1941. Studies on the chemistry of pigments in
          the Lepidoptera, with reference to their
          bearing on systematics, 1: The antho-
          xanthins. *Proceedings of the Royal Ento-
          mological Society of London,* ser. A,
          16:65–90.
  1942. Studies on the chemistry of pigments in
          the Lepidoptera, with reference to their
          bearing on systematics, 2: Red pig-
          ments in the genus *Delias* Hubner. *Pro-
          ceedings of the Royal Entomological Society
          of London,* ser. A, 17:87–92.
  1944a. Studies on the chemistry of pigments in
          the Lepidoptera, with reference to their
          bearing on systematics, 3: The red pig-
          ments of the Papilionidae. *Proceedings of
          the Royal Entomological Society of London,*
          ser. A, 19:92–106.
  1944b. Studies on the chemistry of pigments in
          the Lepidoptera, with reference to their
          bearing on systematics, 4: The classifi-
          cation of the Papilionidae. *Proceedings of
          the Royal Entomological Society of London,*
          ser. A, 19:201–223.
  1945. *Butterflies.* Collins, London.

  1947a. A murexide test for the recognition of
          pterins in intact insects. *Proceedings of
          the Royal Entomological Society of London,*
          ser. A, 22:72–76.
  1947b. Studies on the chemistry of pigments in
          the Lepidoptera, with reference to their
          bearing on systematics, 5: *Pseudopon-
          tia paradoxa* Felder. *Proceedings of the
          Royal Entomological Society of London,* ser.
          A, 22:77–78.
  1971. *Ecological genetics.* 3d ed. Chapman and
          Hall, London.
Fukuda, S., and K. Endo
  1966. Hormonal control of the development
          of seasonal forms in the butterfly, *Poly-
          gonia c-aureum* L. *Proceedings of the Japan
          Academy* 42:1082–1087.
Futuyma, D.
  1986. *Evolutionary biology.* 2d ed. Sinauer,
          Sunderland, Mass.
Fuzeau-Braesch, S.
  1985. Colour changes. In *Comprehensive insect
          physiology, biochemistry, and pharmacology,*
          vol. 9, ed. G. A. Kerkut and L. I. Gil-
          bert, 549–589. Pergamon Press, New
          York.
Garcia-Bellido, A., P. A. Lawrence, and G. Mo-
  rata
  1979. Compartments in animal development.
          Scientific American 241:102–209.
Gehring, W. J., and Y. Hiromi
  1986. Homeotic genes and the homeobox.
          Annual Review of Genetics 20:147–173.
Ghiradella, H.
  1974. Development of ultraviolet-reflecting
          butterfly scales: How to make an inter-
          ference filter. *Journal of Morphology*
          142:395–410.
  1984. Structure of iridescent lepidopteran
          scales: Variations on several themes.
          *Annals of the Entomological Society of
          America* 77:637–645.
  1985. Structure and development of iridescent
          lepidopteran scales: The Papilionodae as
          a showcase family. *Annals of the Entomo-
          logical Society of America* 78:252–264.

Ghiradella, H., and W. Radigan
1976. Development of butterfly scales, 2: Struts, lattices, and surface tension. *Journal of Morphology* 150:279–298.

Ghiradella, H., D. Aneshansley, T. Eisner, R. E. Silberglied, and H. E. Hinton
1972. Ultraviolet reflection of a male butterfly: Interference color caused by thin-layer elaboration of wing scales. *Science* 178:1214–1217.

Gilbert, L. E.
1984. The biology of butterfly communities. In *The biology of butterflies,* ed. R. I. Vane-Wright and P. R. Ackery, 41–54. Academic Press, London.

Gilbert, L. E., H. S. Forrest, T. D. Schultz, and D. J. Harvey
1988. Correlations of ultrastructure and pigmentation suggest how genes control development of wing scales of *Heliconius* butterflies. *Journal for Research on the Lepidoptera* 26:141–160.

Gilbert, S. F.
1985. *Developmental biology.* Sinauer, Sunderland, Mass.

Gloor, H.
1947. Phänokopie-Versuche mit Aether an *Drosophila. Revue Suisse de Zoologie* 54:637–712.

Goldschmidt, R. B.
1935a. Gen und Auszeneigenschaft (Untersuchungen an *Drosophila*), 1: *Zeitschrift für Induktieve Abstammungs- und Vererbungslehre* 69:38–69.

1935b. Gen und Auszeneigenschaft (Untersuchungen an *Drosophila*), 2: *Zeitschrift für Induktieve Abstammungs- und Vererbungslehre* 69:70–131.

1938. *Physiological genetics.* McGraw-Hill, New York.

1940. *The material basis of evolution.* Yale University Press, New Haven, Conn.

1955. *Theoretical genetics.* University of California Press, Los Angeles.

Graham, M. W. R. deV.
1950. Postural habits and colour-pattern evolution in Lepidoptera. *Transactions of the Society for British Entomology* 10:217–232.

Hadorn, E.
1955. Letalfaktoren in ihrer Bedeutung für Erbpathologie und Genphysiologie der Entwicklung. Thieme, Stuttgart, West Germany.

Hafernik, J. E., Jr.
1982. Phenetics and ecology of hybridization in buckeye butterflies (Lepidoptera: Nymphalidae). *University of California Publications in Entomology* 96:1–109.

Haugum, J., and A. M. Low
1979. *A monograph of the birdwing butterflies.* Vol. 2. Scandinavian Science Press, Klampenborg, Denmark.

Henke, K.
1928. Über die Variabilität des Flügelmusters bei *Larentia sordidata* F. und einigen anderen Schmetterlingen. *Zeitschrift für Morphologie und Ökologie der Tiere* 12:240–282.

1933a. Untersuchungen an *Philosamia cynthia* Drury zur Entwicklungsphysiologie des Zeichnungsmusters auf dem Schmetterlingsflügel. *Wilhelm Roux' Archiv für Entwicklungs Mechanik der Organismen* 128:15–107.

1933b. Zur morphologie und Entwicklungsphysiologie der Tierzeichnung. [4 parts.] *Naturwissenschaften* 21:633–640, 654–659, 665–673, 683–690.

1936. Versuch einer vergleichenden Morphologie des Flügelmusters der Saturniden auf entwicklungsphysiologischer Grundlage. *Nova Acta Leopoldina* 18:1–37.

1943. Vergleich und experimentelle Untersuchungen an *Lymantria* zur Musterbildung auf dem Schmetterlingsflügel. *Nachrichten der Akademie der Wissenschaften in Göttingen, Mathematisch-physikalische Klasse,* 1–48.

1944. Über die Determination der Querbindenzeichnung und die Entstehung der Scheinsymmetrie bei der Saturnide *An-*

theraea pernyi Guer. *Biologisches Zentralblatt* 64:98–148.

1946. Über die verschiedenen Zellteilungsvorgange in der Entwicklung des beschuppten Flügelepithels der Mehlmotte *Ephestia kühniella* Z. *Biologisches Zentralblatt* 65:120–135.

1948. Einfache Grundvorgange in der tierischen Entwicklung, 2: Über die Enstehung von Differenzierungsmustern. [3 parts.] *Naturwissenschaften* 35:176–181, 203–211, 239–246.

Henke, K., and G. Kruse
1941. Über Feldgliederungsmuster bei Geometriden und Noctuiden und den Musterbauplan der Schmetterlinge im allgemein. *Nachrichten der Akademie der Wissenschaften in Göttingen, Mathematisch-physikalische Klasse,* 138–197.

Henke, K., and H.-J. Pohley
1952. Differentielle Zellteilungen und Polyploedie bei der Schuppenbildung der Mehlmotte *Ephestia kühniella. Zeitschrift für Naturforschung,* ser. B, 7:65–79.

Hennig, W.
1965. Phylogenetic systematics. *Annual Review of Entomology* 10:97–116.

1966. *Phylogenetic systematics.* University of Illinois Press, Chicago.

Hidaka, T., and H. Takahashi
1967. Temperature conditions and maternal effect as modifying factors in the photoperiodic control of the seasonal form in *Polygonia c-aureum. Annotationes Zoologicae Japonenses* 40:200–204.

Hoffmann, R. J.
1973. Environmental control of seasonal variation in the butterfly *Colias eurytheme,* 1: Adaptive aspects of a photoperiodic response. *Evolution* 27:387–397.

1974. Environmental control of seasonal variation in the butterfly *Colias eurytheme:* Effects of photoperiod and temperature on pteridine pigmentation. *Journal of Insect Physiology* 20:1913–1924.

Horstadius, S.
1973. *Experimental embryology of echinoderms.* Oxford University Press, London.

Huxley, J.
1975. The basis of structural colour variation in two species of *Papilio. Journal of Entomology,* ser. A, 50:9–22.

1976. The coloration of *Papilio zalmoxis* and *P. antimachus* and the discovery of Tyndall blue in butterflies. *Proceedings of the Royal Society, London,* ser. B, 193:44–53.

Inagami, K.
1954. Mechanism of the formation of red melanin in the silkworm. *Nature* 174:1105.

Kauffman, S. A.
1977. Chemical patterns, compartments, and a binary epigenetic code in *Drosophila. American Zoologist* 17:631–648.

Kayser, H.
1985. Pigments. In *Comprehensive insect physiology, biochemistry, and pharmacology,* vol. 10, ed. G. A. Kerkut and L. I. Gilbert, 367–415. Pergamon Press, New York.

Keino, H., and K. Endo
1973. Studies on the determination of seasonal forms in the butterfly, *Araschnia burejana* Bermer. [In Japanese with English summary.] *Zoological Magazine* 82:48–52.

Kingsolver, J.
1983. Thermoregulation and flight in *Colias* butterflies: Elevational patterns and mechanistic limitations. *Ecology* 64:534–545.

1985. Thermoregulatory significance of wing melanization in *Pieris* butterflies (Lepidoptera: Pieridae): Physics, posture, and pattern. *Oecologia* 66:546–553.

1988. Thermoregulation, flight, and the evolution of wing pattern in pierid butterflies: The topography of adaptive landscapes. *American Zoologist* 28:899–912.

Kingsolver, J., and R. K. Moffat
1982. Thermoregulation and the determination of heat transfer in *Colias* butterflies. *Oecologia* 53:27–33.

Kingsolver, J., and D. C. Wiernasz

1990. Analysing color pattern as a complex trait: Wing melanization in pierine butterflies. In *Adaptive coloration in invertebrates,* ed. M. Wicksten. University of Texas Press, Austin. In press.

Klots, A. B.

1933. A generic revision of the Pieridae (Lepidoptera). *Entomologia Americana* 12:139–242.

Koch, P. B.

1985. Die Hormonale Steuerung des Saisondimorphismus von *Araschnia levana* L. (Nymphalidae, Lepidoptera). Thesis, University of Ulm, West Germany.

1987. Die Steuerung der Saisondimorphen Flügelfärbung von *Araschnia levana* L. (Nymphalidae, Lepidoptera) durch ecdysteroide. *Mitteilungen der Deutschen Gesellschaft für Allgemeine und Angewandte Entomologie* 5:195–197.

Koch, P. B., and D. Bückmann

1985. Der Saisondimorphismus von *Araschnia levana* L. (Nymphalidae) in seiner Beziehung zur hormonalen Entwicklungssteuerung. *Verhandlungen der Deutschen Zoologischen Gesellschaft* 78:260.

1987. Hormonal control of seasonal morphs by the timing of ecdysteroid release in *Araschnia levana* L. (Nymphalidae: Lepidoptera). *Journal of Insect Physiology* 33:823–829.

Köhler, W.

1932. Die Entwicklung der Flügel bei der Mehlmotte *Ephestia kühniella* Z., mit besonderer Berücksichtigung des Zeichnungsmusters. *Zeitschrift für Morphologie und Ökologie der Tiere* 24:582–681.

Köhler, W., and W. Feldotto

1935. Experimentelle Untersuchungen über die Modifikabilität der Flügelzeichnung, ihrer Systeme und Elemente in den sensibelen Perioden von *Vanessa urticae* L., nebst einigen Beobachtungen an *Vanessa io* L. *Archiv der Julius Klaus Stiftung für Vererbungsforschung, Sozialan-*

thropologie, und Rassenhygiene 10:313–453.

1937. Morphologische und experimentelle Untersuchungen über Farbe, Form, und Struktur der Schuppen von *Vanessa urticae* und ihre gegenseitigen Beziehungen. *Wilhelm Roux' Archiv für Entwicklungs Mechanik der Organismen* 136:313–399.

Kremen, C.

1987. Metamorphosis of the butterfly, *Precis coenia* (Nymphalidae): Commitment of the imaginal disks and epidermis to pupal development. Ph.D. thesis, Duke University, Durham, N.C.

Kühn, A.

1926. Über die Änderung des Zeichnungsmusters von Schmetterlingen durch temperaturreize un das Grundschema der Nymphalidenzeichnung. *Nachrichten von der Gesellschaft der Wissenschaften zu Göttingen, Mathematisch-physikalische Klasse,* 120–141.

1971. *Lectures on developmental physiology.* 2d ed. Springer-Verlag, New York.

Kühn, A., and M. von Engelhardt

1933. Über die Determination des Symmetriesystems auf dem Vorderflügel von *Ephestia kühniella. Wilhelm Roux' Archiv für Entwicklungs Mechanik der Organismen* 130:660–703.

1936. Über die Determination des Flügelmusters bei *Abraxas grossulariata* L. *Nachrichten von der Gesellschaft der Wissenschaften zu Göttingen, Biologische Klasse* 2:171–199.

1944. Mutationen und Hitzemodifikationen des Zeichnungsmusters von *Ptychopoda seriata* Schrk. *Biologisches Zentralblatt* 64:24–73.

Kuntze, H.

1935. Die Flügelentwicklung bei *Philosamia cynthia* Drury, mit besonderer Berücksichtigung des Geäders der Lakunen und der Tracheensysteme. *Zeitschrift für Morphologie und Ökologie der Tiere* 30:544–572.

Lewis, E. B.
  1963. Genes and developmental pathways. *American Zoologist* 3:33–56.
Lewis, H. L.
  1978. A gene complex controlling segmentation in *Drosophila*. *Nature* 276:565–570.
Lewis, H. L.
  1987. *Butterflies of the world.* Harrison House, New York.
Lewis, J., J. M. W. Slack, and L. Wolpert
  1977. Thresholds in development. *Journal of Theoretical Biology* 65:579–590.
Lewontin, R. C.
  1974. The analysis of variance and the analysis of causes. *American Journal of Human Genetics* 26:400–411.
Linzen, B.
  1974. The tryptophan-ommochrome pathway in insects. *Advances in Insect Physiology* 10:117–246.
Lippert, W., and K. Gentil
  1959. Über lamellare Feinstrukturen bei den Schillerschuppen der Schmetterlinge vom *Urania-* und *Morpho*-typ. *Zeitschrift für Morphologie und Ökologie der Tiere* 48:115–122.
Locke, M., and P. Huie
  1981a. Epidermal feet in insect morphogenesis. *Nature* 293:733–735.
  1981b. Epidermal feet in pupal segment morphogenesis. *Tissue and Cell* 13:787–803.
Magnussen, K.
  1933. Untersuchungen zur Entwicklungsphysiologie des Schmetterlingsflügels. *Wilhelm Roux' Archiv für Entwicklungs Mechanik der Organismen* 128:447–479.
Mallet, J.
  1989. The genetics of warning colour in Peruvian hybrid zones of *Heliconius erato* and *H. melpomene*. *Proceedings of the Royal Society, London,* ser. B, 236:163–185.
Mason, C.
  1926. Structural colors in insects, 1: *Journal of Physical Chemistry* 30:383–395.

  1927a. Structural colors in insects, 2: *Journal of Physical Chemistry* 31:320–354.
  1927b. Structural colors in insects, 3: *Journal of Physical Chemistry* 31:1856–1872.
Maynard Smith, J., R. Burian, S. Kauffman, P. Alberch, J. Campbell, B. Goodwin, R. Lande, D. Raup, and L. Wolpert
  1985. Developmental constraints and evolution. *Quarterly Review of Biology* 60:265–287.
McLeod, L.
  1968. Controlled environment experiments with *Precis octavia* Cram. (Nymphalidae). *Journal for Research on the Lepidoptera* 7:1–18.
McWhirter, K. G., and E. R. Creed
  1971. An analysis of spot placing in the meadow brown butterfly. *Heredity* 21:517–521.
Meinhardt, H.
  1982. *Models of biological pattern formation.* Academic Press, New York.
Meinhardt, H., and A. Gierer
  1974. Applications of a theory of biological pattern formation based on lateral inhibition. *Journal of Cell Science* 15:321–346.
Merrifield, F.
  1890. Systematic temperature experiments on some Lepidoptera in all their stages. *Transactions of the Entomological Society of London,* 131–159.
  1891. Conspicuous effects on the markings and colouring of Lepidoptera caused by exposure of the pupae to different temperature conditions. *Transactions of the Entomological Society of London,* 155–168.
  1892. The effects of artificial temperature on the colouring of several species of Lepidoptera, with an account of some experiments on the effects of light. *Transactions of the Entomological Society of London,* 33–44.
  1893. The effects of temperature in the pupal stage on the colouring of *Pieris napi, Vanessa atalanta, Chrysophanus phloeas,* and

Ephyra punctaria. Transactions of the Entomological Society of London, 55–67.

1894. Temperature experiments in 1893 on several species of *Vanessa* and other Lepidoptera. *Transactions of the Entomological Society of London,* 425–438.

Milkman, R.

1966. Analyses of some temperature effects on *Drosophila* pupae. *Biological Bulletin* 131:331–345.

Miller, J. S.

1987. Phylogenetic studies in the Papilioninae (Lepidoptera: Papilionidae). *Bulletin of the American Museum of Natural History* 186:365–512.

Mitchell, H. K., and L. S. Lipps

1978. Heat shock and phenocopy induction in *Drosophila. Cell* 15:907–918.

Morris, R. B.

1975. Iridescence from diffraction structures in the wing scales of *Callophrys rubi,* the green hairstreak. *Proceedings of the Royal Entomological Society of London,* ser. A, 49:149–154.

Morris, S. J., and B. H. Thomson

1963. The flavonoid pigments of the marbled white butterfly (*Melanargia galathea* Seltz). *Journal of Insect Physiology* 9:391–399.

1964. The flavonoid pigments of the small heath butterfly, *Coenonympha pamphilus* L. *Journal of Insect Physiology* 10:377–383.

Müller, H. J.

1955. Die Saisonformenbildung von *Araschnia levana*—ein photoperiodisch gesteuerte Diapause-Effekt. *Naturwissenschaften* 42:134–135.

1956. Die Wirkung verscheidener diurnaler Licht-Dunkel-Relationen auf die Saisonformenbildung von *Araschnia levana. Naturwissenschaften* 43:503–504.

Munroe, E., and P. R. Ehrlich

1960. Harmonization of concepts of higher classification of the Papilionidae. *Journal of the Lepidopterists' Society* 14:169–175.

Murray, J. D.

1981a. On pattern formation mechanisms for lepidopteran wing patterns and mammalian coat markings. *Philosophical Transactions of the Royal Society, London,* ser. B, 295:473–496.

1981b. A pre-pattern formation mechanism for animal coat markings. *Journal of Theoretical Biology* 88:161–199.

1982. Parameter space for Turing instability in reaction-diffusion mechanisms: A comparison of models. *Journal of Theoretical Biology* 98:143–163.

1989. *Mathematical biology.* Springer-Verlag, New York.

Nardi, J., and S. M. Magee-Adams

1986. Formation of scale spacing patterns in a moth wing, 1: Epithelial feet may mediate cell rearrangement. *Developmental Biology* 116:278–290.

Needham, A. E.

1974. *The significance of zoochromes.* Springer-Verlag, New York.

Nijhout, H. F.

1978. Wing pattern formation in Lepidoptera: A model. *Journal of Experimental Zoology* 206:119–136.

1980a. Ontogeny of the color pattern on the wings of *Precis coenia* (Lepidoptera: Nymphalidae). *Developmental Biology* 80:275–288.

1980b. Pattern formation on lepidopteran wings: Determination of an eyespot. *Developmental Biology* 80:267–274.

1981. The color patterns of butterflies and moths. *Scientific American* 245:145–151.

1984. Colour pattern modification by cold-shock in Lepidoptera. *Journal of Embryology and Experimental Morphology* 81:287–305.

1985a. Cautery-induced colour patterns in *Precis coenia* (Lepidoptera: Pieridae). *Journal of Embryology and Experimental Morphology* 86:191–203.

1985b. The developmental physiology of colour patterns in Lepidoptera. *Advances in Insect Physiology* 18:181–247.

1985c. Independent development of homologous pattern elements in the wing patterns of butterflies. *Developmental Biology* 108:146–151.

1986. Pattern and pattern diversity on lepidopteran wings. *Bioscience* 36:527–533.

1990. A comprehensive model for color pattern formation in butterflies. *Proceedings of the Royal Society, London,* ser. B, 239:81–113.

Nijhout, H. F., and L. W. Grunert

1988. Colour pattern regulation after surgery on the wing disks of *Precis coenia* (Lepidoptera: Nymphalidae). *Development* 102:377–385.

Nijhout, H. F., and D. E. Wheeler

1982. Juvenile hormone and the physiological basis of insect polymorphisms. *Quarterly Review of Biology* 57:109–133.

Nijhout, H. F., and G. A. Wray

1986. Homologies in the colour patterns of the genus *Charaxes* (Lepidoptera: Nymphalidae). *Biological Journal of the Linnean Society* 28:387–410.

1988. Homologies in the colour patterns of the genus *Heliconius* (Lepidoptera: Nymphalidae). *Biological Journal of the Linnean Society* 33:345–365.

Nijhout, H. F., G. Wray, and L. E. Gilbert

1990. An analysis of the phenotypic effects of certain color pattern genes in *Heliconius* (Lepidoptera: Nymphalidae). *Biological Journal of the Linnean Society* 40:357–372.

Oster, G., and P. Alberch

1982. Evolution and bifurcation of developmental programs. *Evolution* 36:444–459.

Othmer, H. G., and E. Pate

1980. Scale-invariance in reaction-diffusion models of spatial pattern formation. *Proceedings of the National Academy of Sciences* 77:4180–4184.

Oudemans, J. T.

1903. Etudes sur la position de repos chez les Lepidopteres. *Verhandelingen der Koninglijke Akademie van Wetenschappen* 10:1–90.

Overton, J.

1966. Microtubules and microfibrils in morphogenesis of the scale cells of *Ephestia kühniella* Zeller. *Zeitschrift für Zellforschung* 63:840–870.

Pansera, M. C. G., and A. M. Arauju

1983. Distribution and heritability of the red raylets in *Heliconius erato phyllis* (Lepid.; Nymph.). *Heredity* 51:643–652.

Patterson, C.

1982. Morphological characters and homology. In *Problems of phylogenetic reconstruction,* ed. K. A. Joysey and A. E. Friday, 21–74. Academic Press, New York.

Paulsen, S., and H. F. Nijhout

1990. Correlation structures in butterfly wing patterns. Submitted for publication.

Pinhey, E. C. G.

1949. *Butterflies of Rhodesia; with a short introduction to the insect world.* Rhodesia Science Association, Salisbury.

Pohley, H.-J.

1959. Über das Wachstum der Mehlmotteflügel unter normalen und experimentellen Bedingungen. *Biologisches Zentralblatt* 78:233–250.

Prota, G., and R. H. Thomson

1976. Melanin pigmentation in mammals. *Endeavour* 35:32–38.

Rachootin, S. P., and K. S. Thomson

1981. Epigenetics, paleontology, and evolution. In *Evolution today,* Proceedings of the Second International Congress on Systematic and Evolutionary Biology, ed. G. G. E. Scudder and J. L. Reveal, 181–193. Hunt Institute for Botanical Documentation, Pittsburgh.

Raup, D. M.

1966. Geometric analysis of shell coiling: General problems. *Journal of Paleontology* 40:1178–1190.

Reinhardt, R.

1969. Über den Einfluss der Temperatur auf den Saisondimorphismus von *Araschnia levana* L. (Lepidopt. Nymphalidae) nach photoperiodischer Diapauseinduktion. *Zoologische Jahrbücher, Physiologie* 75:41–75.

Rendel, J. M.

1967. *Canalisation and gene control.* Academic Press, New York.

Riedl, R.

1978. *Order in living organisms.* Wiley, New York.

Riley, P. A.

1977. The mechanism of melanogenesis. *Symposia of the Zoological Society of London* 39:77–95.

Roth, V. L.

1984. On homology. *Biological Journal of the Linnean Society* 22:13–29.

1988. The biological basis of homology. In *Ontogeny and systematics,* ed. C. J. Humphries, 1–26. Columbia University Press, New York.

Russwurm, A. D. A.

1978. *Aberrations of British butterflies.* Classey, Faringdon, England.

Saul, S. J., and M. Sugumaran

1986. Protease inhibitor controls prophenoloxidase activation in *Manduca sexta. Federation of European Biochemical Societies* 208:113–116.

1987. Protease mediated prophenoloxidase activation in the hemolymph of the tobacco hornworm, *Manduca sexta. Archives of Insect Biochemistry and Physiology* 5:1–11.

Saunders, D. S.

1982. *Insect clocks.* Pergamon Press, Oxford, England.

Schwanwitsch, B. N.

1924. On the groundplan of wing-pattern in nymphalids and certain other families of rhopalocerous Lepidopetra. *Proceedings of the Zoological Society of London,* ser. B, 34:509–528.

1925. On a remarkable dislocation of the components of the wing pattern in a Satyride genus *Pierella. Entomologist* 58:226–269.

1926. On the modes of evolution of the wing-pattern in nymphalids and certain other families of the rhopalocerous Lepidoptera. *Proceedings of the Zoological Society of London,* ser. B, 493–508.

1929a. Evolution of the wing-pattern in Palaeartic Satyridae, 1: Genera *Satyrus* and *Oeneis. Zeitschrift für Morphologie und Ökologie der Tiere* 13:559–654.

1929b. Studies upon the wing-pattern of *Prepona* and *Agrias,* two genera of South American nymphalid butterflies. *Acta Zoologica* 11:289–424.

1929c. Two schemes of the wing pattern of butterflies. *Zeitschrift für Morphologie und Ökologie der Tiere* 14:36–58.

1930. Studies upon the wing-pattern of *Catagramma* and related genera of South American nymphalid butterflies. *Transactions of the Zoological Society of London* 40:105–286.

1931. Evolution of the wing-pattern in Palaeartic Satyridae, 2: Genus *Melanargia. Zeitschrift für Morphologie und Ökologie der Tiere* 21:316–408.

1935a. Evolution of the wing-pattern in Palaeartic Satyridae, 3: Genus *Pararge* and five others. *Acta Zoologica* 16:143–281.

1935b. On some general principles observed in the evolution of the wing-pattern of Palaeartic Satyridae. In *Proceedings of the Sixth International Congress of Entomology (Madrid),* 1–8.

1943. Wing-pattern in papilionid Lepidoptera. *Entomologist* 76:201–203.

1948. Evolution of the wing-pattern in Palaearctic Satyridae, 4: Polymorphic radiation and parallelism. *Acta Zoologica* 29:1–61.

1949. Evolution of the wing pattern in the lycaenid Lepidoptera. *Proceedings of the Zoological Society of London,* ser. B, 119:189–263.

1956a. Color-pattern in Lepidoptera. *Entomologeskoe Obozrenie* 35:530–546.

1956b. Wing pattern of pierid butterflies (Lepidoptera, Pieridae). [In Russian.] *Entomologeskoe Obozrenie* 35:285–301.

Schwanwitsch, B. N., and G. N. Sokolov

1934. On the wing-pattern of the genus *Lethe*. *Acta Zoologica* 15:153–181.

Schwartz, V.

1962. Neue Versuche zur Determination des zentralen Symmetriesystems bei *Plodia interpunctella*. *Biologisches Zentralblatt* 81:19–44.

Scott, J. A.

1986. *The butterflies of North America: A natural history and field guide*. Stanford University Press, Stanford, Calif.

Scriber, J. M., and M. H. Evans

1986. An exceptional case of paternal transmission of the dark form female trait in the tiger swallowtail butterfly, *Papilio glaucus* (Lepidoptera: Papilionidae). *Journal for Research on the Lepidoptera* 25:110–120.

Seybold, W. D., P. S. Meltzer, and H. K. Mitchell

1975. Phenol oxidase activation in *Drosophila*: A cascade of reactions. *Biochemical Genetics* 13:85–108.

Shapiro, A. M.

1968. Photoperiodic induction of vernal phenotype in *Pieris protodice* Boisduval and Le Conte (Lepidoptera: Pieridae). *Wasmann Journal of Biology* 26:137–149.

1973. Photoperiodic control of seasonal polyphenism in *Pieris occidentalis* Reakirt. *Wasmann Journal of Biology* 31:291–299.

1974. Natural and laboratory occurrence of "elymi" phenotypes in *Cynthia cardui* (Nymphalidae). *Journal for Research on the Lepidoptera* 13:57–62.

1976. Seasonal polyphenism. *Evolutionary Biology* 9:259–333.

1978. The evolutionary significance of redundancy and variability in phenotypic induction mechanisms of pierid butterflies (Lepidoptera). *Psyche* 85:275–283.

1979. The life histories of the *autodice* and *sterodice* species-groups of *Tatochila* (Lepidoptera: Pieridae). *Journal of the New York Entomological Society* 87:236–255.

1980a. Canalization of the phenotype of *Nymphalis antiopa* (Lepidoptera: Nymphalidae) from subarctic and montane climates. *Journal for Research on the Lepidoptera* 19:82–87.

1980b. Convergence in pierine polyphenisms (Lepidoptera). *Journal of Natural History* 14:781–802.

1981. Phenotypic plasticity in temperate and subarctic *Nymphalis antiopa* (Nymphalidae): Evidence for adaptive canalization. *Journal of the Lepidopterists' Society* 35:124–131.

1982. Redundancy in pierid polyphenisms: Pupal chilling induces vernal phenotype in *Pieris occidentalis* (Pieridae). *Journal of the Lepidopterists' Society* 36:174–177.

1983a. The genetics of polyphenism and its role in phylogenetic interpretation of the *Tatochila sterodice* species-group (Pieridae) in the Andean-Neantarctic region. *Journal for Research on the Lepidoptera (Supplement)* 25:24–31.

1983b. Testing visual species recognition in *Precis* (Lepidoptera: Nymphalidae) using a cold-shock phenocopy. *Psyche* 90:59–65.

1984a. Experimental studies on the evolution of seasonal polyphenism. In *The biology of butterflies*, ed. R. I. Vane-Wright and P. R. Ackery, 297–307. Academic Press, London.

1984b. The genetics of seasonal polyphenism and the evolution of "general purpose genotypes" in butterflies. In *Population biology and evolution*, ed. K. Wohrmann and V. Loeschke, 16–30. Springer-Verlag, Berlin.

Sheppard, P. M.

1953. Polymorphism, linkage, and the blood groups. *American Naturalist* 87:283–294.

1958. *Natural selection and heredity*. Hutchinson, London.

1963. Some genetic studies of Muellerian mimics in butterflies of the genus *Heliconius*. *Zoologica* 48:145–154.

Sheppard, P. M., J. R. G. Turner, K. S. Brown, W. W. Benson, and M. C. Singer
1985. Genetics and the evolution of Muellerian mimicry in *Heliconius* butterflies. *Philosophical Transactions of the Royal Society, London,* ser. B, 308:33–613.

Shields, O.
1987. Presence of pterin pigment in wings of Libytheidae butterflies. *Journal of Chemical Ecology* 13:1843–1847.

Sibatani, A.
1980. Wing homoeosis in Lepidoptera: A survey. *Developmental Biology* 79:1–18.

1983a. A compilation of data on wing homoeosis in Lepidoptera. *Journal for Research on the Lepidoptera* 22:1–46.

1983b. Compilation of data on wing homoeosis on Lepidoptera: Supplement 1. *Journal for Research on the Lepidoptera* 22:118–125.

1987a. Homoeosis of dorsal and ventral wing surfaces in butterflies. *Journal of Liberal Arts of the Kansai Medical University* 11:11–12.

1987b. Oudemans' principle and its extension in pattern formation on the wings of Lepidoptera (Insecta). *Journal of Liberal Arts of the Kansai Medical University* 11:1–10.

Silberglied, R., and O. R. Taylor
1973. Ultraviolet differences between the sulphur butterflies, *Colias eurytheme* and *C. philodice,* and a possible isolating mechanism. *Nature* 241:406–408.

Smart, P.
1975. *The illustrated encyclopedia of the butterfly world.* Salamander Books, London.

Sokal, R. R., and F. J. Rohlf
1981. *Biometry.* 2d ed. Freeman Press, San Francisco.

Sokolow, G. N.
1936. Die Evolution der Zeichnung der Arctiidae. *Zoologische Jahrbücher, Anatomie* 61:107–238.

Sondhi, K. C.
1963. The biological foundations of animal patterns. *Quarterly Review of Biology* 38:289–327.

Standfuss, M.
1896. *Handbuch der palearctischen Gross-Schmetterlinge für Forscher und Sammler.* G. Fischer, Jena, Germany.

Stossberg, M.
1937. Über die Entwicklung der Schmetterlingsschuppen. *Biologisches Zentralblatt* 57:393–402.

1938. Die Zellvorgange bei der Entwicklung der Flügelschuppen von *Ephestia kühniella* Z. *Zeitschrift für Morphologie und Ökologie der Tiere* 34:173–206.

Süffert, F.
1924a. Bestimmungsfaktoren des Zeichnungsmusters beim Saisondimorphismus von *Araschnia levana-prorsa. Biologisches Zentralblatt* 44:173–188.

1924b. Morphologie und Optik der Schmetterlingsschuppen. *Zeitschrift für Morphologie und Ökologie der Tiere* 1:171–308.

1927. Zur vergleichende Analyse der Schmetterlingszeichnung. *Biologisches Zentralblatt* 47:385–413.

1929a. Die Ausbildung des imaginalen Flügelschnittes in der Schmetterlingspuppe. *Zeitschrift für Morphologie und Ökologie der Tiere* 14:338–359.

1929b. Morphologische Erscheinungsgruppen in der Flügelzeichnung der Schmetterlinge, insbesondere die Querbindenzeichnung. *Wilhelm Roux' Archiv für Entwicklungs Mechanik der Organismen* 120:229–383.

1937. Die Geschichte der Bildungszellen im Puppenflügelepithel bei einem Tagschmetterling. *Biologisches Zentralblatt* 57:615–628.

Sugumaran, M., S. J. Saul, and N. Ramesh
1985. Endogenous protease inhibitors prevent undesired activation of prophenolase in insect hemolymph. *Biochemical and Biophysical Research Communications* 132:1124–1129.

Suomalainen, E., L. M. Cook, and J. R. G. Turner
1971. Chromosome numbers of heliconiine butterflies from Trinidad, West Indies

(Lepidoptera, Nymphalidae). *Zoologica* 56:121–124.

Tauber, M. J., C. A. Tauber, and S. Masaki
  1986. *Seasonal adaptations of insects.* Oxford University Press, New York.

Thompson, D. W.
  1917. *On growth and form.* Cambridge University Press, Cambridge.

Thomson, D. L.
  1926a. The pigments of butterflies' wings, 1: *Melanargia galatea. Biochemical Journal* 20:73–75.
  1926b. The pigments of butterflies' wings, 2: Occurrence of the pigment of *Melanargia galathea* in *Dactylis glomerata. Biochemical Journal* 20:1026–1027.

Toussaint, N., and V. French
  1988. The formation of pattern on the wing of the moth, *Ephestia kühniella. Development* 103:707–718.

Truman, J. W.
  1971. Hour-glass behavior of the circadian clock controlling eclosion of the silkmoth *Antheraea pernyi. Proceedings of the National Academy of Sciences* 68:595–599.

Turner, J. R. G.
  1971a. The genetics of some polymorphic forms of the butterflies *Heliconius melpomene* (Linnaeus) and *H. erato* (Linnaeus), 2: The hybridization of subspecies from Surinam and Trinidad. *Zoologica* 56:125–157.
  1971b. Two thousand generations of hybridization in a *Heliconius* butterfly. *Evolution* 25:471–482.
  1982. How do refuges produce biological diversity? Allopatry and parapatry, extinction, and gene flow in mimetic butterflies. In *Biological diversification in the tropics,* ed. G. T. Prance, 309–335. Columbia University Press, New York.
  1984. Mimicry: The palatability spectrum and its consequences. In *The biology of butterflies,* ed. R. I. Vane-Wright and P. R. Ackery, 141–161. Academic Press, London.

  1986. The genetics of an adaptive radiation: A neo-Darwinian theory of punctuated evolution. In *Patterns and processes in the history of life,* ed. D. M. Raup and D. Jablonski, 183–207. Springer-Verlag, Berlin.

Turner, J. R. G., and J. Crane
  1962. The genetics of some polymorphic forms of the butterflies *Heliconius melpomene* Linnaeus and *H. erato* Linnaeus, 1: Major genes. *Zoologica* 47:141–152.

Umebachi, Y.
  1980. Wing pigments derived from tryptophan in butterflies. In *Biochemical and medical aspects of tryptophan metabolism,* ed. O. Hayaishi, Y. Ishimura, and R. Kido, 117–124. Elsevier, Amsterdam.

Umebachi, Y., and H. Takahashi
  1956. Kynurenine in the wings of papilionid butterflies. *Journal of Biochemistry, Tokyo* 43:73–81.

Vane-Wright, R. I.
  1979. The coloration, identification, and phylogeny of *Nessaea* butterflies. *Bulletin of the British Museum (Natural History), Entomology* 38:27–56.

Waddington, C. H.
  1942. Canalization of development and the inheritance of acquired characters. *Nature* 150:563–565.
  1953. Genetic assimilation of an acquired character. *Evolution* 7:118–126.
  1956a. Genetic assimilation of the bithorax phenotype. *Evolution* 10:1–13.
  1956b. *Principles of embryology.* Allen and Unwin, London.

Wasserthal, L. T.
  1975. The role of butterfly wings in regulation of body temperature. *Journal of Insect Physiology* 21:1921–1930.
  1982. Antagonism between haemolymph transport and tracheal ventilation in an insect wing (*Attacus atlas* L.): A disproof of the generalized model of insect wing circulation. *Journal of Comparative Physiology* 147:27–40.

Watt, W. B.

1964. Pteridine components of wing pigmentation in the butterfly *Colias eurytheme*. *Nature* 201:1326–1327.

1967. Pteridine biosynthesis in the butterfly *Colias eurytheme*. *Journal of Biological Chemistry* 242:565–572.

1968. Adaptive significance of pigment polymorphism in *Colias* butterflies, 1: Variation of melanin in relation to thermoregulation. *Evolution* 22:437–458.

1969. Adaptive significance of pigment polymorphism in *Colias* butterflies, 2: Thermoregulation and photoperiodically controlled melanin variation in *Colias eurytheme*. *Proceedings of the National Academy of Sciences* 63:767–774.

1974. Adaptive significance of pigment polymorphism in *Colias* butterflies, 3: Progress in the study of the "alba" variant. *Evolution* 27:537–548.

Wehrmaker, A.

1959. Modifikabilität und Morphogenese des Zeichnungsmusters von *Plodia interpunctella* (Lepidoptera: Pyralidae). *Zoologische Jahrbücher für Zoologie und Physiologie* 68:425–496.

Weismann, A.

1875. *Studien zur Descendenztheorie.* Engelmann, Leipzig.

1896. New experiments on seasonal dimorphism of Lepidoptera. [7 parts.] *Entomologist* 29:29–253.

Wiernasz, D. C.

1989. Female choice and sexual selection of male wing melanin pattern in *Pieris occidentalis* (Lepidoptera). *Evolution* 43:1672–1682.

Wiernasz, D. C., and J. Kingsolver

1990. Developmental organization of wing melanization pattern in *Pieris* butterflies. Submitted for publication.

Wigglesworth, V. B.

1965. *The principles of insect physiology.* Methuen, London.

Wiley, E. O.

1981. *Phylogenetics: The theory and practice of phylogenetic systematics.* Wiley, New York.

Wolpert, L.

1969. Positional information and the spatial pattern of cellular differentiation. *Journal of Theoretical Biology* 25:1–47.

1971. Positional information and pattern formation. *Current Topics in Developmental Biology* 6:183–224.

Wright, S.

1968. *Evolution and the genetics of populations.* Vol. 1: *Genetics and biometric foundations.* University of Chicago Press, Chicago.

1982. Character change, speciation, and the higher taxa. *Evolution* 36:427–443.

Wuhlkopf, H.

1936. Hitze- und Frostraize ihrer Wirkung auf das Flügelmuster der Mehlmotte *Ephestia kühniella* Z. *Wilhelm Roux' Archiv für Entwicklungs Mechanik der Organismen* 134:209–223.